MW00844571

Visualizing the Invisible

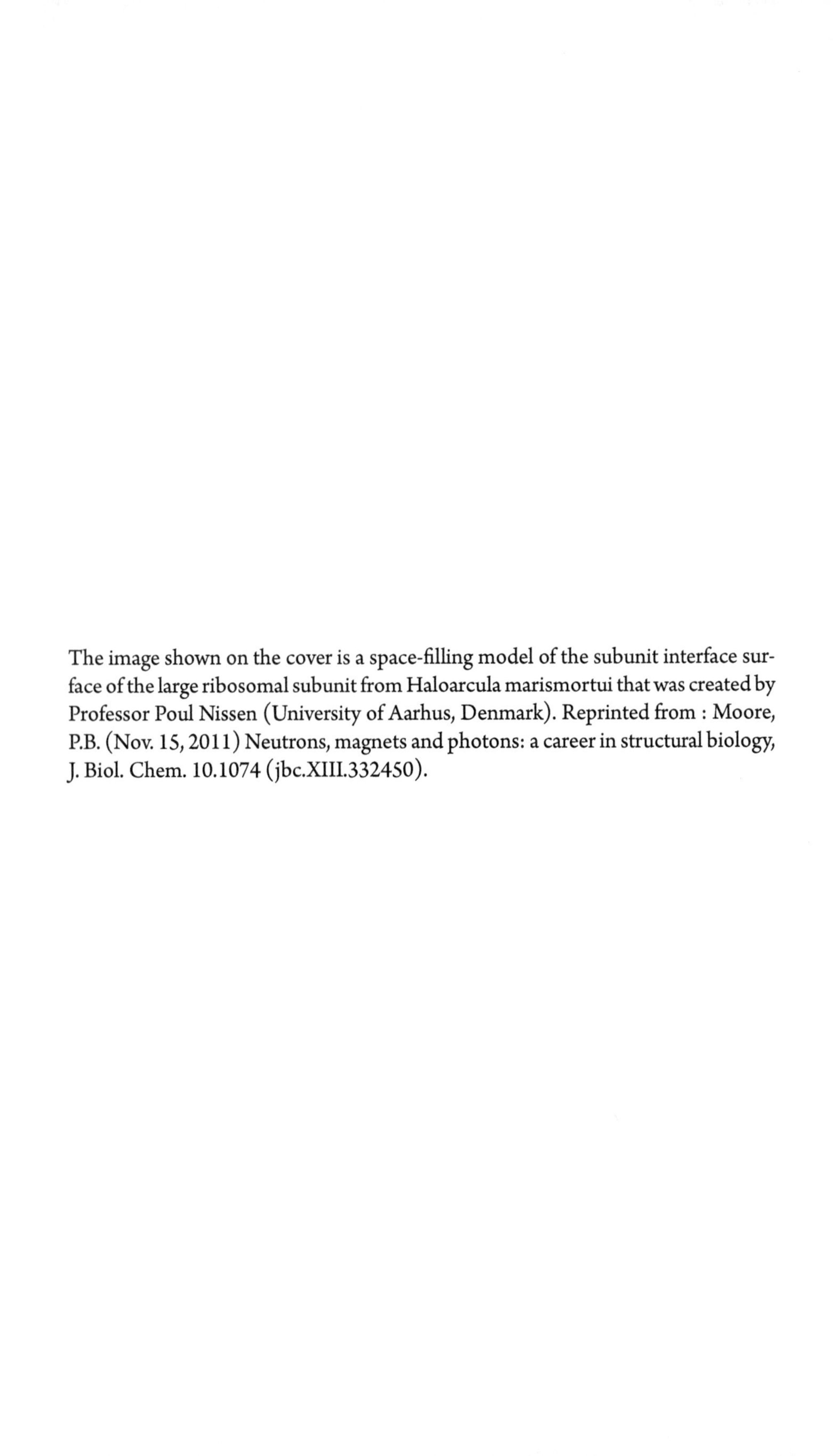

The image shown on the cover is a space-filling model of the subunit interface surface of the large ribosomal subunit from Haloarcula marismortui that was created by Professor Poul Nissen (University of Aarhus, Denmark). Reprinted from : Moore, P.B. (Nov. 15, 2011) Neutrons, magnets and photons: a career in structural biology, J. Biol. Chem. 10.1074 (jbc.XIII.332450).

Visualizing the Invisible: Imaging Techniques for the Structural Biologist

PETER B. MOORE

OXFORD
UNIVERSITY PRESS

OXFORD
UNIVERSITY PRESS

Oxford University Press, Inc., publishes works that further
Oxford University's objective of excellence
in research, scholarship, and education

Oxford New York

Auckland Cape Town Dar es Salaam Hong Kong Karachi
Kuala Lumpur Madrid Melbourne Mexico City Nairobi
New Delhi Shanghai Taipei Toronto

With offices in

Argentina Austria Brazil Chile Czech Republic France Greece
Guatemala Hungary Italy Japan Poland Portugal Singapore
South Korea Switzerland Thailand Turkey Ukraine Vietnam

Published by Oxford University Press Inc.,
198 Madison Avenue, New York, New York 10016

www.oup.com

Library of Congress Cataloging-in-Publication Data
Moore, Peter, 1939–
Visualizing the invisible : imaging techniques for the structural biologist / Peter B. Moore.
p. cm.
Includes bibliographical references and index.
ISBN 978-0-19-976709-0 (hardcover : alk. paper) 1. Ultrastructure (Biology)
2. Molecular structure. 3. Fourier transformations. 4. Imaging systems in biology.
5. Cytology—Experiments. 6. Molecular biology—Experiments. 7. Biology—Experiments. I. Title.
QH324.M66 2012
571.6'33—dc23 2011027000

1 3 5 7 9 8 6 4 2

Printed in the United States of America
on acid-free paper

CONTENTS

PREFACE

It has long been clear that the physiologies of organisms would remain a mystery until their internal structures were fully understood. Thus, the description of human anatomy that Vesalius published in 1543 (*De humani corporis fabrica* [On the Fabric of the Human Body]) is a landmark not only in the history of medicine, but also in the history of biology. However, by the late 17th century, it had become evident that tools Vesalius had relied on, the sharp knife and the naked eye, could not provide all the structural information needed to understand biological function. In 1665, using a primitive compound microscope, Robert Hooke (1635–1702) discovered that tissues are assemblies of entities too small to see that he termed "cells". A few years later, using a microscope of superior design, Anton van Leeuwenhoek (1632–1723) discovered an entire universe of microscopic organisms, the existence of which had been previously unsuspected. During the 19th century, chemists learned that organisms contain an amazing variety of organic molecules, most of which are found nowhere else in nature, and it became obvious that life is absolutely dependent on these molecules, many of which have physico chemical properties that still have the power to amaze us. Thus, by 1900, it was evident that any biologist interested in biological function would have to understand the structures of organisms at length scales beginning at the meter–centimeter length scale of the gross anatomist and extending all the way down to the angstrom (10^{-10} m) length scale of the biochemist.

Most of what is known today about the structure of organisms at length scales smaller than about 1 mm has been generated by experimental techniques that produce magnified images of biological materials, and the impact on the biological sciences of the discoveries that continue to be made using techniques is as large today as it has ever been. This book derives from materials I developed over several years to support a one-term course on biological imaging that I taught to graduate and advanced undergraduate students at Yale University. Thus, it is no accident that the number of chapters in this book is the same as the number of weeks in a semester, and that each chapter is short enough so that its substance can be transmitted to students in roughly one week's worth of class time.

My motivation for teaching this course was my conviction that anyone seriously interested in biology must understand how the microscopic structures are visualized. The biological literature is filled with microscopic images of one kind or another. Absent a sound understanding of how they are generated, how is one to evaluate the validity of the conclusions drawn from them? In addition, at some stage in their careers, most biological scientists will find themselves using at least one of the microscopic imaging techniques described in this book. When that day comes, their need to understand will be even greater.

Biological imaging is a field in ferment today. Novel approaches for imaging biological structures are being developed all the time, and powerful new ways are being found to use old technologies such as light microscopy and X-ray scattering/diffraction. The reason a single textbook can be written that covers all these techniques is that, different as they may seem, they are all applications of a handful of physical principles. Once those principles are grasped, the rest is easy. Readers will be introduced to those principles and will be shown many examples of how they are used to solve biological problems. The hope is that readers will emerge with an understanding of imaging robust enough to make them discerning critics of the relevant literature who are prepared to do structural research of their own.

The presentation of the material in this book reflects my visceral dislike of the ex cathedra pronouncements about physical theory that are found all too often in today's textbooks, even those intended for advanced students. I am comfortable with scientific subjects only when (I think) I understand their relationships to fundamental physical and chemical principles. For that reason, this book includes many derivations that are intended to make those connections clear, but because I know that my enthusiasm for derivations is not universally shared, the more complicated of them are sequestered in appendices that appear at the end of each chapter. They can be bypassed by those willing to take on faith the relationships they explain.

The book is intended for readers whose scientific education includes enough biology and biochemistry to know what cells and proteins are, and who have taken (at least) a year of chemistry and a year of physics with calculus. However, it is *not* assumed that the reader remembers it all. In addition, it will quickly become obvious that imaging is best addressed using a mathematical operation called the *Fourier transform*, an aspect of mathematics that is seldom mentioned in first-year calculus. Readers familiar with Fourier transforms will find this book easier to read than those who are not, but it was written with the Fourier-naïve reader in mind. The relevant aspects of that branch of mathematics are presented in the text.

Textbooks are seldom generated out of whole cloth. In the writing of this book, as in the teaching of the course on which it is based, I constantly consulted a handful of books, which I mention here to acknowledge the intellectual debt I owe their authors. For questions about the physics of X-ray diffraction, I find *The Optical Principles of the Diffraction of X-Rays* by R. W. James (Cornell University Press, 1962) exceptionally useful. (Historical note: In 1914, as a freshly minted college graduate, James joined Shackleton's Imperial Trans-Antarctic Expedition, and thus participated in one of the greatest adventures of the twentieth century, or indeed, of any other century.) *Optical Physics*, third edition by S. G. Lipson, H. Lipson, and D. S. Tannhauser (Cambridge University Press, 1995) has also proven helpful. R. N. Bracewell's text on Fourier transforms ("*The Fourier Transform and Its Applications*", McGraw-Hill, 2000) is an excellent source of information not only about the mathematics of the Fourier transform, but also about its practical applications.

For electron microscopy, *Three-Dimensional Electron Microscopy of Macromolecular Assemblies* by Joachim Frank (Oxford University Press, 2006) is exceedingly useful. Finally, I would be remiss not to acknowledge my debt both to *Principles of Optics* by Max Born and Emil Wolf, a classic now in its seventh edition (Cambridge University Press, 1999), and to the *International Tables for Crystallography, Volume F: Crystallography of Biological Macromolecules* (M. G. Rossman & E. Arnold eds., Kluwer Academic Publishers, 2001), which is a collection of essays written by experts that deal with all aspects of macromolecular crystallography. Additional references are provided at the end of each chapter, where appropriate.

I owe a special debt of gratitude to an old friend, Professor David J. DeRosier of Brandeis University, who bravely agreed to read and comment on the chapters in this book that deal with microscopy. Those chapters are much the better for his input. I also thank Dr. Gregor Blaha for his suggestions for improving the opening chapters of this book. Finally, I gratefully acknowledge the students whose responses to my pedagogical efforts over the years helped shape this book.

New Haven, Connecticut
April 2011

Peter B. Moore

NOTES FOR THE READER

Those of you who need to be reminded of the trigonometry you have forgotten, or of the integrals you encountered in freshman calculus that you no longer know how to evaluate, will find the information you seek in *Standard Mathematical Tables and Formulae* (CRC Press). Innumerable editions of this book have been published, and any one of them will do. Much of the material in that book also appears in the *Handbook of Physics and Chemistry*, another CRC Press publication. Finally, never forget the awesome power of Google.

The mathematics you require, beyond what a year or two of college calculus provides, will be supplied in the text. As for the physics, most of it too will be presented in the text, as needed, not only to help you recall material you may have forgotten, but also because many first-year courses in physics do not cover all the topics that are important here, for example, optics.

Almost all the phenomena discussed in this text can be understood adequately using the concepts of classical, 19th-century, physics. They will be described that way because in most cases there is little to be gained conceptually by treating them otherwise, and it is much simpler if we do not. Most of the time, classical reasoning will lead us to conclusions that are at least qualitatively correct, which is all we will usually require. Quantum mechanics would have to be used to obtain quantitatively accurate descriptions of many of the phenomena discussed here, and from time to time, quantum mechanical concepts will be invoked, but only at a level that should cause no discomfort to someone who has had a year of chemistry and a year of physics.

A number of conventions are adhered to in the text.

1. Words or expressions that you are encouraged to add to your vocabulary are printed in **boldface** when they first appear in the text.
2. In mathematical expressions, symbols that represent quantities, either real or complex, are *italicized*. Symbols that stand for functions are never italicized, even when their arguments are not indicated in the context in which they appear. Vectors are always represented by **boldfaced** symbols.
3. Meter-kilogram-second (i.e., SI) units of measure are used throughout. This is particularly important for equations describing electrical and magnetic phenomena. In SI units, Coulomb's law is:

$$\mathbf{F} = (1/4\pi\epsilon_o)(q_1 q_2/r_{12}^2)\mathbf{1}_r,$$

where ε_o is the permittivity of a vacuum ($= 8.854 \times 10^{-12}\ \mathrm{kg}^{-1}\ \mathrm{m}^{-3}\ \mathrm{s}^4\ \mathrm{A}^2$), q_1 and q_2 are the charges on particles 1 and 2 in coulombs; r_{12} is the distance between the two particles in meters; and $\mathbf{1}_r$ is a vector of unit length that joins particle 1 to particle 2. **F**, is a force in newtons. In expressions that deal with magnetic phenomena, the permeability of free space, μ_o, will often appear ($= 4\pi \times 10^{-7}$ m kg s^{-2} A^{-2}). [NB: $(1/\mu_o\epsilon_o)^{1/2} = c$, the speed of light in a vacuum.]

4. As was pointed out in the preface, some aspect of the mathematics we need, as well as many complicated derivations are presented in appendices that appear at the end of the chapters to which they are relevant.
5. Scattered about the text there are italicized sections called *comments*. Their purpose is usually to connect the material being presented with other aspects of science that you may know about, or in some cases, to add some historical color.
6. A small number of questions will be found at the end of each chapter, the purpose of which is to encourage you to think about the material presented.

Instructors can find the answers to end-of-chapter problems at the book's companion website, located at www.oup.com/us/visualizingtheinvisible. Several of the figures in this book are black and white renderings of colored originals. The original, colored versions of those illustrations may be found in the website referenced above. The figures in question are: 7.2, 8.4, 10.3, 10.7, 13.6 and 13.7.

Visualizing the Invisible

PART ONE

Fundamentals

1

On the Scattering of Electromagnetic Radiation by Atoms and Molecules

The ultimate goal of the structural biologist is to determine the three-dimensional arrangement of atoms and molecules in living organisms. Some of that information is there for the taking once the chemical composition of a biological specimen has been determined, thanks to what chemists have learned about the covalent structures of biopolymers and other organic molecules over the years. To get beyond that, structural biologists must do experiments of their own, and conceptually, at least, the experiments they usually do are remarkably simple. Radiation of some sort is made to pass through a sample that may be as complex as a piece of living tissue or as simple as a protein solution, and the radiation that emerges is analyzed to determine how it was affected by its passage through the sample. Beams of photons (e.g., visible light and X-rays) are often used for this purpose, but beams of electrons and neutrons are also employed.

When a quantum of radiation passes through an object there are four possible outcomes: (1) it passes through unaltered, or (2) it gets absorbed, or (3) it gets scattered **inelastically**, which implies that both its trajectory and its energy change, or (4) it gets scattered **elastically**, which means that its trajectory changes, but not its energy. The probabilities of these outcomes depend on the structure of the sample, as well as on the kind of radiation used and its energy.

Absorption and inelastic scattering measurements provide information about the differences in energy of the states that are possible for some substance. Experiments aimed at delineating energy differences usually entail measurement of the dependence of the outcomes of radiation experiments on the energy of the radiation used, and the data are often displayed as spectra (singular: spectr*um*), which are plots that show how the magnitude of some effect varies with the energy of the radiation used to elicit it. Consequently, this sort of research is called *spectroscopy*. Although the spectroscopic properties of a substance are a consequence of its three-dimensional structure, it is usually hard to learn much about the three-dimensional structure of biological materials from their spectra because the spectroscopic properties of substances are often dominated by groups of atoms that account for only a small fraction of their total mass. (Nuclear magnetic resonance spectroscopy is the great exception in this regard.)

The interactions between radiation and substances that structural biologists care about most are those that result in elastic scattering, because the elastic scattering properties of materials depend on the locations of *all* of the atoms they contain. Furthermore, the relationship between the structure of a sample and the elastic scatter it produces is often so straightforward that magnified two- or three-dimensional images of its structure can be recovered from that scatter. Images can be obtained from scattered radiation either directly, using some form of microscopy, or indirectly, using any one of the several techniques that depend on the analysis of scattering data, for example, X-ray crystallography.

Even though absorption plays an important role in some of the experimental techniques discussed in this book, they are all elastic scattering techniques, first and foremost. Thus, it is appropriate to begin by reviewing the way atoms, molecules, and, by extension, entire assemblies of molecules scatter radiation elastically. We will start by considering the scattering of electromagnetic radiation—light and X-rays—because it is the type of radiation most often used by structural biologists.

1.1 WHAT IS ELECTROMAGNETIC RADIATION?

From the classical point of view, electromagnetic radiation has been well understood since the second half of the 19th century. Oscillating electric fields induce oscillating magnetic fields, and vice versa. Thus, if a charged particle is made to oscillate, the electric field associated with it will oscillate in sympathy, and that will induce an oscillating magnetic field in the surrounding volume that generates its own oscillating electric field, and so on. Electromagnetic disturbances propagate through a vacuum at the speed of light, c ($= 3 \times 10^8\ \mathrm{ms}^{-1}$).

A sinusoidally oscillating electric field, $\mathbf{E}(\mathbf{r},\ t)$, and an associated, sinusoidally oscillating magnetic induction, $\mathbf{B}(\mathbf{r},\ t)$, that obey the following pair of equations:

$$\begin{aligned} \mathbf{E}(\mathbf{r},t) &= E_o \sin\left[2\pi\left(\mathbf{r}\cdot\mathbf{1}_z\right)/\lambda - \omega t\right]\mathbf{1}_x \\ \mathbf{B}(\mathbf{r},t) &= (E_o/c)\sin\left[2\pi\left(\mathbf{r}\cdot\mathbf{1}_z\right)/\lambda - \omega t\right]\mathbf{1}_y \end{aligned} \tag{1.1}$$

would constitute a simple electromagnetic wave that causes radiant energy to flow in the $+z$ direction (see figure 1.1). E_o is the maximum value $|\mathbf{E}(\mathbf{r},t)|$ attains, and its units are volts/meter. $\mathbf{1}_x$, $\mathbf{1}_y$, and $\mathbf{1}_z$ are unit vectors in the $+x$, $+y$, and $+z$ directions, respectively. The vector from the origin of the coordinate system to the point in space where $\mathbf{E}$ and/or $\mathbf{B}$ are being observed is $\mathbf{r}$. The angular frequency of the oscillating electric field, ω, is measured in radians per second, and thus $\omega = 2\pi\upsilon$, where υ is the frequency of the oscillating field measured in hertz. λ is the wavelength of the radiation (in meters), and if the light is propagating through a vacuum,

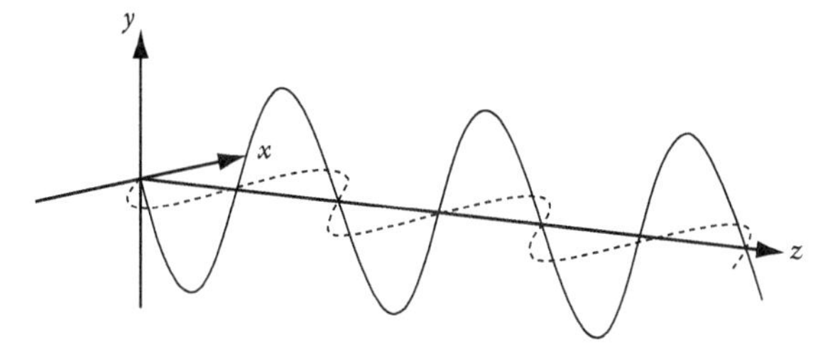

Figure 1.1 A sinusoidal electromagnetic wave that transports energy in the $+z$ direction. The electric field vector lies in the x-z plane (dotted profile). The magnetic field vector is in the y-z plane (solid profile).

$\lambda \upsilon = c$, where c is the speed of light. The quantity (E_o/c) has the dimensions of webers/meter^2 and is thus a magnetic induction. Because it is also the maximum value attained by $|\mathbf{B}(\mathbf{r}, t)|$, it is often designated "B_o". The flux of energy transported by this wave in the z direction, that is, its intensity, is $(1/2\mu_o c)E_o{}^2$. The dimensions of that flux are J m^{-2} s^{-1} or, equivalently, Wm^{-2}.

Comment: Visible light has wavelengths around 500 nm, which implies frequencies around 6×10^{14} Hz. X-rays, the other form of electromagnetic radiation routinely used by structural biologists, have wavelengths around 10^{-10} m, or 1 angstrom, and thus frequencies around 3×10^{18} Hz.

Comment: We will constantly be interested in the "dot products" of vectors. If **a** *is a vector that in some Cartesian coordinate system can be written $(a_x\mathbf{1}_x + a_y\mathbf{1}_y + a_z\mathbf{1}_z)$, and* **b** *is a vector that can be described in the same coordinate system as $(b_x\mathbf{1}_x + b_y\mathbf{1}_y + b_z\mathbf{1}_z)$, then $\mathbf{a} \cdot \mathbf{b} = a_xb_x + a_yb_y + a_zb_z$. NB:It is also true that $\mathbf{a} \cdot \mathbf{b} = |\mathbf{a}||\mathbf{b}| \cos\theta$, where θ is the angle between the two vectors. ($|\mathbf{a}|$ is the length, or magnitude, of vector* **a***.)*

It is clear from equations (1.1) that at $t = 0$, $\mathbf{E} = \mathbf{B} = 0$, and $d\mathbf{E}/dt$ and $d\mathbf{B}/dt$ are both negative at the origin (i.e., the point where $\mathbf{r} = 0$) of the axis system chosen. If some other point in space had been picked as the origin, or some other time had been taken as $t = 0$, a phase increment, which might be called Φ, would have had to be added to the argument of the sine functions to make oscillations measured in that new time/coordinate system correspond to the oscillations described in equations (1.1). For example, if the origin of the wave were moved from (0, 0, 0) to $(0, 0, \lambda/2)$, and the time $t = 0$ remained the same, the argument of the sine term in equation (1.1) would have to be altered from $[2\pi(\mathbf{r} \cdot \mathbf{1}_z)/\lambda - \omega t]$ to $[2\pi(\mathbf{r} \cdot \mathbf{1}_z)/\lambda - \omega t + \pi]$. Using elementary trigonometry, one can show that $\sin[2\pi(\mathbf{r} \cdot \mathbf{1}_z)/\lambda - \omega t + \pi] = -\sin[2\pi(\mathbf{r} \cdot \mathbf{1}_z)/\lambda - \omega t] = \sin[\omega t - 2\pi(\mathbf{r} \cdot \mathbf{1}_z)/\lambda]$. Thus, the wave specified by equations (1.1) can be described mathematically in many different, but equivalent ways, and not only is there is no way for an observer stationed at some fixed point in space to tell whether the time-dependent component of the argument of this particular wave is increasing with time—rotating counterclockwise—or decreasing—rotating clockwise—it is immaterial except in one regard. When solving problems, or carrying out derivations involving electromagnetic waves, it is *essential* to treat their time dependences consistently.

Four additional points need to be made about the electromagnetic wave described by equations (1.1).

1. It is perfectly **polarized**. This means that the orientations of its electric and magnetic field vectors are fixed. Not all electromagnetic waves behave this way. Some are circularly polarized, which means that instead of oscillating up and down in one direction, the electric vector of the light wave remains constant in magnitude but rotates in the (x, y) plane, tracing out a helical trajectory in space as the wave advances. Exponential notation provides a convenient way of describing circularly polarized light (see appendix 1.1). For example, we might write $\mathbf{E}(\mathbf{r}, t) = E_o \exp[i(2\pi(\mathbf{r} \cdot \mathbf{1}_z)/\lambda - \omega t)]$ for such a wave, taking the real part of this function to be the $\mathbf{1}_x$ component of its electric field, and its imaginary part to be the component of its electric field in the $\mathbf{1}_y$ direction. Interpreted this way, $E_o \exp(i[2\pi(\mathbf{r} \cdot \mathbf{1}_z)/\lambda - \omega t])$ describes a circularly polarized electromagnetic wave that transports energy in the $+z$ direction, and its polarization is rotating about the z-axis in a clockwise direction as z increases. Furthermore, a beam of plane-polarized light can be thought of as the vector sum of two beams of circularly polarized light rotating in opposite directions.

In the laboratory, one often works with **unpolarized** radiation. The (net) electric vector of unpolarized radiation varies randomly in orientation (and length) in the (x, y) plane over times that are short compared to experimental measurement times, and so, on average, it has no preferred direction. Unpolarized light can be thought of as the superposition of two electromagnetic waves that are polarized in directions at right angles to each other in the (x, y) plane and that differ slightly in frequency and vary in $t = 0$, so that sum of their two electric vectors varies erratically with time. The energy delivered by an unpolarized beam is the sum of the energies delivered by its two components taken separately. Unpolarized light is produced by radiation sources much larger than single atoms that are operating under conditions such that the phase and electric vector orientation of the radiation emitted by any point in the source are unrelated to those of the radiation emitted by any other. The filament of an incandescent lamp produces unpolarized radiation.

2. For the wave described by equations (1.1), the surfaces in space on which the value of the quantity $[2\pi(\mathbf{r} \cdot \mathbf{1}_z)/\lambda]$ is the same at any instant in time are planes, and hence, this particular electromagnetic wave is a **plane wave**. The surfaces of constant $[2\pi(\mathbf{r} \cdot \mathbf{1}_x)/\lambda]$ associated with electromagnetic radiation are called **wave fronts**, and they need not be planes. The wave fronts of the radiation produced by a pointlike source are spherical. Lasers can generate plane waves, but most of the radiation that experimentalists describe as "plane-polarized" is really a small portion of a spherical wave front gener-

ated by a point source that is far from the place where observations are being made. Whatever their shapes, wave fronts travel through a vaccum at the speed of light.

3. The wave described by equations (1.1) is **monochromatic**, which means that only a single frequency (and wavelength) is associated with the flow of radiant energy it describes. Perfectly monochromatic radiation is never encountered in practice. Often the radiation that experimentalists described as "monochromatic" is obtained by using a device like a grating or prism to select radiation having a narrow range of frequencies from a continuum that is generated by a **white** source. (The sun is the prototypical white light source.) Laser light is the most perfectly monochromatic light commonly available to the experimentalist, but even it is not perfectly so.

4. The radiation described by equation (1.1) is perfectly **coherent**, both spatially and temporally. "Perfectly coherent" means that if you were to determine the values of **E** (or **B**), and its time derivative for that wave at some point in space and some instant in time, you would be able to predict its value at all other locations in space for the rest of time with perfect certainty. Not only is real radiation never perfectly monochromatic, it is never perfectly coherent either. (Laser light is highly coherent both spatially and temporally, just as it is highly monochromatic.)

The degree of coherence of the radiation used for structural investigations is important because many of the imaging techniques used by biologists depend on the interference/diffraction effects that arise when substances scatter radiation coherently, and those effects can be observed *only* if the radiation used to elicit them is appropriately coherent to begin with.

1.2 ATOMS ARE ELECTRICALLY POLARIZED BY ELECTROMAGNETIC RADIATION

Electromagnetic radiation interacts strongly with atoms and molecules because they are composed of electrically charged particles. At the right frequencies, it can even cause changes in the electronic configurations of atoms and molecules that lead to the absorption (or emission) of electromagnetic energy. However, the interactions of primary concern here are those that occur at frequencies often far removed from the frequencies at which atoms and molecules absorb (or emit) electromagnetic energy efficiently because those interactions produce most of the elastic scattering on which structure determinations depend.

An atom is (usually) an electrically neutral assembly of negatively charged electrons surrounding a positive nucleus. Prior to its exposure to electromagnetic radiation, the atom will usually be in its ground state, and the center of gravity of its (time-averaged) electron distribution will coincide with the position of its nucleus. When an atom is exposed to a static electric field, its electrons and its nucleus experience forces that push them in opposite directions, and the center of gravity of the atom's electron distribution will no longer coincide with its nucleus. An atom that has been distorted this way is (electrically) **polarized**, and polarized objects have **dipole moments.** The dipole moment, $\boldsymbol{\mu}$, of an assembly consisting of a positive charge, $+\delta q$, and a negative charge, $-\delta q$, separated by a distance, **d**, is:

$$\boldsymbol{\mu} = \delta q \mathbf{d}. \quad (1.2)$$

By convention, $\boldsymbol{\mu}$ points toward the positive end of the dipole.

Experiment shows that if the strengths of the electric fields to which an atom is exposed are modest, the dipole moments induced in them will be proportional to field strength, which is to say that $\boldsymbol{\mu}$ will be related to the inducing field as follows:

$$\boldsymbol{\mu} = \alpha \varepsilon_o \mathbf{E}. \quad (1.3)$$

The proportionality constant in this equation, α, is the (static) **polarizability** of the atom in question, and its value depends on the identity of that atom. (Similar equations can be used to describe the response of molecules to external electric fields.) If the electric field that is inducing a dipole moment in some atom or molecule is changing with time, the dipole moment of that atom or molecule will vary in magnitude and direction in sympathy with the external electric field.

Equation (1.3) might be taken as implying that a single number can be used to characterize the electrical response of an atom or molecule to external electric fields, but this is not the case. Polarizabilities depend strongly on the frequency of the inducing field (see section 3.6). In addition, polarizabilities are not necessarily scalar quantities. The dipole moment induced in an asymmetric molecule will always have a large component in the direction of the inducing field, but it need not be parallel to it. If it is not, α will be a tensor, rather than a scalar. However, for spherically symmetric objects such as isolated atoms, life is simple: induced dipole moments are always parallel to inducing fields. So this is where our discussion will begin.

1.3 OSCILLATING DIPOLES EMIT ELECTROMAGNETIC RADIATION

The dipole moments induced in atoms by electromagnetic radiation oscillate in sympathy with that radiation because the electrons (and nuclei) they contain are constantly being accelerated by the ever-changing electric field of the incident radiation. Maxwell's equations show that electromagnetic radiation is generated whenever electrical charges are accelerated, and the equation that describes the electric field of the electromagnetic radiation produced by a sinusoidally oscillating electric dipole, $\mathbf{E}_s(R, \varphi, t)$, may be found in any textbook on electricity and magnetism. At distances from the oscillating dipole that are large compared to $|\mathbf{d}|$,

$$\mathbf{E}_s(R, \varphi, t) = -\left(1/4\pi c^2\right)\left(\omega^2 \alpha / R\right) \sin(\varphi)\, E_o \exp\left[-i\omega(t - R/c)\right] \mathbf{1}_\varphi. \quad (1.4)$$

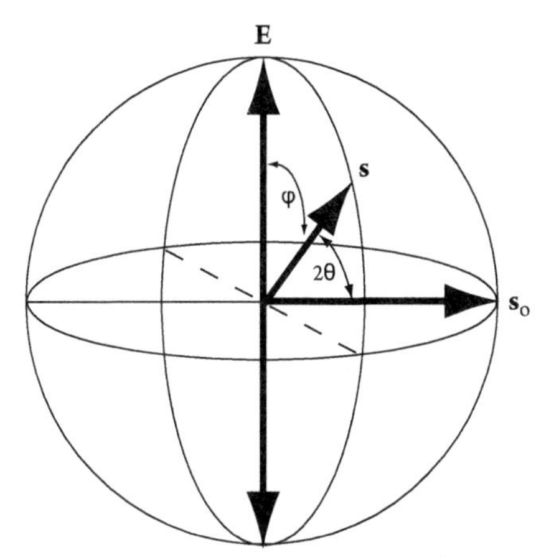

Figure 1.2 Definitions of the angles and vectors used to describe the scattering of electromagnetic radiation by single atoms.

Figure 1.2 shows how the angles and directions relevant to this equation are defined. **E** is the direction of the electric vector of the incident electromagnetic radiation. The origin of the coordinate system is the center of the radiating dipole, and the sphere shown has unit radius (e.g., a radius of 1 m). The unit vector that points toward the location where the scatter is to be measured is **s**, while the unit vector normal to the wave fronts of the incident radiation is $\mathbf{s}_0$, and φ is the angle between **E** and **s**. $\mathbf{1}_{\varphi}$ is a unit vector perpendicular to **s** that points toward the "south pole" of the sphere. It is parallel to the circle of longitude that connects the distal ends of **E** and **s**. R is the distance from the atom to the point where the scattered radiation is being observed. [Note that the time and space dependences of the radiation in question are written in exponential form instead of using sines and cosines. It makes no difference in the end. (See appendix 1.1.) Note also that $i\omega(t - R/c) = i(\omega t - 2\pi R/\lambda)$.]

Equation (1.4) has four properties worthy of comment.

1. Of the two angles specified in figure 1.2, only one, the polar angle, φ, affects the amplitude of the electric field of the scattered radiation, $|\mathbf{E}_s(R, \varphi, t)|$. Thus, for any given value of R, $|\mathbf{E}_s(R, \varphi, t)|$ is the same for all **s** that have the same polar angle.

2. Because the radiant energy scattered by an oscillating dipole is proportional to the time-averaged value of $\mathbf{E}_s{}^2$, the $(1/R)$ dependence of $\mathbf{E}_s(R, \varphi, t)$ ensures that the intensity of the scattered radiation will obey the inverse square law, as expected.

3. The radiation an atom emits because of its interactions with an incident beam of radiation is directed away from the atom radially in *all* directions. Thus, the energy in the incident beam must be less on the far side of the atom

than it is on the near side, and the energy lost from the beam must correspond to the energy scattered by the atom.

4. If the polarizability of the entity scattering electromagnetic radiation is a positive number, which it need not be (see section 3.8), the scattered field will be in phase with the inducing field, which means that in the forward direction (i.e., the $2\theta = 0$ direction), at all distances from the atom, $\mathbf{E}_s$ will point in the same direction as the electric field vector of the radiation that was not scattered by the atom. (NB: When $\varphi = 90°$, as it is when $\mathbf{s} = \mathbf{s}_o$, $\mathbf{1}_\varphi$ points down.)

Comment: As will be explained at greater length in section 3.6, displacements of electrons account for almost all of the dipole moment induced in atoms by external electric fields. Consequently, it is oscillations in the positions of electrons, not oscillations in the positions of nuclei that account for essentially all the electromagnetic radiation scattered by atoms.

Under most circumstances, the quantity of concern in scattering experiments is not $\mathbf{E}_s(R, \varphi, t)$, but rather it is $\mathbf{E}_{\mathrm{rel}}(R, \varphi, t)$, which is defined as follows:

$$\begin{aligned} \mathbf{E}_{\mathrm{rel}}(R,\varphi,t) &= \mathbf{E}_s(R,\varphi,t)/E_o \exp\left[-i\omega\left(t - r/c\right)\right] \\ &= -\left(1/4\pi c^2\right)\left(\omega^2\alpha/R\right)\sin\varphi\,\mathbf{1}_\varphi. \end{aligned} \tag{1.5}$$

By now, you may be wondering why has nothing been said about the effect of oscillating magnetic fields on atoms. The reason magnetic effects have been ignored is that they contribute little to the scattering of concern here. Magnetic fields induce circular motions in charged particles, and time-varying circular motions usually do not engender the time-varying electric dipole moments that would emit electromagnetic radiation. The only circumstances under which the magnetic fields associated with electromagnetic radiation make a difference to chemists and biologists are: (1) when light interacts with chiral molecules, and (2) when the magnetic properties of biological materials are being investigated directly. Magnetic effects account for the circular dichroism/optical rotatory dispersion spectra of chiral molecules, and both electron spin resonance and nuclear magnetic resonance experiments are specifically designed to exploit the interaction of atoms with the magnetic components of electromagnetic radiation.

1.4 THE ELECTRONS IN ATOMS AND MOLECULES SCATTER X-RAYS AS THOUGH THEY WERE UNBOUND

Equation (1.5) describes the way electromagnetic radiation is scattered by atoms and molecules when the wavelength of that radiation is large compared to their linear dimensions. Life would be simpler if the polarizability of a molecule was the sum of the polarizabilities of its constituent atoms under these circumstances, but it is not. Molecular polarizabilities have to be measured directly. However, if the frequency of the electromagnetic radiation being scattered is so high that the energy of its photons greatly exceeds the energy needed to ionize atoms and molecules, which is always true when wavelengths are comparable to atomic dimensions, the physics of molecular scattering becomes simpler. To a usefully good first approximation, all the electrons in an atom or molecule respond to very short wavelength electromagnetic radiation (i.e., to X-rays) the way they would if each was an isolated entity, unbound to any atom or molecule. For this reason, the electric field that corresponds to the X-rays scattered by an atom or molecule is the vector sum of the electric fields of the X-rays scattered by each of its electrons independently.

Comment: Atomic and molecular ionization energies are typically tens of electron volts (eV), the energy of a photon of visible (e.g., $\lambda = 500$ nm) light is about 2.5 eV, and the energy of a 0.1-nm wavelength X-ray photon is about 12,000 eV.

The equation that describes the elastic scattering of X-rays by a (single) free electron was first derived by J. J. Thomson, the discoverer of the electron:

$$\mathbf{E}_s(R,\varphi,t) = \left(q_e^2/4\pi\varepsilon_0 m_e c^2\right)(\sin\varphi/R)\,E_o \exp\left[-i\omega\,(t - R/c)\right]\mathbf{1}_\varphi. \qquad (1.6)$$

In this equation, q_e is the charge of the electron; m_e is its mass; and the incident X-rays are polarized as shown in figure 1.2. (NB: This scatter is 180° out of phase with respect to the inducing radiation.)

1.5 THE SCATTERING OF X-RAYS BY MOLECULES DEPENDS ON ATOMIC POSITIONS

When using equation (1.6) to compute the way atoms or molecules scatter X-rays, allowance must be made for the fact that the electrons in atoms and molecules are not all at the same location in space, and the distances between them can be substantial compared to the wavelength of X-rays. (NB: The wavelength of light is much larger than the linear dimensions of almost all biological macromolecules.) The (average) distribution of electrons in an atom or molecule can be described using an **electron density distribution** function, $\rho(\mathbf{r})$, the units of which are electrons per cubic meter. Some point within a molecule, say the nucleus of one of its atoms, is chosen arbitrarily as the origin of a molecular coordinate system, and the positions of all other points in the molecule relative to that origin are indicated by $\mathbf{r}$. $\rho(\mathbf{r})$ is the probability of finding an electron in a volume of differential size at $\mathbf{r}$. At distances that are far from the molecule compared to its linear dimensions, the electric field due to the scatter from the volume element at $\mathbf{r}$ will be:

$$\left(q_e^2/4\pi\varepsilon_0 m_e c^2\right)(\sin\varphi/R)\,E_o\ \exp\left[-i\omega(t - R/c)\right]\rho\,(\mathbf{r})\,dV\,\mathbf{1}_\varphi,$$

where dV is the differential element of volume.

Comment: If the electronic wave function of an atom or molecule is known, $\Psi(\mathbf{r})$, $\rho(\mathbf{r})$ can easily be computed because $\rho(\mathbf{r}) = [\Psi(\mathbf{r})]^2$.

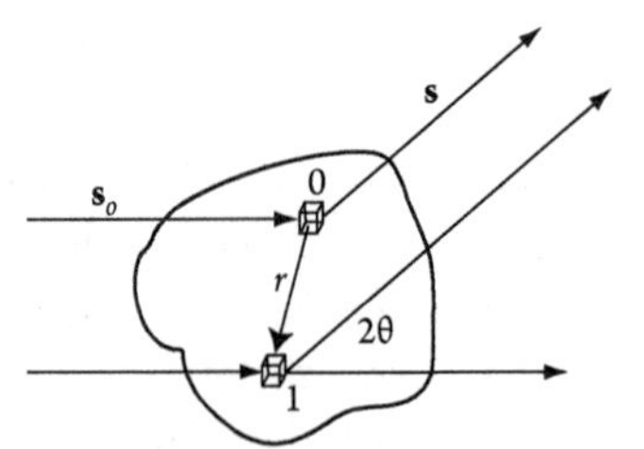

Figure 1.3 Scattering from a single molecule. The irregular outline represents the boundaries of a macromolecule. The cubes shown are (differentially) small volumes within that molecule.

Now suppose that a single molecule is irradiated by a perfectly polarized, perfectly planar wave of perfectly monochromatic X-rays (see figure 1.3). As before, we will designate the unit vector normal to the wave fronts associated with this beam $\mathbf{s}_o$, and the unit vector that points from the molecule to the detector is designated $\mathbf{s}$. Suppose also that the scattering produced by the molecule is measured using a detector that is so far from the molecule that all of that the radiation it scatters appears to come from a single point in space. When this is so, $\mathbf{s}$ is (effectively) the same vector for the scatter produced all the volume elements in the molecule. The scattering observed when these geometric conditions are fulfilled is called **Fraunhofer scattering** (or diffraction).

Because the entire molecule is illuminated, the electric fields of the waves scattered from all parts of the molecule will add vectorially at the detector. This means that the field at the detector, $\mathbf{E}_d$, will be the integral over all volume elements of $\mathbf{E}_i dV$, where $\mathbf{E}_i$ is the electric field generated by scattering from the ith volume element in the molecule. In order to add these field contributions correctly, we need to work out the differences in the distances traveled by the radiation scattered from different parts of the molecule as it progresses from the source to the detector.

As is evident in figure 1.3, the distance from the source to volume element 0 in the molecules and then on to the detector is not the same as the corresponding distance from the source to the detector via volume element 1. If the vector from volume 0, which we have picked arbitrarily as the origin of the molecule's internal coordinate system, to volume 1 is designated, $\mathbf{r}$, then the path involving volume 1 will be longer than the path involving volume 0 by an amount $[\mathbf{r} \cdot (\mathbf{s}_o - \mathbf{s})]$. This means that the scattered wave from volume 1 will have a phase that is more positive than that of the scattered wave from volume 0 by an amount equal to $\{+(\omega/c)[\mathbf{r} \cdot (\mathbf{s}_o - \mathbf{s})]\}$ or, alternatively, $\{+(2\pi/\lambda)[\mathbf{r} \cdot (\mathbf{s}_o - \mathbf{s})]\}$ because $c = (\omega/2\pi)\lambda$. Thus, at the detector, the total electric field of the scattered radiation originating in the two volume elements will be:

$$\mathbf{E}_{0+1} = \left(q_e^2/4\pi\varepsilon_o m_o c^2\right)(\sin\varphi/R)\,E_o\,\exp\left[-i\omega\,(t - R/c)\right]$$
$$\times\left[\rho\,(0) + \rho\,(\mathbf{r})\exp\left(i\,(2\pi/\lambda)\left[\,\mathbf{r}\cdot(\mathbf{s}_o - \mathbf{s})\right]\right)\right]dV\mathbf{1}_\varphi$$

[The effects that molecular-scale differences in distance have on the [sin(φ)/R] term in this equation are so small that they can be ignored (see question 4).]

When writing equations like the preceding one, it is often is often convenient to represent its ($\mathbf{s}_o - \mathbf{s}$) component using a new vector, **S**, which is called the **scattering vector**, $\mathbf{S} = (\mathbf{s} - \mathbf{s}_o)/\lambda$. When this substitution is made, and the equation generalized to include all the volume elements in the molecule, rather than just two, one finds that the electric field at the detector will be:

$$\mathbf{E}_d = \left(q_e^2/4\pi\varepsilon_o m_e c^2\right)(\sin\,\varphi/R)\,E_o \exp\,[-i\omega\,(t - R/c)]$$
$$\left[\int \rho\,(\mathbf{r}) \exp\,(-2\pi\mathbf{r}\cdot\mathbf{S})\,dV\right]\mathbf{1}_\varphi. \quad (1.7)$$

(The integration in equation (1.7) is over the entire volume of the molecule.) Note that:

$$|\mathbf{S}| = (2\,\sin\,\theta)\,/\lambda, \quad (1.8)$$

where θ is half the scattering angle, 2θ, which is the angle between **s** and $\mathbf{s}_o$ (see question 1).

Rather than deal with equation (1.7) in its entirety, it is often convenient to work with $\mathbf{E}_{\mathrm{rel}}$, which is obtained from equation (1.7) the same way equation (1.5) is obtained from equation (1.4), namely by dividing by $E_o\exp[-i\omega(t - R/c)]$:

$$\mathbf{E}_{\mathrm{rel}} = \left(q_e^2/4\pi\varepsilon_o m_e c^2\right)(\sin\,\varphi/R)\left[\int \rho\,(\mathbf{r}) \exp\,(-2\pi i\mathbf{r}\cdot\mathbf{S})\,dV\right]\mathbf{1}_\varphi. \quad (1.9)$$

Equation (1.9) accurately describes the way atoms/molecules scatter X-rays provided two conditions hold. First, as already stated, the detector must be very far from the sample compared to the linear dimensions of the objects of interest in the sample, a condition that is usually easily met. The second condition is subtler and can be more difficult to deal with experimentally. Implicit in the derivation of equation (1.9) is the assumption that the amount of radiation scattered by the sample/molecule is so small that the intensity of the beam does not change appreciably as it penetrates the sample. It is also assumed that the probability that radiation might be scattered more than once before it exits the specimen is so small that the effects of multiple scattering can be ignored. The two go hand in hand. If a sample is so thin that it will scatter only, say, 2% of the incident radiation once, then no more than 2% of that 2% will be scattered twice. If the sample is so thick that scatter of radiation that has already been scattered once must be taken into account, equation (1.9) must be replaced by a more complicated expression that contains additional terms, the quantitative importance of which increases with sample thickness. When the conditions under which a scattering experiment is conducted are such that only the lead term in that more comprehensive equation need be taken into account (i.e., when equation (1.9) is "good enough"), the **first Born approximation** is said to hold.

1.6 RADIATION DETECTORS MEASURE ENERGY, NOT FIELD STRENGTH

Most devices for detecting electromagnetic radiation, including the human eye, do not measure the strengths of the electric fields (or magnetic fields) associated with electromagnetic waves directly. It is not that electric/magnetic field strengths cannot be measured, but simply that it is hard to measure the instanteous value of quantities that change sign every 10^{-15} s or so. What most electromagnetic radiation detectors do instead is measure the amount of radiant energy that reaches them, and because it takes a finite length of time for any detector to respond, their outputs are invariably proportional to a time-averaged value of the incident energy, where the lengths of times over which that averaging takes place are very long compared to 10^{-15} s.

We know that the energy flux associated with a beam of electromagnetic radiation that has a maximum field strength of E_o is $(1/2\mu_o c)E_o{}^2$, and it should be pointed out here that if $\mathbf{E}$ is a complex quantity, as it is likely to be, then $\mathbf{E}_o{}^2 = \mathbf{E}^2 = \mathbf{E}\mathbf{E}^*$, where $\mathbf{E}^*$ is the complex conjugate of $\mathbf{E}$.

Comment: The complex conjugate of any quantity is obtained by replacing i with -i everywhere in that quantity. Hence, $(A + iB)^* = A - iB$.

Thus, devices that measure molecular scattering will yield numbers that are proportional to an energy flux, I_d, that obeys the following equation:

$$I_d(\mathbf{S}) = (1/2\mu_o c)\left(q_e^2 E_o \sin\varphi/4\pi\varepsilon_o m_e c^2 R\right)^2 \times \left[\iint \rho(\mathbf{r})\,\rho(\mathbf{r}')\exp\left(-2\pi i\left(\mathbf{r}-\mathbf{r}'\right)\cdot\mathbf{S}\right)dV\,dV'\right]. \tag{1.10}$$

For light scattering, as opposed to X-ray scattering, the equivalent expression is:

$$I_d(\mathbf{S}) = (1/2\mu_o c)(\omega^2 E_o \sin\varphi/4\pi c^2 R)^2 \left[\iint \alpha(\mathbf{r})\alpha\left(\mathbf{r}'\right)\exp\left(-2\pi i\left(\mathbf{r}-\mathbf{r}'\right)\cdot\mathbf{S}\right)dV\,dV'\right]. \tag{1.11}$$

In this case, the differential element of volume is a cube the edges of which are, say, a tenth the wavelength of light, which is a volume much larger than most macromolecules. The polarizabilities assigned to volume elements in this equation are thus bulk polarizabilities, not molecular polarizabilities.

1.7 IF THE RADIATION BEING SCATTERED IS UNPOLARIZED, THE POLARIZATION CORRECTION DEPENDS ONLY ON SCATTERING ANGLE

The only reason structural biologists measure $I_d(\mathbf{S})$ is to find out what the integral in square brackets in equation (1.10) (or (1.11)) looks like, and the two would be proportional to each other if equation (1.10) (and all of it relatives) did not include a $\sin^2(\varphi)$ contribution that reflects the effect that the polarization of the incident beam has on scattered intensities. Obviously, measured scattering data can be corrected for the effect that polarization has on intensities by dividing $I_d(\mathbf{S})$ by the appropriate value of $\sin^2(\varphi)$ for every value of **S**, and for that reason, $\sin^2(\varphi)$ is often called a **polarization correction.** (NB: The division called for will be problematic whenever φ is close to 0° or 180°.)

Scattering/diffraction experiments are sometimes done using radiation that is unpolarized, but contrary to what might be expected, data measured using unpolarized radiation must be corrected for polarization. In appendix 1.2, it is shown that the polarization correction for data obtained with unpolarized radiation is $(1/2)(1 + \cos^2 2\theta)$ rather than $\sin^2(\varphi)$. [NB: Happily, unlike $\sin^2(\varphi)$, $(1/2)(1 + \cos^2 2\theta)$ is never 0.]

1.8 THE COHERENCE LENGTH OF THE RADIATION USED IN SCATTERED EXPERIMENTS AFFECTS THE ACCURACY WITH WHICH I_d CAN BE MEASURED

Everyone who studies molecular structure by techniques that depend on the analysis of scattering profiles needs to obtain data that correspond as accurately as possible to:

$$\left[\iint \rho(\mathbf{r})\, \rho(\mathbf{r}') \exp\left(-2\pi i(\mathbf{r} - \mathbf{r}') \cdot \mathbf{S}\right) dV\, dV'\right].$$

We now know that scattering experiments will yield such data provided they have been designed so that: (1) the first Born approximation holds; (2) the scattering measured is Fraunhofer scatter; (3) the radiation used is monochromatic; and (4) the wave fronts of that radiation are planar. It is usually easy to arrange experiments so that the first two conditions are met, but the radiation sources available never produce radiation that is perfectly monochromatic and that has perfectly planar wave fronts. The deviations of the radiation used for a scattering experiment from perfection have consequences that must be attended to.

Suppose the radiation source used for some scattering experiment produces planar radiation that is perfectly polarized, but that there are two frequencies represented in the beam, not one. As always, the electric field vector of scattered radiation reaching the detector at any instant, **E**, will be the sum of the vectors that result from scattering at both wavelengths:

$$\mathbf{E} = \mathbf{E}_1 \exp\left[i(2\pi z/\lambda_1 - \omega_1 t)\right] + \mathbf{E}_2 \exp\left[i\left(2\pi z/\lambda_2 - \omega_2 t + \Phi\right)\right],$$

where both $\mathbf{E}_1$ and $\mathbf{E}_2$ can be obtained using equation (1.7), and a phase factor Φ is inserted to acknowledge that the two components of the radiation may not have had the same phase when they passed the origin of the axis system at $t = 0$. The intensity of the energy falling on the detector, I, will be proportional to EE^*, and thus:

$$I(t) = \mathbf{E}_1^2 + \mathbf{E}_2^2 + 2|E_1||E_2| \cos\left[2\pi z(1/\lambda_1 - 1/\lambda_2) - (\omega_1 - \omega_2)t + \Phi\right].$$

What the detector will measure is not the instantaneous value of $I(t)$, but rather its value averaged over some period of time, T. Clearly if $T >> 2\pi/(\omega_1 - \omega_2)$, the contribution made by the cosine part of $I(t)$ will be very small compared to contribution made by its $\left(\mathbf{E_1}^2 + \mathbf{E_2}^2\right)$ part, and the energy the detector registers will be the sum of what it would have registered if the intensities of the two rays had been measured separately. In other words, it will correspond to $\left(\mathbf{E_1}^2 + \mathbf{E_2}^2\right)$, rather than to $(\mathbf{E_1} + \mathbf{E_2})^2$.

As we saw earlier, X-rays have frequencies of the order of 3×10^{18} Hz. Thus, if the detector used for some experiment makes measurements every 10^{-6} s, and the frequencies of the two components of the beam differ by more than a few parts in 10^{12}, the outcome will be the one just described. The energies contributed by scattering at the two wavelengths will add.

A corollary of these arguments is that the output of any scattering experiment that is done with polychromatic radiation will be the appropriately weighted sum of the outputs that would have been obtained if series of experiments had been done using perfectly monochromtic radiation at frequencies spanning the range of concern and the data summed. If the spatial distributions of the energy scattered by molecules were independent of the wavelength of the radiation being scattered, it would make little practical difference whether the radiation generating the scattering was monochromatic or not, but this is not the case. As we already know, $|\mathbf{S}| = 2 \sin \theta/\lambda$, and, therefore, if the orientation of some molecule is held constant and its $I_d(\mathbf{S})$ explored as a function of $|\mathbf{S}|$ in any given direction, it will be found that $I_d(\lambda_1, \mathbf{S})$ corresponds to $I_d(\lambda_2, \mathbf{S})$ when $\sin \theta_2/\sin \theta_1 = \lambda_2/\lambda_1$. Thus, as judged by scattering angle, molecular scattering profiles expand as wavelengths increase, but they do so in a nonlinear manner. Hence, the scattering profile measured for some molecule using nonmonochromatic radiation is an appropriately weighted superposition of monochromatic scattering profiles that are *not* congruent.

The superposition effect just described makes scattering profiles measured with nonmonochromatic radiation smeared versions of the profiles one is trying to measure. Because $|\mathbf{S}| = 2 \sin \theta/\lambda$, $d|\mathbf{S}|/d\lambda = -|\mathbf{S}|/\lambda$. Thus, if $I_d(\mathbf{S})_{\text{measured}}$ is to correspond reasonably well to $I_d(\mathbf{S})_{\text{ideal}}$, the range of wavelengths in the beam used to measure $I_d(\mathbf{S})$, $\Delta\lambda$, must be such that $[(dI_d(\mathbf{S})/d|\mathbf{S}|)(-|\mathbf{S}|/\lambda)\Delta\lambda]$ is small for all $\mathbf{S}$ of interest. To put it another way, the separations between the peaks and valleys in $I_d(\mathbf{S})_{\text{ideal}}$ must be large compared to $(|\mathbf{S}|/\lambda)\Delta\lambda$ if they are to be measured well, and the larger the value of $|\mathbf{S}|$, the more monochromatic the radiation must be to ensure that this is so.

1.9 MEASUREMENT ACCURACY ALSO DEPENDS ON TRANSVERSE COHERENCE LENGTH

The concept of **transverse coherence** is also easy to understand. You should remember from first-year physics that when two parallel slits are illuminated by a plane wave of monochromatic light, an interference fringe is seen. In theory, the transverse coherence length of a beam of light can be measured experimentally by using it to illuminate two parallel slits and then varying the distance between slits.

The transverse coherence length of the radiation is the separation between the slits beyond which interference fringes are no longer seen. For perfectly monochromatic, perfectly planar waves (equation (1.1)), the transverse coherence length is infinite.

Transverse coherence is determined by the geometry of the apparatus used to conduct a scattering experiment. As figure 1.4 shows, if both the area of the radiation source and the area of the sample are finite, the directions of the normals to the wave fronts of the light originating at the source that pass through the sample will vary, and, hence, the direction that corresponds to $2\theta = 0$ will not be the same for those wave fronts either. If we define the on-axis direction as the $2\theta = 0$ direction for the whole system, the range of angles of the wave front normals that reach the sample will be $\pm(1/2r)(d_1 + d_2)$. (Remember that for small angles, tan $2\theta \approx 2\theta$.) Thus, features in the scattering profile that are closer together in the scattering profile than $\sim (1/r\lambda)(d_1 + d_2)$ will be poorly represented in $\mathrm{I_d}(\mathbf{S})_{\mathrm{measured}}$ due to a blurring/smearing effect similar to that which would be seen if the radiation used were not monochromatic. If the sample is a single molecule, $d_2 << d_1$, and $(d_1/r\lambda)$ becomes the criterion that determines how close together peaks and valleys in $\mathrm{I_d}(\mathbf{S})$ can be and not be obscured in $\mathrm{I_d}(\mathbf{S})_{\mathrm{measured}}$ due to blurring. $(r\lambda/d_1)$ has the dimensions of distance, and to within some arbitrary constant that is determined by the details of the definition used for transverse coherence length, that distance is the transverse coherence length of a given experimental setup.

When the two slits illuminated by some beam of radiation are further apart than the transverse coherence length of the apparatus being used to examine their scatter, the interference patterns generated by the different rays of light passing through the sample add in the $\left(E_1{}^2 + E_2{}^2\right)$ sense in such a way that peaks of one pattern superimpose on the troughs of another, which eliminates the intensity ripple that is the hallmark of interference. It should also be evident that longitudinal coherence length effects have the same result; they too can cause an addition of (ideal) slit interference patterns that make peaks superimpose on troughs. What distinguishes

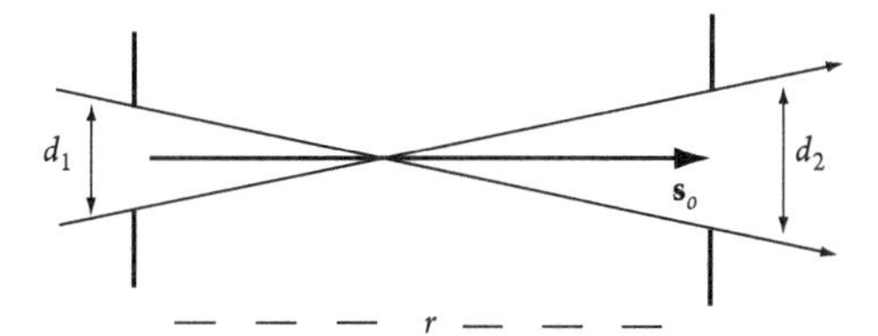

Figure 1.4 The geometry that determines transverse coherence widths. $\mathbf{s}_o$ is a unit vector that points down the axis of the instrument being used to measure scattering. The gap at the left of the diagram represents the physical size of the source of the radiation that is illuminating the sample, the physical dimensions of which are indicated by the gap on the right side of the diagram. The distance from the source to the sample is r.

these two types of coherence effects is that coherence length effects get worse as $|\mathbf{S}|$ increases but transverse coherence effects do not.

Finally, it is important to note that in this discussion of transverse coherence length, we have assumed that the size of the entry aperture of the device being used to measure the radiation scattered is such that its angular width measured in units of $|\mathbf{S}|$ is small compared with the distances between whatever distinctive features there may be in the scattering profile of interest on the face of the detector. If this is not so, then the effective transverse coherence length of the experimental apparatus will be less than that specified by $\pm(1/2r)(d_1 + d_2)$.

PROBLEMS

1. Show that $|\mathbf{S}| = 2 \sin \theta/\lambda$.
2. Nuclear magnetic resonance experiments are commonly done using electromagnetic radiation that has a frequency of 500 MHz. What is the wavelength of this radiation?
3. Consider figure 1.3. Suppose that the angle between $\mathbf{s}_o$ and $\mathbf{r}$ is 210°, and the angle between $\mathbf{s}$ and $\mathbf{r}$ is 30°. Suppose also that $|\mathbf{r}|$ is 15 Å. What will the value of $(\mathbf{s} - \mathbf{s}_o) \cdot \mathbf{r}$ be?
4. Assume that the distance from the reference volume in figure 1.3 to the place where the radiation scattered from it is to be measured is 0.2 m. Using the result obtained from problem 3, show that the impact of the difference in the distance of that volume from the detector and the second volume from the detector on the $(1/R)$ component of equation (1.9) will be so small that it can be ignored.
5. In radians, what does an angle of 1° correspond to?
6. Suppose that the structure of an object that has a maximum linear dimension of 100 Å is to be studied using X-rays that have a wavelength of 1 Å. If the distance between the X-ray source to be used and the sample is 1 m, how large can the diameter of the source be and still deliver a beam to the sample that has a transverse coherence adequate for this investigation?
7. Using series expansions, demonstrate that $\exp(ix) = \cos(x) + i \sin(x)$.

APPENDIX 1.1 EXPONENTIAL NOTATION, COMPLEX NUMBERS, AND ARGAND DIAGRAMS

Sines and cosines appear naturally in most of the equations used to describe scattering processes, but they are inconvenient to use when algebra has to be done with them. The mathematics can often be made simpler by replacing those trigonometric functions with their exponential equivalents, which is easy to do because $\exp(i\theta) = \cos\theta + i \sin\theta$. Thus:

$$\sin\theta = (1/2i)\left[\exp(i\theta) - \exp(-i\theta)\right]$$
$$\cos\theta = (1/2)\left[\exp(i\theta) + \exp(-i\theta)\right].$$

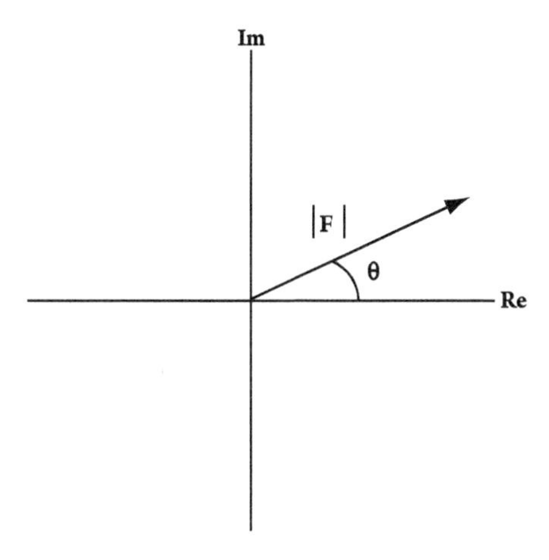

Figure 1.1.1 An Argand diagram.

[NB: $i = (-1)^{1/2}$.]

It is important to note that functions of the form $A\exp(i[2\pi(\mathbf{r}\cdot\mathbf{1}_z)/\lambda - \omega t])$ will satisfy any differential equation that (also) has solutions of the form $A\cos[2\pi(\mathbf{r}\cdot\mathbf{1}_z)/\lambda - \omega t]$ or $A\sin[2\pi(\mathbf{r}\cdot\mathbf{1}_z)/\lambda - \omega t]$ because linear combinations of functions that individually satisfy differential equations will also satisfy those same equations, and Maxwell's equations, which are the source of all expressions that have to do with electromagnetic radiation, are all differential equations. Thus, we could write the **E** part of equation (1.1) as $\mathrm{Re}[E_o\exp(i[2\pi(\mathbf{r}\cdot\mathbf{1}_z)/\lambda - \omega t])]\mathbf{1}_y$, where Re means "the real part of".

In addition, it is often helpful to represent complex quantities, such as $A + iB$, graphically using an Argand diagram as in figure 1.1.1. On an Argand diagram, a complex number is a point, the abscissa of which is A, and ordinate of which is B. This method for depicting complex numbers immediately suggests another way to express such numbers. If the number depicted were thought of as the point at the end of the arrow—the vector—in the diagram shown, then $|\mathbf{F}|$, the length of that vector would be $(A^2 + B^2)^{1/2}$, θ will be $\tan^{-1}(B/A)$, and $(A + iB)$ could be written as $|\mathbf{F}|\exp(i\theta)$. We will often write complex quantities that way, and we note here that the quantity θ in expressions of this sort is often referred to as a "phase".

APPENDIX 1.2 THE POLARIZATION CORRECTION FOR UNPOLARIZED RADIATION

Figure 1.2.1 shows the coordinate system we will use to analyze the polarization correction problem. In equation (1.4), φ is the angle between the direction of polarization of the inducing **E** field, and the direction from the middle of the induced dipole to the point of observation, **s**. The unit vector $\mathbf{s}_o$, which represents the direction of the incoming beam of radiation points in the $+z$ direction, and the y- and z-axes have been set so that **s** lies in the (x, z) plane. In this axis system, as in any other,

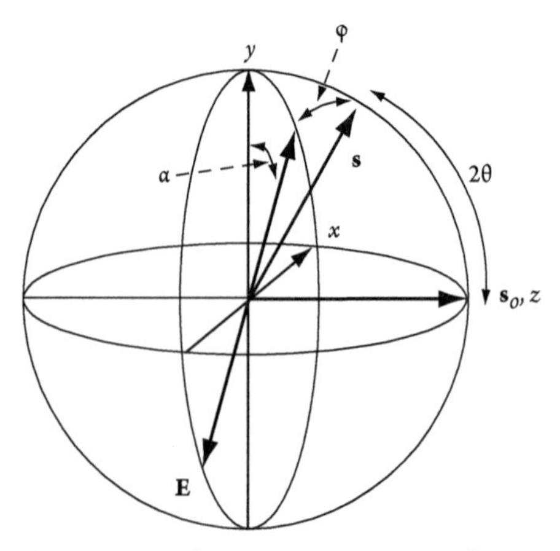

Figure 1.2.1 Geometric parameters relevant to the determination of the polarization correction.

$|\mathbf{1}_E \times \mathbf{s}| = \sin\,\varphi$, where $\mathbf{1}_E$ is a unit vector in the direction of polarization of the inducing **E** field. It is also true that in this particular axis system:

$$\mathbf{1}_E = \cos\alpha\,\mathbf{1}_x - \sin\alpha\,\mathbf{1}_y,$$

$$\mathbf{s} = \sin(2\theta)\,\mathbf{1}_x + \cos(2\theta)\,\mathbf{1}_z.$$

In these equations, $\mathbf{1}_x$, $\mathbf{1}_y$, and $\mathbf{1}_z$ are unit vectors in the $+x$, $+y$, and $+z$ directions, respectively, and α is the angle we will use to specify the orientation of $\mathbf{1}_E$ in the (x, y) plane.

The cross product $(\mathbf{1}_E \times \mathbf{s})$ can be written as a 3×3 determinant, and when that determinant is evaluated, one finds that:

$$\mathbf{1}_E \times \mathbf{S} = -\sin(\alpha)\cos(2\theta)\mathbf{1}_x + \cos(\alpha)\cos(2\theta)\,\mathbf{1}_y + \sin(\alpha)\sin(2\theta)\,\mathbf{1}_z.$$

Thus:

$$(\mathbf{1}_E \times \mathbf{s})^2 = \sin^2\varphi = [\sin(\alpha)\cos(2\theta)]^2 + [\cos(\alpha)\cos(2\theta)]^2 + [\sin(\alpha)\sin(2\theta)]^2,$$

and using simple trigonometric identities, it can be shown that this expression is equivalent to:

$$\sin^2\varphi = \cos^2(2\theta) + [\sin(\alpha)\sin(2\theta)]^2.$$

Because the average value of $\sin^2(\alpha)$ over the interval $\alpha = 0$ to $\alpha = 2\pi$ is $1/2$, it follows that the average value of $\sin^2\varphi$ over that same interval must be

$(\cos^2(2\theta) + [\sin^2(2\theta)/2])$, which with a little more trigonometric manipulation, can be rearranged to give:

$$\langle \sin^2 \varphi \rangle = (1/2)\left[1 + \cos^2(2\theta)\right].$$

Comment: If there are two vectors, $\mathbf{a}$ $(= a_x \mathbf{1}_x + a_y \mathbf{1}_y + a_z \mathbf{1}_z)$, *and* $\mathbf{b}$ $(= b_x \mathbf{1}_x + b_y \mathbf{1}_y + b_z \mathbf{1}_z)$, *then their cross product,* $\mathbf{a} \times \mathbf{b}$, *is:*

$$\mathbf{a} \times \mathbf{b} = \begin{vmatrix} \mathbf{1}_x & \mathbf{1}_y & \mathbf{1}_z \\ a_x & a_y & a_z \\ b_x & b_y & b_z \end{vmatrix} = \left(a_y b_z - a_z b_y\right) \mathbf{1}_x + (a_z b_x - z_x b_z)\mathbf{1}_y + (a_x b_y - a_y b_x)\mathbf{1}_z.$$

Furthermore:

$$|\mathbf{a} \times \mathbf{b}| = |\mathbf{a}|\,|\mathbf{b}| \sin\theta,$$

where θ *is the angle between the two vectors.*

2

Molecular Scattering and Fourier Transforms

In chapter 1, we derived an expression for the relative magnitude of the electric field of the radiation scattered when X-rays encounter molecules (equation (1.9)):

$$\mathbf{E}_{\text{rel}} = \left(q_e^2/4\pi\epsilon_o m_e c^2\right)(\sin\varphi/R)\left[\int \rho(\mathbf{r})\exp(-2\pi i\mathbf{r}\cdot\mathbf{S})dV\right]\mathbf{1}_{\varphi}.$$

$\mathbf{E}_{\text{rel}}$ is the product of a first term that depends on physical constants, the polarization of the incident radiation, and the location where the scattered field is to be observed and a second term that depends only on the structure of the scattering molecule and its orientation relative to the incident beam. The latter part of $\mathbf{E}_{\text{rel}}$, the part of the equation inside the square brackets, is the **structure factor** or **form factor** of the molecule in question. It is the same as $\mathbf{E}_{\text{rel}}$ (molecule)/$\mathbf{E}_{\text{rel}}$ (free electron) (see equation (1.6)), and it specifies the way the radiation scattered by that molecule varies with scattering angle and molecular orientation. As promised in chapter 1, the structure factor of a molecule depends on the location of every one of its electrons.

If the structure of a molecule is known, it is easy to compute its structure factor, but that is not the problem structural biologists usually confront. The data produced by many of their experiments provide information about structure factors, rather than structures; hence, the following question often arises. What can be learned about the structure of a molecule (i.e., about $\rho(\mathbf{r})$) if all that is known about the molecule is its structure factor, or something related to its structure factor? How can the information about $\rho(\mathbf{r})$ embedded in the structure factor of a molecule be recovered from it? The answer depends on circumstances, and one of the goals of this book is to spell out what can be done.

In this chapter, we will explore the mathematical relationship that exists between structures and structure factors. We will discover that the structure factor of a molecule is the Fourier transform of its scattering density distribution function. Therefore, structural biologists need to understand Fourier series and Fourier transforms.

2.1 F(S) IS A FUNCTION OF THREE ANGULAR VARIABLES

We will often find it convenient to write the structure factor of a molecule as $F(\mathbf{S})$, where, by definition:

$$F(\mathbf{S}) \equiv \int \rho(\mathbf{r}) \exp(-2\pi i \mathbf{r} \cdot \mathbf{S}) dV. \tag{2.1}$$

$F(\mathbf{S})$, which is a function of a three-dimensional function, $\rho(\mathbf{r})$, is itself three-dimensional. It is a function of three angles. When scattering data are collected from a molecule held in a fixed orientation in the incident beam, one is effectively "looking at" the molecule from a single point of view. To fully characterize *any* structure in three dimensions, it must be examined from *all* points of view, which means that scattering data must be collected when the molecule is in every possible orientation with respect to the incident beam of radiation, and at every orientation, its scattering must be measured at all possible scattering angles. If the instrument used to collect scattering data resembles the one shown in figure 2.1, one can easily understand why $F(\mathbf{S})$ is three-dimensional. It takes two angles to specify the orientation of an object (e.g., latitude and longitude), and the third angle is the scattering angle, 2θ, which is the rotational angle of the arm on which the detector is mounted in the instrument shown in figure 2.1.

Comment: You may find it odd that $F(\mathbf{S})$ is three-dimensional. As humans, we are used to perceiving the world in three dimensions, and the "instruments" that enable us to do so, our eyes, generate two images of whatever we are looking at

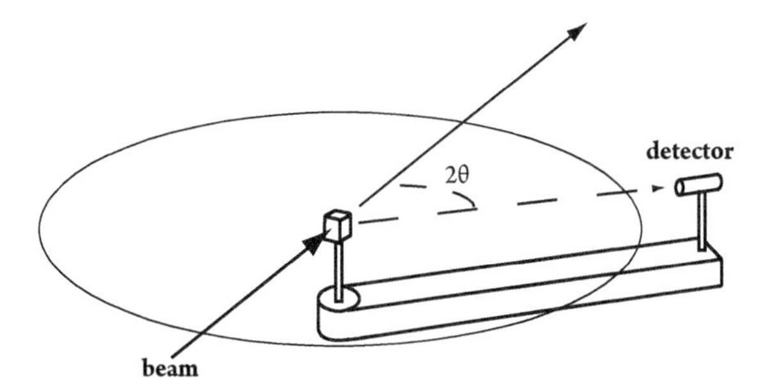

Figure 2.1 Schematic diagram of a simple device for measuring the radiation scattered by a sample. The arm connecting the sample and the detector rotates around an axis perpendicular to the incident beam of radiation that passes through the sample. The sample is mounted on a device (not shown) that allows it to be oriented in all possible ways with respect to the incident beam.

from slightly different points of view. By some means that are still not understood, our brains automatically convert the small differences between those images into a single, three-dimensional image of our surroundings. Because this is so, you might think that two views of any object would be enough to determine its structure in three dimensions. However, most of the light we see is light that has been scattered off the surfaces of the objects, and two views of a surface from different points of view can tell you a lot about its basically two-dimensional shape. Two views are not enough if you need to understand the arrangement of components inside the surface of a three-dimensional object.

2.2 FOURIER SERIES ARE A USEFUL WAY TO REPRESENT STRUCTURES

In the early 19th century, Joseph Fourier (1768–1830) invented the method for representing functions as sums of sine and cosine terms now called the **Fourier series**. He used them to solve the differential equations he encountered as he developed his theory of heat conduction, which is one of the monuments of the physics of his time. Fourier series are useful in many other contexts, including this one. For simplicity, we will develop the theory of Fourier series and Fourier transforms in one dimension, but the conclusions we will reach are easily extended into three.

Fourier's assertion is/was that any well-behaved function, g(x), that has a value different from 0 only over some finite interval in space, such as within the interval that starts, say, at $x = 0$ and ends at $x = d$, can be accurately represented as a sum of sine and cosine functions as follows:

$$\mathbf{g}(x) = a_0/2 + \sum_{1}^{\infty} \left[a_n \cos(2\pi nx/d) + b_n \sin(2\pi nx/d) \right]. \qquad (2.2)$$

What distinguishes the Fourier series representation of one function from that of any other function over the same interval are its coefficients, a_i and b_i, and so a prescription must be provided for evaluating them.

To obtain a_m, multiply both sides of equation (2.2) by $\cos(2\pi mx/d)$ and integrate from 0 to d. Equation (2.2) then becomes:

$$\int_0^d \mathrm{g}(x) \cos(2\pi mx/d)\, dx = \int_0^d (a_o/2) \cos(2\pi mx/d) dx$$

$$+ \sum_{n=1}^{\infty} \int_0^d \Big[a_n \cos(2\pi nx/d) \cos(2\pi mx/d)$$

$$+ b_n \sin(2\pi nx/d) \cos(2\pi mx/d) \Big] dx.$$

If $m = 0$, all the integrals inside the sum will be 0, and the a_o integral will be $(a_o/2)d$. Thus:

$$a_o = (2/d)\int_0^d g(x)dx.$$

For all values of $m > 0$, the sine-cosine part of the integral inside the sum will always be 0. (If you are unsure of this, consult any table of integrals, or a calculus textbook.) As far as the cosine–cosine terms are concerned, all terms where $n \neq m$ are 0 too, and the $m = n$ term will be $(a_m/2)d$. The b_n terms are found in a similar way. Thus:

$$a_n = (2/d)\int_0^d g(x)\cos(2\pi nx/d)\,dx,$$

and:

$$b_n = (2/d)\int_0^d g(x)\sin(2\pi nx/d)\,dx.$$

What is being evaluated in all of these integrals is the degree to which $g(x)$ resembles a particular sine or cosine function, and it is easy to show that the values of the coefficients produced by the equations just given are "best" values in the sense that they minimize the sum of the squares of the point-to-point differences between $g(x)$ and a particular sine or cosine function over the interval from 0 to d (see problem 2).

Fourier series have a curious property. The series just discussed has nonzero values outside the interval 0 to d, and in fact, $g(x) = g(x + d) = g(x + 2d) = g(x - d)$, etc. In short, it repeats in x, which the function it is supposed to represent, $g(x)$, does not. However, if all we care about is what happens between 0 and d, it makes no difference. Another way to put it is that if $g(x)$ *did* happen to repeat every d, the coefficients of the Fourier series that represents that repeating function would be proportional to those obtained for a single repetition of $g(x)$. This property of Fourier series will have important consequences when we come to consider the way crystals diffract radiation.

Fourier series can be also written using exponentials. This is done by replacing the sines and cosines in equation (2.2) with their exponential equivalents (see appendix 1.1), and then gathering terms. If we write $c_n = (1/2)(a_n - ib_n)$ and $c_{-n} = (1/2)(a_n + ib_n)(= c_n{}^*)$, then:

$$g(x) = \sum_{-\infty}^{+\infty} c_n \exp(2\pi inx/d), \text{ and : } c_n = (1/d)\int_0^d g(x)\exp(-2\pi inx/d)\,dx. \tag{2.3}$$

The initial response of the physicists and mathematicians of Fourier's day to his assertion that functions can be accurately represented using sine/cosine series was skeptical because most functions do not resemble $\sin(x)$ or $\cos(x)$ in any way. However, their concerns were addressed in the years that followed, and the bottom line is the following. Any function of limited spatial extent that describes an object or phenomenon that could possibly exist in nature can be represented as a Fourier series. From a practical, physical point of view, Fourier was right.

2.3 IN THE LIMIT OF $d = \infty$, THE FOURIER SERIES BECOMES THE FOURIER TRANSFORMATION

What happens as the interval over which $g(x)$ is nonzero increases in size? This issue can be explored by considering a function $G(n/d)$, which will be defined as:

$$G(n/d) \equiv \int_0^d g(x) \exp(-2\pi inx/d)dx.$$

From what has just been said, this equation implies that:

$$g(x) = (1/d) \sum_{-\infty}^{\infty} G(n/d) \exp(2\pi inx/d)$$

(see equation (2.3)). Now (n/d) is an interesting quantity that we have not have encountered before. It is a **spatial frequency**; it is a number of sinusoidal oscillations per unit distance. We will call it s. Furthermore, the absolute magnitude of the interval in spatial frequency between $s = n/d$ and $s = (n \pm 1)/d$, which we will call δs, is clearly $(1/d)$. So we can write:

$$g(x) = \sum_{-\infty}^{\infty} G(s) \exp(2\pi isx)\delta s,$$

and thus in the limit as $d \to \infty$, it must be true that:

$$g(x) = \int_{-\infty}^{\infty} G(s) \exp(2\pi isx)ds. \quad \textbf{(2.4)}$$

Furthermore, it is easy to show that if equation (2.4) is true, then:

$$G(s) = \int_{-\infty}^{\infty} g(x) \exp(-2\pi isx)dx. \quad \textbf{(2.5)}$$

Any pair of functions that bear the same relationship to each other as $G(s)$ and $g(x)$ are **Fourier mates**, and sometimes we will use $FT(f(x))$ to write functions such as $G(s)$, or use $FT^{-1}(G(s))$ in place of $g(x)$. The integral operations that interconvert $g(x)$ and $G(s)$ are called **Fourier** and **inverse-Fourier transformations.**

When $g(x)$ is Fourier-transformed, the limits of integration are $-\infty$ to $+\infty$, but if $g(x)$ happens to differ from 0 only over some finite region, the representation of $g(x)$ produced by inverse Fourier transformation of $G(s)(= FT(g(x)))$ will be 0 everywhere outside that region, as it should be. Thus, unlike Fourier series, Fourier transforms are not "accidentally" periodic.

Unhappily, the definition used for the Fourier transformation is not the same everywhere in the literature. In some documents, it is:

$$FT(g(x)) = \int g(x) \exp(2\pi isx)dx,$$

but in other documents, such as this one, it is:

$$FT(g(x)) = \int_{-\infty}^{\infty} g(x) \exp(-2\pi isx)dx.$$

In the context of scattering experiments, the sign of the argument for the exponential is determined by the way one chooses to describe the way the phases of electromagnetic waves evolve with time. If $E \propto \exp(-i\omega t)$, then the function inside the integral should be $\exp(-2\pi isx)dx$, as it is here. If phases rotate counterclockwise, then the function inside the integral should be $(2\pi isx)$. The two definitions are related. If:

$$G(s) \equiv \int g(x) \exp(-2\pi isx)dx,$$

then:

$$\int g(x) \exp(2\pi isx)dx = G(-s).$$

Physically, it makes no difference which definition is used for the Fourier transform. What *does* matter, however, is internal consistency. If the forward Fourier transformation is defined the way we are defining it here, then waves must advance with time as $\exp(-i\omega t)$, and:

$$g(x) = FT^{-1}(FT(g(x))) = \int FT(g(x)) \exp(2\pi isx)ds.$$

If the other convention is adopted, then the signs in all the exponent arguments must change also.

2.4 THE GREAT EXPERIMENT

It has doubtless already dawned on you that $F(\mathbf{S})$ (equation 2.1) is the three-dimensional Fourier transform of $\rho(\mathbf{r})$ and, hence, that $F(\mathbf{S})$ and $\rho(\mathbf{r})$ are Fourier mates. Thus, it must be true that:

$$\rho(\mathbf{r}) = \int F(\mathbf{S}) \exp(2\pi i \mathbf{r} \cdot \mathbf{S}) dV_s. \tag{2.6}$$

The integral in this equation runs over all of spatial frequency space, or, operationally, over the entire region of reciprocal space where $F(\mathbf{S})$ is measurably different from 0. Thus, when you measure scattering and/or diffraction patterns, you are exploring $F(\mathbf{S})$ experimentally because $I_d(\mathbf{S})$ is proportional to $|F(\mathbf{S})|^2$.

Now suppose you actually did measure $F(\mathbf{S})$ experimentally for some specimen, which means measuring $F(\mathbf{S})$ at all possible scattering angles and with the specimen in all possible orientations relative to the incident beam of radiation. When you were done, all you would have to do to recover $\rho(\mathbf{r})$, the three-dimensional scattering density distribution of that specimen, is to compute the inverse Fourier transform of the measured data. In essence, this is the Great Experiment in structural biology, and one way or another, it is the experiment many structural biologists spend their time doing.

The Great Experiment can be difficult. As noted in chapter 1, radiation detectors do not measure the amplitudes and phases of the electric fields of electromagnetic radiation, instead they measure energy intensities, which are proportional to $|F(\mathbf{S})|^2$ (i.e., the amplitude of $F(\mathbf{S})$ squared). In addition, $F(\mathbf{S})$ is generally a complex number, and thus $|F(\mathbf{S})|^2 = F(\mathbf{S})F^*(\mathbf{S})$. To understand why this makes a difference, write $F(\mathbf{S})$ either as $A + iB$, or, equivalently, as $|F(\mathbf{S})| \exp(i\varphi)$, where φ is the **phase** of $F(\mathbf{S})(= \tan^{-1}(B/A))$. Then, $F(\mathbf{S})F^*(\mathbf{S}) = A^2 + B^2 = |F(\mathbf{S})|^2$. Thus, intensity measurements tell you what $|F(\mathbf{S})|$ is, but they do not tell you how to divide $|F(\mathbf{S})|$ into its real (A) and imaginary (B) parts, or, equivalently, what its phase is. This loss of information, which is intrinsic to the measurement process, is the source of what crystallographers refer to as the **phase problem**. As we will see, it does not absolutely prevent you from using measured diffraction data to determine molecular structures, but it can make the process difficult.

The phase problem gets solved in two entirely different ways. In some cases, it can be bypassed altogether using instruments that do not measure $|F(\mathbf{S})|^2$ at all, but instead make the radiation scattered by specimens Fourier transform itself. Devices of this sort, which are called **microscopes**, produce magnified images of $\rho(\mathbf{r})^2$ (see section 10.2) rather than images of $[FT(\rho(\mathbf{r}))]^2$. The uglier alternative is to accept the loss of phase information that occurs when scattering/diffraction patterns are measured directly and proceed from there. As we shall see, the phase information lost when diffraction data are recorded can be recovered experimentally, or, in some cases, computationally. Once it has been, it is easy to compute $\rho(\mathbf{r})$, and if you are computing $\rho(\mathbf{r})$, you can represent it graphically at any level of magnification you like. Thus, the product can be a magnified image of some object that is not all that different from what a microscope operating with the same radiation might have produced.

Before examining either of these two approaches to solving the phase problem, it makes sense to spend some time looking carefully at the formal properties of the Fourier transform because once its properties are understood, it will be easy to the answer many of the practical questions that arise in imaging experiments. A modest investment in Fourier transform theory now will pay big dividends later.

2.5 THE SHIFT THEOREM LEADS TO A SIMPLE EXPRESSION FOR THE SCATTERING OF MOLECULES

Suppose $G(s)$ is the Fourier transform of some function $g(x)$, the origin of which is located at $x = 0$. What will the Fourier transform be of the function obtained when the origin of $g(x)$ is moved from $x = 0$ to $x = \Delta$? Call the translated version of $g(x)$ "$g'(x)$": $g'(x) = g(x - \Delta)$. (Clearly at x, $g'(x)$ has the same value as $g(x)$ at $(x - \Delta)$.) The Fourier transform of g' is:

$$\int_{-\infty}^{\infty} g'(x)\exp(-2\pi ixs)dx = \int_{-\infty}^{\infty} g(x - \Delta)\exp(-2\pi ixs)dx.$$

Making the substitutions: $u = x - \Delta$, and $du = dx$ one finds that:

$$\int_{-\infty}^{\infty} g(x - \Delta)\exp(-2\pi ixs)dx = \int_{-\infty}^{\infty} g(u)\exp\left[-2\pi i(u + \Delta)s\right]du$$

$$= \exp(-2\pi i\Delta s)\int_{-\infty}^{\infty} g(u)\exp(-2\pi ius)du$$

$$= \exp(-2\pi\,\Delta s)G(s).$$

Thus:

$$\mathrm{FT}(g(x - \Delta)) = \exp(-2\pi i\Delta s)\mathrm{FT}(g(x)). \tag{2.7}$$

This equation is the **shift theorem**, and it is easily extended into three dimensions:

$$\mathrm{FT}(\rho(\mathbf{r} - \boldsymbol{\Delta})) = \exp(-2\pi i\boldsymbol{\Delta}\cdot\mathbf{S})\mathrm{FT}(\rho(\mathbf{r})) = \exp(-2\pi i\boldsymbol{\Delta}\cdot\mathbf{S})F(\mathbf{S}).$$

This theorem imples that no information about the location of a molecule can be recovered from its scattering pattern, $I(\mathbf{S})$, beyond the fact that the molecule was in the radiation beam. The reason is that the information about its location is contained in the phases of the radiation it scatters, and there is no phase information in $I(\mathbf{S})$. This assertion is easy to prove. No matter what the value of $\boldsymbol{\Delta}$, in other words, no matter where the molecule is:

$$I(\mathbf{S}) = \exp(-2\pi i\boldsymbol{\Delta}\cdot\mathbf{S})F(\mathbf{S})\left[\exp(-2\pi i\boldsymbol{\Delta}\cdot\mathbf{S})\,F(\mathbf{S})\right]^*$$

$$= \exp(-2\pi i\boldsymbol{\Delta}\cdot\mathbf{S})\exp(2\pi i\boldsymbol{\Delta}\cdot\mathbf{S})F(\mathbf{S})F^*(\mathbf{S}) = F(\mathbf{S})F^*(\mathbf{S}) = I(\mathbf{S})$$

Another way of putting it is that the phase information in a transform is basically information about where things are located in space. It is important to note that the same is *not* true about a molecule's rotational orientation relative to the incident beam; scattering patterns are highly sensitive to rotational orientation because when molecules rotate, their transforms rotate with them.

The shift theorem provides a convenient way to specify molecular transforms that is particularly useful in the context of X-ray scattering or diffraction. A single, isolated atom has an electron density distribution, the dimensions of which are of the same order as the wavelength of X-rays (i.e., angstroms), and, therefore, in isolation, each atom has a nontrivial structure factor, $f(\mathbf{S})$, that one can either measure or calculate from first principles because:

$$f_{\text{atom}}(\mathbf{S}) = \int \Psi^2(\mathbf{r})\exp(-2\pi i\mathbf{r}\cdot\mathbf{S})dV_s, \tag{2.8}$$

where $\Psi(\mathbf{r})$ is the wave function that describes the electronic structure of the atom in its ground state (usually). [$\Psi(\mathbf{r})$ can be computed quantum mechanically, and $\Psi^2(\mathbf{r})$, is the same as $\rho(\mathbf{r})$.] Fortunately, even when they are covalently bonded to their neighbors in molecules, the time-averaged distributions of electrons around atoms are sufficiently close to spherical in symmetry, so that $f_{\text{atom}}(\mathbf{S})$ depends only weakly on orientation, in other words, $f_{\text{atom}}(\mathbf{S}) \approx f(|\mathbf{S}|)$. The $f(|\mathbf{S}|)$ functions of atoms are called **atomic structure factors**, or **atomic form factors**. The scattering factor of the *i*th atom in some molecule is usually written f_i, f_i being implicitly a function of $|\mathbf{S}|$, or if there is some reason to call attention to that dependence, one might write $f_i(|\mathbf{S}|)$ NB: the expression for atomic form factors given in equation 2.8 is the scatter an isolated atom would give divided by Thomson's expression for the scattering of a single electron (equation 1.6).

The scattering factors of the atoms of all the elements may be found in the *International Tables for Crystallography*. Figure 2.2 shows what the scattering factor for carbon looks like. At $\mathbf{S} = 0$, as figure 2.2 confirms, the scattering factor of any (neutral) atom is equal its atomic number: $f_{\text{atom}}(0) = Z_{\text{atom}}$ (see equation (2.8)).

Because the scattering factors of atoms are (largely) independent of chemical context, a usefully accurate expression for the structure factor of a molecule that contains N atoms can be obtained using the shift theorem:

$$F_{\text{mol}}(\mathbf{S}) = \sum_{1}^{N} f_i(|\mathbf{S}|)\exp(-2\pi i\mathbf{r}_i\cdot S), \tag{2.9}$$

where $\mathbf{r}_i$ specifies the position of the nucleus of *i*th atom in the molecule measured in any coordinate system for the molecule that is convenient, and f_i is its atomic scattering factor.

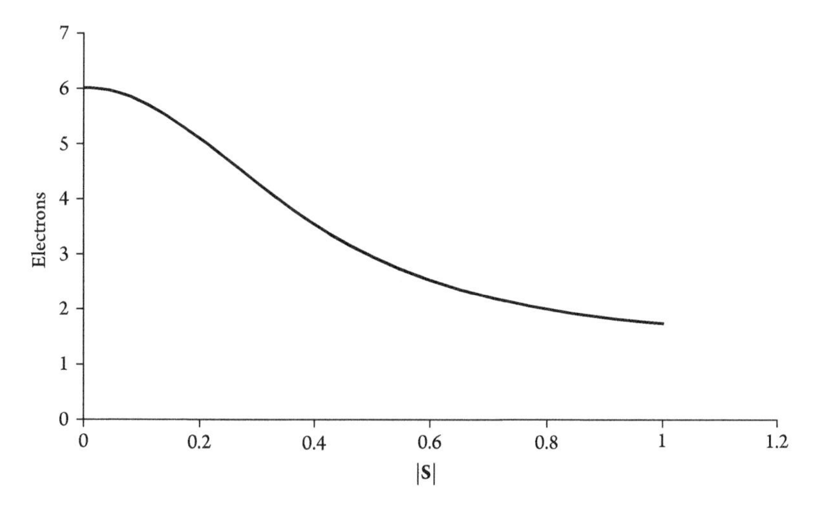

Figure 2.2 The scattering factor of carbon. $|\mathbf{S}|(= 2 \sin \theta/\lambda)$ is given in $Å^{-1}$.

2.6 THE SCALING THEOREM: BIG THINGS IN REAL SPACE ARE SMALL THINGS IN RECIPROCAL SPACE

In discussions of functions and their Fourier transforms, the phrases **real space** and **reciprocal space** often appear. Real space is the world in which real objects exist, and reciprocal space is the world of their Fourier transforms. Distances in the space occupied by $\rho(\mathbf{r})$ (i.e., real space) are measured in meters, whereas in the space of $\mathrm{FT}(\rho(\mathbf{r}))$, distances are measured in units of meters^{-1}. (As pointed out earlier, any quantity that has the dimensions of length^{-1} is a spatial frequency.) Hence, dimensions in the two spaces are reciprocally related.

There is another reason for describing the two spaces this way. Suppose a function exists in real space, $g(x)$, the transform of which is $G(s)$. How will $G(s)$ relate to the Fourier transform of a function, $g'(x)$, which is just like $g(x)$ except that $g'(x) = g(ax)$? Clearly $g'(x)$ is a stretched ($a < 1.0$) or shrunken ($a > 1.0$) version of $g(x)$. Again, one writes:

$$G'(s) = \int g'(x) \exp(-2\pi ixs)dx = \int g(ax) \exp(-2\pi ixs)dx,$$

and by making the substitution $u = ax$, one obtains:

$$G'(s) = (1/a) \int g(u) \exp\Big[-2\pi iu(s/a)\Big]du = (1/a)G(s/a). \qquad \mathbf{(2.10)}$$

The message conveyed by equation (2.10) is that if $g(x)$ is expanded, say by stretching it out by a factor of 2 ($a = 0.5$), $G'(s) = 2g(2s)$. In other words, the shape of the transform of the expanded function will have shrunk in reciprocal space by a factor of 2 compared to that of the unexpanded function, but its amplitude at corresponding points will have increased by a factor of 2. Thus, functions that extend over wide expanses in real space have transforms that extend over narrow ranges in reciprocal space, and vice versa. Big in one space means small in the other.

2.7 THE SQUARE WAVE AND THE DIRAC DELTA FUNCTION

At this point, it will be helpful to examine the Fourier transform of a specific, one-dimensional function, namely the square wave, $\mathrm{sq}(x)$. It is defined is as follows: $\mathrm{sq}(x) = b$ from $x = -a/2$ to $x = +a/2$, and $\mathrm{sq}(x) = 0$ everywhere else. The transform of $\mathrm{sq}(x)$, $\mathrm{Sq}(s)$ is easy to compute:

$$\mathrm{Sq}(s) = b \int_{-a/2}^{+a/2} \exp(-2\pi ixs)dx,$$

and when that integral is evaluated, we find:

$$\mathrm{Sq}(s) = b\ \sin(\pi as)/(\pi s). \quad \textbf{(2.11)}$$

[For the record, if $a = b = 1$, then $\mathrm{Sq}(s) = \sin(\pi s)/(\pi s)$, a function often referred to in the literature as **sinc**(s).] Figure 2.3 shows what $\mathrm{sq}(x)$ and $\mathrm{Sq}(s)$ look like. $\mathrm{Sq}(s)$ has a central maximum that is flanked on both sides with oscillating features that die away as $|s|$ increases.

Imagine a family of square waves, all centered on $x = 0$, that have the property that $(ab) = 1$. Clearly, as a decreases, b will increase, and as it does so, the square wave will look more and more like a tall spike centered on $x = 0$. At the same time, the central peak of $\mathrm{Sq}(s)(= \sin(\pi as)/(\pi as))$ will get broader and broader. In the limit, as a goes to 0, the width of the square wave will become indistinguishable from 0, and its height will increase to ∞, but its area will remain 1.0. Furthermore, in the limit, $\mathrm{Sq}(s)$ will be 1.0 for all values of s because $(\sin(0)/0) = 1.0$, as you learned in first-year calculus. In short, the central peak of $\mathrm{Sq}(s)$ will have become infinitely wide. The limiting member of the family of real space functions just discussed, that is, the square wave of area 1.0 that has 0 width, is called the **Dirac delta function,** $\delta(x)$.

Comment: The Dirac delta function can be arrived at in other ways. For example, we could equally well have considered the family of Gaussians of unit area that differ in width. The member of that family that has a width of 0 is also $\delta(x)$, and no matter how it is defined, FT($\delta(x)$)=1.0 everywhere in reciprocal space.

The Dirac delta function has some interesting properties:

$$\int_{-\infty}^{\infty} \delta(x)g(x)dx = g(0); \int_{-\infty}^{\infty} \delta(x-u)g(x)dx = g(u).$$

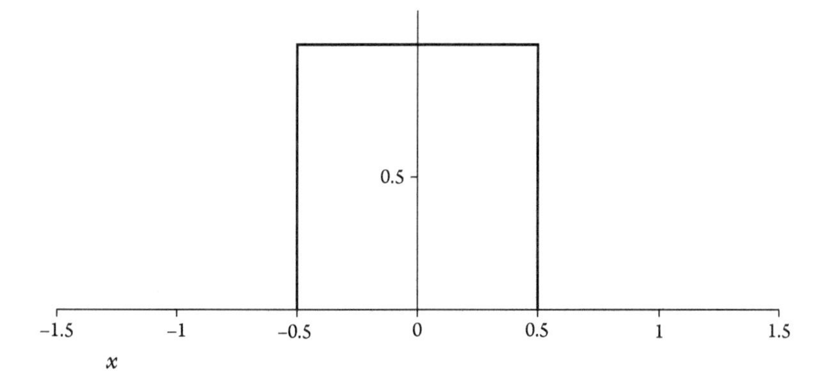

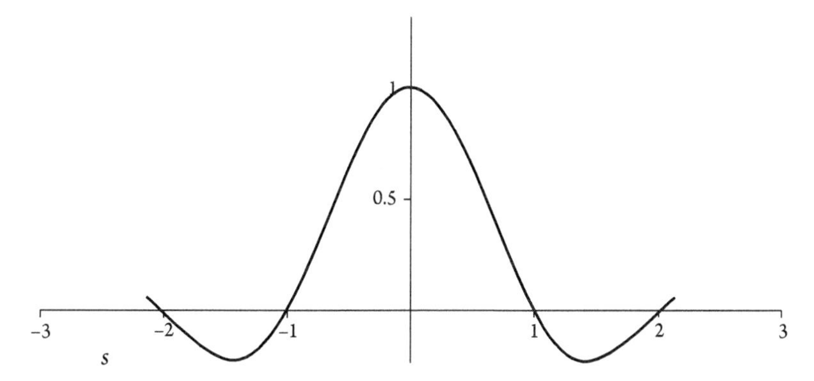

Figure 2.3 The square wave of width 1.0 and height 1.0 and its Fourier transform.

The reason these equations are valid is that because the peak of $\delta(x)$ is infinitesimally wide, the value of $g(x)$ does not change as it is traversed, and outside the peak, $\delta(x) = 0$. Therefore:

$$\int_{-\infty}^{\infty} \delta(x)g(x)dx = g(0) \int_{-\infty}^{\infty} \delta(x)dx = g(0).$$

Similarly:

$$g(x)\delta(x-a) = g(a)\delta(x-a).$$

Using delta functions and the shift theorem, it is easy to work out the Fourier transforms of $\cos(2\pi x/d)$ and $\sin(2\pi x/d)$. The shift theorem tells us that the inverse Fourier transform of $\delta(s-(1/d))$ is $[\exp(+2\pi ix/d)\mathrm{FT}^{-1}(\delta(s))]$, but $\mathrm{FT}^{-1}(\delta(s)) = 1.0$, therefore, $\mathrm{FT}^{-1}(\delta(s-1/d)) = \exp(2\pi ix/d)$. (Note the sign change in the shift theorem formula used here. It results when FT^{-1} is the operation of concern rather than FT.) It follows that:

$$\begin{aligned}\mathrm{FT}(\cos(2\pi x/d)) &= (1/2)(\delta(s-1/d)+\delta(s+1/d)),\\ \mathrm{FT}(\sin(2\pi x/d)) &= (1/2_i)(\delta(s-1/d)-\delta(s+1/d)).\end{aligned} \tag{2.12}$$

2.8 MULTIPLICATION IN REAL AND RECIPROCAL SPACE: THE CONVOLUTION THEOREM

It often happens that the function whose transform is required, namely $p(x)$, can be written as a product of two other functions, that is, $p(x) = g(x)h(x)$, the Fourier transforms of which are already known. A specific example would be the cosine wave that begins at $x = -a/2$ and ends at $x = +a/2$, which can be written as the product of a cosine function that extends from $x = -\infty$ to $x = +\infty$, or $\cos(2\pi x/d)$, and the square wave of amplitude 1.0 that extends from $x = -a/2$ and ends at $x = +a/2$. We know what the Fourier transformations of both these functions are separately. What is the Fourier transform of their product? In this instance, because the two functions are so simple, it is easy to compute the Fourier transform of their product directly, but if we take a more general approach, insights will emerge that can be taken advantage of when Fourier integrals are harder to evaluate directly.

Suppose there are two functions, $g(x)$ and $h(x)$ whose transforms are $G(s)$ and $H(s)$, respectively. The Fourier transform of the function of interest, $(g(x)h(x))$, $P(s)$, will be:

$$P(s) = \int_{-\infty}^{\infty} g(x)h(x)\exp(-2\pi ixs)\,dx.$$

Now replace $g(x)$ with $\mathrm{FT}^{-1}(G(s'))$. (Remember: It does not matter what the working variable in a definite integral is called if the range of integration is $-\infty$ to $+\infty$.) Once that is done, one finds that:

$$\begin{aligned}P(s) &= \int_{-\infty}^{\infty}\left[\int_{-\infty}^{\infty} G(s')\exp(2\pi is'x)\,ds'\right]h(x)\exp(-2\pi ixs)\,dx\\ &= \int_{-\infty}^{\infty} G(s')\,ds'\left[\int_{-\infty}^{\infty} h(x)\exp(-2\pi i(s-s'))\,dx\right]\end{aligned}$$

Now the integral in square brackets is clearly $H(s)$ evaluated at $(s - s')$. Therefore,

$$P(s) = \mathrm{FT}(g(x)h(x)) = \int_{-\infty}^{\infty} G(s')H(s-s')\,ds'. \tag{2.13a}$$

and, using the similar reasoning, it is easy to show that:

$$\mathrm{FT}^{-1}(\mathrm{G}(s)\mathrm{H}(s)) = \int_{-\infty}^{\infty} \mathrm{g}(x')\mathrm{h}(x - x')dx'. \qquad \textbf{(2.13b)}$$

The integral operations called for on the right sides of both equations (2.13a) and (2.13b) are called **convolutions**, and what we have just shown is that the Fourier (or inverse Fourier) transform of a product of functions in one space is the convolution of their individual Fourier transforms (or inverse Fourier transforms) in the other space. This statement is the **convolution theorem.**

Convolutions appear so often in mathematical physics that there is a shorthand notation for them. For example, equation (2.13a) might often be written:

$$\mathrm{P}(s) = \mathrm{FT}(\mathrm{g}(x)\mathrm{h}(x)) = \mathrm{G}(s) * \mathrm{H}(s).$$

Note that convolution both commutes, that is, $\mathrm{P}(s) = \mathrm{G}(s){*}\mathrm{H}(s) = \mathrm{H}(s){*}\mathrm{G}(s)$, and associates, that is, $\mathrm{F}(s){*}(\mathrm{A}(s) + \mathrm{B}(s)) = \mathrm{F}(s){*}\mathrm{A}(s) + \mathrm{F}(s){*}\mathrm{B}(s)$.

What do convolutions correspond to physically? Why should we care about them? A simple example may help explain why convolutions are useful. Consider the convolution of $\delta(s - 1/d)$ with $\sin(\pi as)/(\pi s)$. That integral can be written as:

$$\delta(s - 1/d) * \sin(\pi as)/\pi s = \int \delta(s - s' - 1/d)(\sin(\pi as')/\pi s')ds'.$$

If $s = 1/d$, then the integral will be nonzero only at $s' = 0$, and there its value will be $[\sin(0)/0] = 1.0$. As other values are entered for s, one discovers that the convolution under consideration is $\sin(\pi a(s - 1/d))/(\pi(s - 1/d))$, which is a $\sin(x)/x$ function centered at $s = 1/d$ instead of at $s = 0$. Thus, the convolution of any function, $\mathrm{g}(x)$, with $\delta(x - a)$ is $\mathrm{g}(x - a)$; it has the effect of moving the function's origin to a new location, an idea we will make incessant use of later on in our discussions of image formation.

Equipped with the convolution theorem, we can also deal with the Fourier transformation of a truncated cosine wave. Because the truncated cosine wave can be written as the product of a cosine wave of infinite extent with a square wave, the transform of the truncated cosine wave must be the convolution of the transforms of those two functions, both of which we already know. Thus:

$$\begin{aligned}&(1/2)\,(\delta(s - 1/d) + \delta(s + 1/d)) * \sin(\pi as)/\pi s = \\ &\qquad (1/2)\left[\sin\left(\pi a(s - 1/d)\right)/\pi(s - 1/d) + \sin(\pi a(s + 1/d))/\pi(s + 1/d)\right].\end{aligned}$$

Thus, the difference between the transform of the infinite cosine wave and the transform of a truncated cosine wave is easily described in words, even if it looks messy on paper (see figure 2.4). The transform of a cosine wave of wavelength d

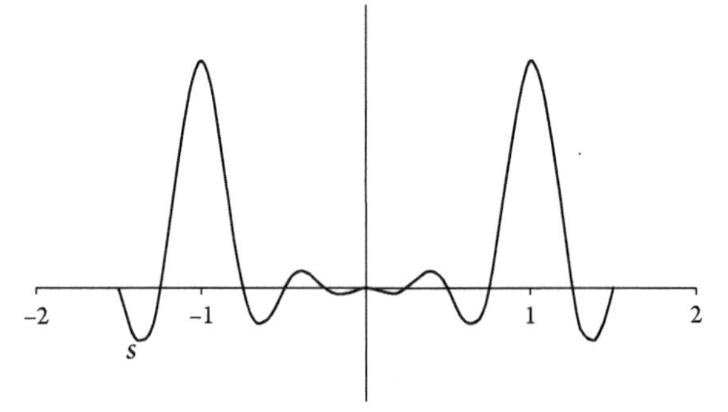

Figure 2.4 The Fourier transform of a cosine wave of finite length. The wavelength of the cosine wave of is 1 in arbitrary units, and it is truncated after 8 wavelengths. Only the region between $s = -1.5/\text{wavelength}$ and $s = +1.5/\text{wavelength}$ is shown, and the vertical scale is arbitrary.

that has infinite extent is the sum of two delta functions: one centered at $-(1/d)$ and the other at $+(1/d)$. The transform of the truncated cosine wave of wavelength d is also the sum of two functions centered at $-(1/d)$ and $+(1/d)$, but the functions are $\sin(x)/x$ functions, not delta functions.

Comment: The convolution theorem explains the uncertainty principle in quantum mechanics. As you will remember, moving particles correspond to de Broglie waves, the wavelengths of which equal Planck's constant divided by their momenta. The de Broglie wave of a particle in a stationary state that has some particular momentum, but that is completely delocalized in some one-dimensional space can thus be represented as a cosine function that extends through that space indefinitely. As we already know, the Fourier transform of such a wave has two spatial frequencies associated with it, $+p/h$ *and* $-p/h$*, and both are precisely defined because* $g(s)$ *for an infinite cosine wave is the sum of two Dirac delta functions. If a particle having that momentum is confined to a region that extends from* $-a/2$ *to* $+a/2$*, the Fourier transform of its wave function will be two* $\sin(\pi as)/(\pi as)$ *functions, one centered at* $-p/h$ *and the other at* $+p/h$*, as we just demonstrated. Thus, there will be a range of spatial frequencies associated with such a particle, not just two specific frequencies:* $+p/h$ *and* $-p/h$*. The width of that range can be conveniently characterized using the frequency difference between the center of either of these two* $\sin(\pi as)/(\pi as)$ *distributions and the point where it first becomes 0. That spatial frequency difference,* Δs*, is* $(1/a)$*, where* a *is the width of the region in which the particle is confined. If we call that width* Δx*, then* $\Delta s = 1/\Delta x$*. Because,* Δs *also equals* $\Delta p/h$*,* $\Delta x \Delta p = h$*, which is Heisenberg's uncertainty principle.*

In the versions of the Heisenberg equation that appear in text books, h *is multiplied by a constant, the value of which depends on the convention used to describe the width of the central maximum of a* $\sin(\pi as)/(\pi as)$ *function; it is physically inconsequential. Thus the uncertainty principle is not some abstruse principle of quantum mechanics that cannot be connected to the rest of physics, it is instead a straightforward consequence of the wavelike properties of moving particles. Uncertainty principles can be formulated for all kinds of waves.*

2.9 INSTRUMENT TRANSFER FUNCTIONS AND CONVOLUTIONS

Convolution has innumerable applications in technology. For example, it can be used to describe the performance of the amplifier in a sound reproduction system. The input received by the amplifier is a voltage that varies with time, $\mathrm{v}(t)$, which might be the output from a microphone or a tape deck, and what the designer of the amplifier will certainly have intended is that it produce an output signal, $\mathrm{V}(t)$, that has the property that $\mathrm{V}(t)/\mathrm{v}(t)$ be a constant larger than 1.0, which we might call G, the gain of the amplifier. The question for the high-fidelity buff, as well as the electrical engineer, is how similar $\mathrm{V}(t)$ actually is to $G\mathrm{v}(t)$, making due allowance for the fact that there is a slight time delay connected with the process that converts $\mathrm{v}(t)$ into $\mathrm{V}(t)$?

One way this question can be addressed experimentally is by measuring the response of the amplifier to strong voltage pulses of short duration. If the total electrical energy in a series of input pulses is maintained constant but their duration shortened, it will invariably be found that once their duration becomes short enough, the output of the amplifier will become independent of both the shape and duration of the pulse. As far as the amplifier is concerned, signals that short (or shorter) are Dirac delta functions.

It is equally certain that the response of the amplifier to short, sharp signals of this sort will *not* be a delta function. The reason is that the onset of input signals like these is very abrupt, and it takes time for the circuits in the amplifier to respond; they have "inertia". The cessation of such an input signal is also abrupt, but once the circuits of the amplifier have been excited by some input, it takes time for them to quiet down after the input drops to 0. The output produced by the amplifier in response to an impulsive input, $\mathrm{O}(t)$, is its **response function** or its **instrument transfer function.**

What will the output signal of that amplifier be if the input is $\mathrm{v}(t)$, instead of $\delta(t)$? One can think of $\mathrm{v}(t)$ as a succession of closely spaced delta functions each having an integrated area equal to $\mathrm{v}(t)dt$. Thus, the contribution made to the output signal measured at t by the input delta function that entered the device at a time t' prior to t, must be: $\mathrm{v}(t - t')\mathrm{O}(t')dt$. It follows that the total output observed at t must be:

$$V(t) = \int_0^T \mathrm{v}(t - t')\mathrm{O}(t')dt',$$

where T is the time interval over which $\mathrm{O}(t')$ differs significantly from 0. In short, the output of the amplifier will be the convolution of its input with its response function.

The implications of this equation for the performance of the amplifier are best understood by computing the Fourier transforms of v(t) and O(t). The reciprocal space of both transforms is frequency space because frequency, υ, equals $1/t$. If v(t) varies in time, its Fourier transform is likely to have appreciable amplitude over a wide range of frequencies. In the best of all possible worlds, the instrument transfer function of the amplifier, O(t), would be a delta function having an area equal to G. If it were, its transform would be G at all frequencies; the product of the two transforms would be proportional to GFT(v(t)); V(t) would be perfectly proportional to v(t); and the amplifier would have perfect fidelity. However, for any real amplifier, O(t) is bound to be a peaklike function of finite width in time; hence, its transform will die off as υ increases. This means that the relative amplitudes of high-frequency components of V(t) will be weaker than the relative amplitudes of the high-frequency components of v(t) because FT(V(t)) = FT(v(t))FT(O(t)).

The reason this happens is easy to understand. There is a limit to how fast an amplifier can respond to changes in input signal, and no amplifier can track oscillations in v(t) the frequencies of which are the reciprocal of the duration of O(t), or higher, $>1/T$. It will tend to smooth those oscillations, which is the equivalent of reducing their amplitude in reciprocal space. Hence, the amplitudes of high-frequency components will be reduced in V(t) compared to what they are in v(t), and beyond some frequency, components present in v(t) may be missing from V(t) altogether. By the way, it is not guaranteed that FT(O(t)) will be a real function, and

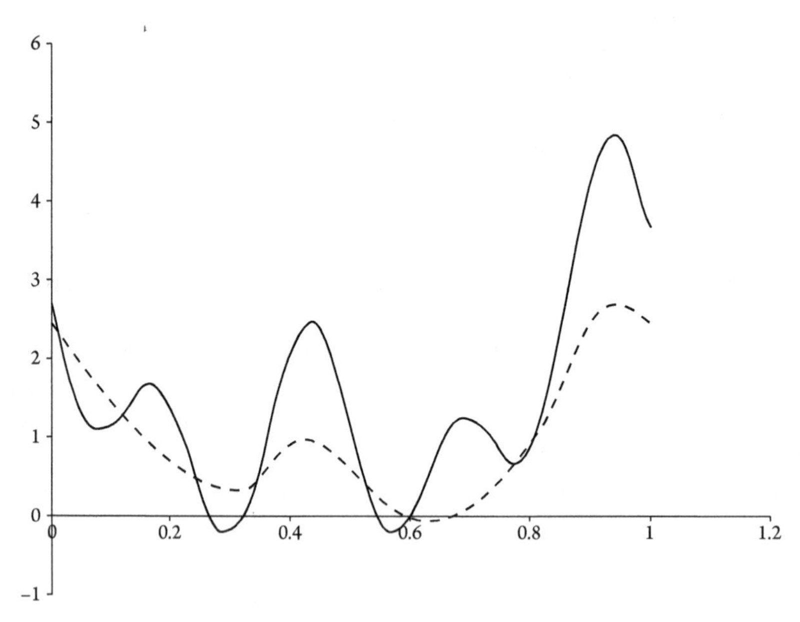

Figure 2.5 The effect of instrument transfer functions on instrument output. The input signal is: $\mathrm{v}(t) = 1 + \cos(2\pi\nu t + \varphi_1) + \cos(4\pi\nu t + \varphi_2) + \cos(8\pi\nu t + \varphi_4)$ (solid line; only one full repeat of the input is shown.) The units of time on the horizontal axis are multiples of $1/\nu$. The dotted profile is the output an instrument would produce in response to v(t) if its instrument transfer function were a square wave that has a (full) width of $0.222/\nu$.

if it is not, then the relative phases of the different components of v(t) will also be distorted by the amplifier, and that could be a problem too. Fortunately for the hi-fi buff, the (youthful) human ear cannot hear sound waves that have frequencies above roughly 20,000 Hz. (The human ear has an O(t) too!). This gives audio engineers some room to work with; O(t) does not have to be a perfect delta function in order for their amplifiers to satisfy their customers.

Figure 2.5 shows what would happen if v(t) were the sum of a constant and three sinusoidal waves, the frequencies of which are ν, 2ν, and 4ν, and the transfer function of the amplifier processing v(t) was a square wave that is almost as wide as a full period of the highest frequency wave, that is, $0.222/\nu$. The signal averaging caused by O(t) would reduce the amplitudes of the two lowest frequency waves only modestly, by factors of 0.92 and 0.71, but would almost eliminate the contribution of the 4ν wave to V(t). Its amplitude would be reduced by a factor of 0.12.

The performance of any kind of linear device that generates an output in response to an input can be described using convolutions, and one can learn a lot about the performance of all such instruments using the convolution theorem. Later on, we will use instrument transfer functions to describe the performance of microscopes.

2.10 THE AUTOCORRELATION THEOREM

As our discussion of scattering, diffraction, and imaging proceeds, we will occasionally be interested in the Fourier transform of the product of some function with its own complex conjugate, that is, $\mathrm{FT}(g(x)g^*(x))$. From the convolution theorem, we know that this transform must be the convolution of the transforms of $g(x)$ and $g^*(x)$, that is, $\int G(s')H(s-s')ds'$, where $G(s) = \mathrm{FT}(g(x))$, and $H(s) = \mathrm{FT}(g^*(x))$. The only issue remaining is the relationship between $\mathrm{FT}(g^*(x))$ and $G(s)$, which is easily sorted out because if $G(s) = \int g(x)\exp(-2\pi i x s)dx$, then:

$$G^*(s) = \int g^*(x)\exp(2\pi i x s)dx,$$

as we already know, and thus:

$$G^*(-s) = \int g^*(x)\exp(-2\pi i x s)dx = \mathrm{FT}(g^*(x)).$$

Therefore:

$$\mathrm{FT}(g(x)g^*(x)) = \int G(s')G^*(s'-s)ds' = A(s). \qquad \textbf{(2.14)}$$

Equation (2.14) is called the **autocorrelation theorem** because functions such as $A(s)$ are called **autocorrelation functions**. Starting with this result, it is easy to show that:

$$\mathrm{FT}(g(x)g^*(x)) = \int G^*(s')G(s'+s)ds' = A(s),$$

and that:

$$\mathrm{FT}(G(s)G^*(s)) = \int g(x')g(x'-x)dx'.$$

For convenience, functions such as $A(s)$ are often normalized by dividing them by $A(0)$ so that their values at $s = 0$ will be 1.0. Autocorrelation functions are important in many areas of the physical sciences, including crystallography, as we will see.

The only difference between the autocorrelation theorem equation (equation (2.14)), and the convolution theorem equation (equation (2.13a)) is a sign change in the argument for G^*, which may look inconsequential, but is not. $A(0)$, the value of $A(s)$ when $s = 0$, is simply the product of $G(s)$ with its own complex conjugate, integrated over all of space, which is necessarily going to be a positive number. When s is not 0, $G(s)$ is the overlap that exists between $G(s)$ and $G^*(s)$ when the origin of one of them has been shifted relative to the origin of the other by an amount equal to s. Thus, $A(s)$ describes the way the overlap of a function and its own complex conjugate varies with the offset between their origins.

Some autocorrelation functions are easy to understand qualitatively, and they are not too hard to compute. For example, it is obvious that the autocorrelation function of $\cos(2\pi x)$ must be 1.0 for all integer values of x, -1.0 when x is a half integer, and 0 half way between. Thus, it is not too hard to persuade oneself that the autocorrelation function of $\cos(2\pi x)$ might be $\cos(2\pi x)$, which indeed it is (see question 9).

2.11 RAYLEIGH'S THEOREM

It is but a short step from the autocorrelation theorem to Rayleigh's theorem. The autocorrelation theorem tells us that:

$$\mathrm{FT}(g(x)g^*(x)) = \int g(x)g^*(x)\exp(-2\pi ixs)dx = \int G(s')G^*(s'-s)ds'.$$

Thus, if $s = 0$, then, clearly:

$$\int g(x)g^*(x)dx = \int G(s')G^*(s')ds'. \qquad \textbf{(2.15)}$$

Equation (2.15) is Rayleigh's theorem. It can be thought of as the equivalent of the principle of the conservation of energy for image processing, and its derivation is so simple it looks like a slight of hand trick.

Ignoring proportionality constants, Rayleigh's theorem implies that for X-ray scattering/diffraction experiments:

$$\int \rho(\mathbf{r})^2 dV_r = \int \mathrm{I}(\mathbf{S}) dV_s,$$

where, as usual, $\rho(\mathbf{r})$ is the electron density distribution of an object, and the real space integral is over the entire volume occupied by the object. $\mathrm{I}(\mathbf{S})$ is the intensity of the radiation scattered by the object at $\mathbf{S}$ $(= \mathrm{F}(\mathbf{S})\mathrm{F}^*(\mathbf{S}))$, and the reciprocal space integral covers the volume in reciprocal space that contains the $\mathrm{I}(\mathbf{S})$ data that contributed to the image represented by $\rho(\mathbf{r})$.

The physical significance of this equation becomes clear when $\rho(\mathbf{r})$ is replaced by $(\langle\rho\rangle + \delta(\mathbf{r}))$, where $\langle\rho\rangle$ is the average value of $\rho(\mathbf{r})$ within the region of space occupied by the object it represents, the volume of which is V, and $\delta(\mathbf{r})$ is the difference between $\rho(\mathbf{r})$ and $\langle\rho\rangle$ at $\mathbf{r}$. Writing $\rho(\mathbf{r})$ that way, one discovers that:

$$\mathrm{F}(\mathbf{S}) = \int (\langle\rho\rangle + \delta(\mathbf{r})) \exp(-2\pi i \mathbf{r} \cdot \mathbf{S}) dV_r,$$

and thus:

$$\mathrm{I}(\mathbf{S}) = \iint [\langle\rho\rangle^2 + \delta(\mathbf{r})\langle\rho\rangle + \delta(\mathbf{r}')\langle\rho\rangle + \delta(\mathbf{r})\delta(\mathbf{r}')] \exp(-2\pi i(\mathbf{r} - \mathbf{r}') \cdot \mathbf{S}) dV_r dV_r'.$$

Because $\int \delta(\mathbf{r}) dV_r = 0$, at $\mathbf{S} = 0$:

$$\mathrm{I}(0) = \langle\rho\rangle^2 V^2.$$

Using that same reasoning, it is easy to show that:

$$\int \rho(\mathbf{r})^2 dV_r = \langle\rho\rangle^2 V + \int \delta^2(\mathbf{r}) dV_r.$$

Thus, Rayleigh's theorem is equivalent to the following equation:

$$\int \delta^2(\mathbf{r}) dV_r = \int \mathrm{I}(\mathbf{S}) dV_s - \mathrm{I}(0)/V.$$

The message conveyed by this equation is that $\mathrm{F}(0)$ [or, equally, $\mathrm{I}(0)$] is determined by the total amount of material (e.g., the total number of electrons) contained in the object responsible for that scattering, and that the distribution of that material in space determines all the rest of $\mathrm{F}(\mathbf{S})$ [or $\mathrm{I}(\mathbf{S})$]. Now $(1/V) \int \delta(\mathbf{r})^2 dV_r$ is the variance of the electron density distribution function of the object; in other words, the square of the difference between $\rho(\mathbf{r})$ and $\langle\rho\rangle$ averaged over the volume occupied by the object. Thus, the intensity of the scattering pattern integrated over the relevant volume of reciprocal space, less a correction for forward scatter, is the volume of the object times the variance of its electron density distribution.

Comment: $I(0)$ can never be measured directly because forward scatter cannot be distinguished from the portion of incident beam that was not scattered/diffracted by the specimen under investigation. Thus, in practice, Rayleigh's theorem can be stated in an even simpler form:

$$\int \delta^2(\mathbf{r})dV_r = \int I(\mathbf{S})dV_s,$$

where $I(\mathbf{S})$ refers now to the intensities that can be measured.

The implications of this theorem can be illustrated by considering the set of two-dimensional structures that can be generated by placing 16 cubes of uniform size and unit density in a square box that is 4 cubes long on each edge. No matter how the cubes are arranged, $I(0)$ will have the same value: 256 [$=$ (number of cubes)^2], and $I(0)$/area will be 16. However, if the cubes are placed in the box so that they form a layer 1 cube thick, then within the box, the variance of the electron density distribution will be 0, and all the scatter the object produces will be forward scatter. If the cubes are distributed so that the box contains 8 stacks of 2 cubes each, then the integral of the variance of the electron density distribution within the area will be 16, because $\delta(\mathbf{r})^2 = 1$ at each position in the box. The integral of the scattered intensity over all of reciprocal space will have increased to 32, and there will be some scatter in directions other than the forward direction. In the ultimate extreme, if the cubes are arranged to make a single stack 16 cubes high somewhere in the box, the integrated value of the variance will rise to 240, and the integrated scattering intensity will have its maximum possible value (256). Thus, Rayleigh's theorem says that the more Dirac delta–like the features in a structure, the bigger the average value of the density fluctuations will be in that structure, the larger and more numerous the Fourier coefficients it will take to describe its electron density distribution adequately, and the bigger $\int I(\mathbf{S})dV_S$ will be, everything else being equal.

Rayleigh's theorem also has something to say about amplifiers. The power associated with a Dirac delta function, the ideal $O(t)$ for an amplifier, is infinite because $FT(\delta(t))$ is G at all frequencies. The implication is that high-power amplifiers will cope with sudden changes in $v(t)$ better than low-power amplifiers will. Hence, the real reason why the high-fidelity buff should prefer amplifiers with high-power ratings is not that they can be used to shatter eardrums, which they certainly can, but rather that they can amplify the signals fed into them more accurately.

We will prove some additional theorems about Fourier transforms as we proceed, but we now have what we need to analyze what happens when radiation passes through condensed phase samples, which is the topic of Chapter 3.

PROBLEMS

1. Show that if equation (2.4) is true, then equation (2.5) must be true also.
2. In section 2.2 it is asserted that the coefficients of the Fourier series representation of a function using the formulas given there are optimal in a statistical sense in that they minimize the following integral, I:

$$I = \int_0^d (f(x) - a_n \cos(2\pi nx/d))^2 \, dx.$$

Prove that this is so by demonstrating that when a_n is set so that I has the smallest value it possibly can:

$$a_n = (2/d) \int_0^d f(x) \cos(2\pi nx/d) dx,$$

which is the formula given in the text for a_n. (It is easy to do the same for b_n.)

3. What is the (three-dimensional) Fourier transform of a cube the sides of which have a length of d? (Hint: Think convolution theorem.)
4. Consider the triangle-shaped function that obeys the following equations:

$$x < -1 : y = 0,$$
$$-1 \leq x \geq 0 : y = (1 + x),$$
$$0 \leq x \geq +1 : y = (1 - x),$$
$$x > +1 : y = 0.$$

What is the Fourier transform of that function? Hints: (1) A function that has the property that $f(x) = f(-x)$ is said to be "even". When the Fourier transforms of even functions are computed, what happens to $i \sin(x)$ part of the integral? (2) Use a table of integrals if you cannot remember how to evaluate the integral that is called for.
5. What is the relationship between the Fourier transform of some function and the Fourier transform of its derivative? It is not hard to work this out using the shift theorem and the primitive definition of the derivative, namely that in the limit as Δx goes to 0:

$$(f(x + \Delta x) - f(x))/\Delta x = df/dx.$$

The relationship you are being asked to derive here is the **derivative theorem.**
6. We know that $FT^{-1}(FT(g(x))) = g(x)$. What is $FT(FT(g(x)))$ equal to?

3

Scattering by Condensed Phases

A More Sophisticated Look at Atomic and Molecular Scattering

Before embarking on a discussion of the many ways information about the three-dimensional structure of biological materials is obtained from the radiation they scatter, it is important to develop an understanding of the physics of scattering that is more sophisticated than what chapters 1 and 2 provide. In both of those chapters, the objects responsible for the scattering considered were assumed to be microscopic objects, such as isolated macromolecules, that are somehow—magically—held in fixed, but controllable orientations in a beam of radiation. In reality, almost all biological scattering experiments are done using liquid or solid samples that are big enough to see (and handle). This simple fact has consequences that need to be examined not only because our exploration of them will deepen our understanding of scattering, but also because it will give us the language we need to describe the way matter scatters radiation that is not electromagnetic.

3.1 THE FORWARD SCATTER FROM MACROSCOPIC SAMPLES IS 90° OUT OF PHASE WITH RESPECT TO THE RADIATION THAT INDUCES IT

Almost no high angle scattering is observed when light passes through bulk samples of a transparent substance such as glass or water; they would not be transparent otherwise. Nevertheless, the molecules in those materials scatter radiation the same way isolated molecules do, and that scatter has measurable effects on the light that passes through such materials. For example, consider what happens when a thin slab of glass is placed a beam of visible light, the wave fronts of which are parallel to its surface (figure 3.1). For simplicity, we will assume that the light is monochromatic and plane-polarized, that the beam has a circular cross-sectional profile of modest

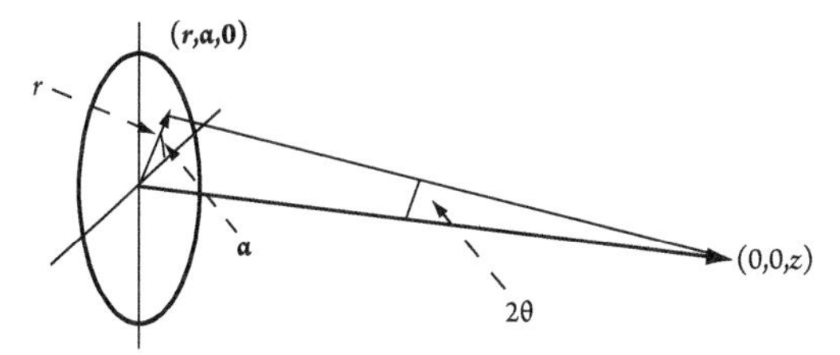

Figure 3.1 Geometric parameters used for computing the forward scatter from a sample of finite size. Note that relative to the on-axis position where the scattered radiation is to be observed, (0, 0, z), the diameter of the circle representing the illuminated area of the sample is grossly exaggerated to make it easier to see how positional variables are defined. The direction the incident radiation is traveling is indicated by the heavy arrow that connects (0, 0, 0) to (0, 0, z).

radius, r_{max}, that might be of the order of 1 mm, and that the transverse coherence length of the beam exceeds its diameter.

As always, the incident beam will induce a dipole moment in every volume element of the sample it illuminates that is proportional to the polarizabilities of those volume elements. Because these induced dipole moments will oscillate in sympathy with the incident radiation, some of the radiation passing through the sample will be scattered. Assuming the polarizability per unit volume is the same everywhere in the sample, which it must be if the sample is to be transparent, at distances from the sample that are very large compared to the diameter of the incident beam, (i.e., in the Fraunhofer limit), the distribution of scattered radiation will be proportional to $|\mathrm{FT}(\text{illuminated volume})|^2$, as we already know. In addition, given the reciprocal relationship that exists between sizes in real space and sizes in reciprocal space, if the illuminated volume is millimeters in size and the wavelength of the incident radiation is that of visible light, $|\mathrm{FT}(\text{illuminated volume})|^2$ will be vanishingly small at scattering angles greater than a few hundredths of a degree (see question 1). Thus, experimentally, the light scattered by the illuminated portion of the sample will be almost perfectly superimposed on the light that was not scattered as it passed through the sample.

Even though it is impossible to measure the radiation scattered by such a sample directly, the electric field vector of that scattered radiation, $\mathbf{E}_{sz}$, is an important quantity, and the equation that describes it is (see appendix 3.1):

$$\mathbf{E}_{sz} = (\omega/2c)\alpha' d\tau \exp(i\pi/2)[\exp(i\omega r_{max}^2/2cz) - 1]\mathbf{E}_o \exp\Big[-i\omega(t - z/c)\Big], \tag{3.1}$$

where α' is the polarizability per unit volume of the sample. This equation is valid provided the distance from the sample to the point of measurement is large enough so that equation (1.4) holds for every volume element in the slab (i.e., is not less than hundreds of thousands of angstroms), but it need not be hugely larger than the diameter of the beam, say 10 cm.

As is usually the case, it is not the absolute magnitude of the electric field generated by scattering in the glass slab that is of interest, but rather its magnitude relative to that of the electric field that would have been seen at (0, 0, z) if the glass slab had not been present, E_{rel}. The field that would have been seen there in the absence of the glass slab can be computed using Huygen's principle, which requires that each point inside the volume that the glass slab would have occupied be considered the origin of a spherical wave of radiation, the field strength of which is $[i(\omega/4\pi Rc)\mathbf{E}_o \exp(-i\omega(t - R/c))]$, where R is the distance from that point to (0, 0, z). (The factor of i in this expression emerges from the integrations that have to be done to evaluate the way Huygen's spherical wavelets propagate; it would be more trouble than it is worth to justify it here.) The field at (0, 0, z) is the vector sum of the fields produced by all the emitting points, and it can be evaluated using the approach outlined in appendix 3.1. The result is quoted in many textbooks of physics and optics:

$$\mathbf{E}(0,0,z) = [\exp(i\omega r_{\max}^2/2cz) - 1]\mathbf{E}_o\exp(-i\omega(t - z/c)). \tag{3.2}$$

Comment: Equation (3.2) implies that downstream of a circular aperture that is illuminated by plane parallel light, the light intensity on axis will be proportional to $[1 - \cos(\omega r_{max}{}^2/2cz)]^2$, *which means that for all* $z = (\omega r_{max}{}^2/4\pi nc)$, *where* n *is an integer, the light intensity will be dead 0, even though it is **not** zero for other values of* z. *This counterintuitive property of light beams that have passed through small circular apertures was confirmed experimentally in the 19th century.*

*When a similar analysis is done of the inverse situation, namely the propagation of plane-parallel electromagnetic radiation beyond an opaque disk, theory predicts that there will be a bright spot in the middle of the disk's shadow at **all** values of* z, *which is, if anything, even more counterintuitive. In the early 19th century, Poisson pointed out that the wave theory of light absolutely requires that this be so, and because bright spots of this kind had never been reported, he argued that the wave theory of light had to be wrong. Thus, Arago's experimental demonstration (ca. 1818) that the shadows of appropriately illuminated opaque disks actually do have bright spots at their centers silenced many of those who doubted the wave theory of light.*

E_{rel} is obtained by dividing equation (3.1) by equation (3.2), and when this operation is carried out, another surprising result emerges:

$$E_{rel} = (\omega/2c)\alpha'\exp(i\pi/2)d\tau. \tag{3.3}$$

The surprise is the $\exp(i\pi/2)$ term in the equation. Assuming that α' is positive (see the following comment), the presence of this term in this equation indicates that the phase of the forward scatter produced by a liquid or solid sample is 90° *behind* the phase of the radiation that induced it instead of being the same as the phase of that radiation, which is what happens when microscopic objects scatter light. This fact has important consequences, as we shall see.

Comment: In chapter 1, we arbitrarily decided to represent electromagnetic waves using equations in which amplitudes are proportional to $\exp(-i\omega t)$*, which means that on Argand diagrams, amplitude vectors rotate clockwise as time advances. The addition of a positive quantity to the argument of* $\exp(-i\omega t)$ *corresponds to a counterclockwise rotation; hence, the conclusion that the phase of* E_{rel} *lags that of the incident radiation. [NB: Had we decided to write expressions for electromagnetic waves in which amplitudes are proportional to* $\exp(+i\omega t)$*, the signs of the expressions that emerged would all have been different, but the arguments made here in would still have led us to conclude that the phase of* E_{rel} *lags that of the incident radiation.]*

The expression for E_{rel} describes what the scattering produced by an illuminated disk of glass having a radius of 1 mm will look like on axis at a distance of 10 cm from the disk. At that distance, the disk will certainly not look like a point source of light. For the record, scattering in the closer-than-infinitely-far regime is called **Fresnel scattering** to distinguish it from Fraunhofer scattering, which is the kind of scattering treated in chapters 1 and 2.

3.2 SCATTERING ALTERS THE PHASE OF *ALL* THE RADIATION THAT PASSES THROUGH A TRANSPARENT SAMPLE

Suppose the sample of interest is a liquid or solid that consists of identical molecules each having a polarizability of α. If there are N of them per unit volume, then from equation (3.3), the forward scatter of a thin slab of such a sample that has a thickness of $d\tau$ will be:

$$\mathbf{E}_{sz} = (\omega/2c)N\alpha\exp(i\pi/2)\mathbf{E}_o\, d\tau,$$

and to first order, the total electric field at $(0, 0, z)$, which is the sum of the unscattered incident field plus the scattered field will be:

$$\mathbf{E}_{\text{total}} = [1 + i(\omega/2c)N\alpha d\tau]\mathbf{E}_o.$$

Comment: It is important to realize that this equation implies that the energy in the beam beyond the sample, which is proportional to $|E_{total}|^2$*, is* ***greater*** *than the energy in the beam upstream of the sample,* $|E_o|^2$*. Assuming that the molecules in the sample are all in their ground states to begin with, which is our standard assumption, this is a violation of the first law of thermodynamics. However, in the thin-sample limit, the violation is so small that it can be ignored.*

If the sample is of finite thickness, say n thin layers deep, but still thin, then $\mathbf{E}_{\text{total}} = [1 + i(\omega/2c)N\alpha d\tau]^n\mathbf{E}_o$. Because $[1 + i(\omega/2c)N\alpha d\tau]^n \approx \exp[i(\omega/2c)N\alpha\Delta\tau]$, where $\Delta\tau$ is defined as $nd\tau$, it follows that:

$$\mathbf{E}_{\text{total}} \approx \exp[i(\omega/2c)N\alpha\Delta\tau]\mathbf{E}_o. \tag{3.4}$$

Thus, the scattering that occurs in condensed phase samples alters the phase of the on-axis radiation that emerges from them.

3.3 PHASE CHANGES ARE INDISTINGUISHABLE FROM VELOCITY CHANGES

Two points need to be made about equation (3.4). First, even though the arguments we used to obtain equation (3.4) are based on an analysis of the passage of radiation through transparent samples, they are valid for materials that are not transparent. If the polarizability of a sample varies with location, that is, if $\alpha = \alpha(\mathbf{r})$, it will not be transparent because it will scatter radiation in all directions. However, $\alpha(\mathbf{r})$ can be written as $(\langle\alpha\rangle + \delta(\mathbf{r}))$, where $\langle\alpha\rangle$ is the average value of $\alpha(\mathbf{r})$, and $\delta(\mathbf{r})$ is the difference between $\alpha(\mathbf{r})$ and $\langle\alpha\rangle$ at $\mathbf{r}$. Because the scatter a sample gives in the forward direction reports only on $\langle\alpha\rangle$, that is the polarizability that should be used in equation (3.4) when discussing the on-axis radiation that passes through samples that are not transparent. The second point is much more important. Because of the phase lag that occurs when a beam of radiation passes through a bulk sample, the length of time it takes for a particular wave front in the incident beam to reach $(0,\ 0,\ z)$ is *greater* when there is material in the path than it would have been if there was not, which is exactly what would be observed if the speed of light inside bulk samples was *less* than the speed of light in a vacuum.

The **refractive index** of a substance, n, is c/v, where v is the velocity of light in that substance. A formula for the relationship between polarizability and refractive index can be obtained by writing an expression for the difference in the time it takes for the radiation to pass through a slab of thickness dz at a speed of v and the time it takes to travel the same distance at a speed of c. The phase difference caused by the passage of radiation through the sample is the product of this time difference and ω. When that phase difference is equated to the phase shift given by equation (3.4), the following formula emerges:

$$n = 1 + N\alpha/2, \tag{3.5}$$

or, alternatively:

$$\alpha = (2/N)(n - 1).$$

Equation (3.5) is not fully accurate because its derivation takes no account of the fact that the strength of the electric field experienced by molecules inside a condensed phase sample is not the same as the strength of the field applied to that sample from the outside, because of polarizability effects. (It is usually less.) When the difference between the net internal field and the external field is properly dealt with, a more complicated expression emerges:

$$\alpha = (3/N)(n^2 - 1)/(n^2 + 2). \quad (3.6)$$

Equation (3.6) is called the **Clausius-Mossotti equation**, or sometimes the **Lorentz-Lorenz equation**, and the set of constants that multiply its $(n^2 - 1)/(n^2 + 2)$ component depend on the definition used for α. Now $n^2 = (c/v)^2 = (1/\mu_o\varepsilon_o)/(1/\mu\varepsilon)$, where μ is the permeability of the sample, and ε is its permittivity (see Notes for the Reader). However, μ is only slightly different from μ_o for most materials; hence, equation (3.6) is also often written with its $(n^2 - 1)/(n^2 + 2)$ part replaced by the expression $[(\varepsilon/\varepsilon_o) - 1]/[(\varepsilon/\varepsilon_o) + 2]$. (NB: the ratio $(\varepsilon/\varepsilon_o)$ is the **dielectric constant** of a substance.)

Dimensional analysis of equation (1.3) shows that α is a volume, and using equation (3.6), we can now understand what that volume is. Because N is the number of molecules per unit volume, $1/N$ is the average volume occupied by a single molecule in a liquid or solid sample. That volume multiplied by $[3(n^2 - 1)]/(n^2 + 2)$, which is a number that is usually not far from 1.0, is the polarizability. Thus, the polarizability of an atom or molecule should be comparable to its van der Waals volume, which it often is. For example, the static polarizability of xenon is about 40×10^{-30} m^3; its polarizability at optical frequencies is about 52×10^{-30} m^3; and its van der Waals volume is about 43×10^{-30} m^3.

Comment: Refractive indices ordinarily vary over a small range. For example, the refractive of a vacuum is 1.0, whereas that of water is 1.33. Thus, $(n^2 + 2)/3$ is generally not far from 1.0, and so $n \approx (1 + N\alpha)^{1/2}$, which if $N\alpha$ is small compared to 1.0, is $\sim (1 + N\alpha/2)$. Thus, equation (3.5) can be thought of as a first approximation version of equation (3.6) that is useful when N^{-1} is much greater than α, which it always is for gases, but it never is for liquid or solid samples.

Two additional points need to be made. First, it has long been recognized that the refractive indices of transparent substances vary with frequency, and that means that $\alpha = \alpha(\omega)$. Second, the change in velocity of light observed when light passes from a vacuum into a transparent liquid or solid medium is due entirely to a change in wavelength. When electromagnetic radiation moves through substances, the atomic/molecular dipole moments it induces in those substances all oscillate at the same frequency as that of the inducing electromagnetic radiation. Thus, in a medium having an index of refraction of n, the wavelength of light that has a wavelength of λ in vacuo is λ/n.

3.4 POLARIZABILITIES DO NOT HAVE TO BE REAL NUMBERS

Equation (3.4) has another implication we need to consider. If α were a complex quantity, that is, if $\alpha = a + ib$, rather than the real number we have assumed it to be so far, then equation (3.4) would become:

$$\mathbf{E}_{\text{total}} \approx \exp\left[i(\omega/2c)Na\Delta\tau\right]\exp\left[-(\omega/2c)Nb\Delta\tau\right]\mathbf{E}_o. \quad (3.7)$$

Thus, the amplitude of the electric field of the radiation beam passing through a sample that has complex polarizability would diminish exponentially as it penetrates the sample, and because energy is proportional to $|\mathbf{E}|^2$, this would mean that the sample was absorbing electromagnetic energy. Transparent materials that are colored do absorb light energy, of course, and turning the argument just made on its head, this must imply that the polarizabilities of colored substances are complex. Moreover, $b \geq 0$, because if it were less than 0, the beam of light would be *gaining* energy as it passed though the sample, which would violate the conservation of energy. Not only that, b must depend strongly on wavelength because colored substances would not appear colored unless the probability of their absorbing electromagnetic radiation varied significantly with wavelength. Thus, in general: $\alpha(\omega) = \mathrm{a}(\omega) + i\mathrm{b}(\omega)$.

Comment: Equation (3.7) indicates that the absorption of radiant energy is proportional to the exponent of a negative quantity that is the product of a coefficient that depends on the nature of the molecules present, b; their number concentration, N; and the thickness of the sample, $\Delta\tau$. Thus, the absorption component of equation (3.7) has exactly the same functional form as the Lambert-Beer law, which is the phenomenological equation chemists and biochemists use to analyze the absorption of light by materials. It follows that the quantity b, must be proportional to the quantity spectroscopists call the molar extinction coefficient (see question 2).

3.5 ATOMIC POLARIZATION EFFECTS ARE SMALL

As a first step toward developing a sense of how the frequency dependences of $\mathrm{a}(\omega)$ and $\mathrm{b}(\omega)$ arise, it is useful to estimate the distances that charged particles move when an atom or molecule is exposed to bright light. This issue can be addressed using equations (1.2) and (1.3) once we decide what the word "bright" means when applied to electromagnetic radiation. Most people would agree that direct sunlight is "bright", and at the earth's surface, the portion of the solar energy flux that is visible to the human eye amounts to $\sim$ 1 kWm^{-2}, which implies an $|\mathbf{E}_o|$ of $\sim$ 1,000 V/m. At optical frequencies, the polarizability of xenon is about 50×10^{-30} m^3 (see section 3.3); thus, the dipole moment induced in a xenon atom by an electromagnetic field of that amplitude would be $\sim 4.5 \times 10^{-37}$ Cm. This is the dipole moment a xenon atom would display if one of its electrons were displaced from its average, ground-state position by $\sim 3 \times 10^{-18}$ m, which is a distance eight orders of magnitude smaller than the diameter of that atom ($\sim 4 \times 10^{-10}$ m)! Thus, this simple calculation suggests that the electron distributions in atoms and molecules (usually) barely quiver when they are exposed to light of ordinary intensity.

3.6 THE FREQUENCY DEPENDENCE OF POLARIZABILITIES CAN BE ADDRESSED CLASSICALLY

The small size of the changes in electron positions that occur when atoms and molecules are exposed to electromagnetic radiation at frequencies far from those at which they absorb makes it easy to develop a classical theory that explains the frequency dependence of α. Most of the time, most atoms and molecules are found in their ground electronic states, which implies that their nuclei and electrons are arranged so that they have the lowest possible energy. The energy of such a system must rise if any of its components are displaced from their ground-state positions, and if the displacement is small, a simple relationship will exist between the energy of the system and those displacements.

If the ground-state energy of an atom is $U(0)$, and the position of one of its electrons is displaced from its average, ground-state position by an amount δ in the x direction, then the dependence of its energy on displacement can be written as a Taylor series:

$$U(\delta) = U(0) + (\partial U/\partial x)_0\delta + (1/2)(\partial^2 U/\partial x^2)_0\delta^2 + \ldots$$

But $(\partial U/\partial x)_0$ must be 0 because otherwise $U(0)$ would not be a minimum. Furthermore, by assumption, δ is very small, so it must be true that $U(\delta) \approx U(0) + (1/2)k\delta^2$, where $k = (\partial^2 U/\partial x^2)_o$. The magnitude of the force experienced by the displaced electron, F, will be the negative of the first derivative of U, and hence $F = -k\delta$. Thus, the displaced electron will experience a force that pushes it back toward its equilibrium position, and the further the electron is displaced from its equilibrium position, the greater the restoring force will be.

Comment: The linear dependence of restoring forces on displacements explains why dipole moments are proportional to applied electric fields (equation (1.3)). The electrical force an electron experiences when the atom/molecule to which it belongs is exposed to an external field of E is $-q_eE$, and this force will cause it to move away from its equilibrium position in the direction the electric field requires until $\delta = -(q_eE/k)$, at which point the restoring force, $-k\delta$, will be equal to and opposite to the external electric force, and its motion will cease. The contribution to the dipole moment of the atom/molecule made by that electron will thus be $(q_e{}^2/k)E$, which is a quantity proportional to E. [This relationship implies that $\alpha = (q_e{}^2/\varepsilon_o k)$.]

The motions of any component of a multicomponent system that has been displaced from its equilibrium position by a small amount, and then released so that it can respond freely to the restoring force obeys the following differential equation:

$$F = ma = m(d^2x/dt^2) = -kx,$$

which is easily solved. One possible solution is:

$$x(t) = A\ \sin[(k/m)^{1/2}t],$$

where $(k/m)^{1/2} = \omega_o = 2\pi\upsilon$; υ being the frequency of the oscillation, and A is its amplitude, in this case, the distance the particle was from its equilibrium position at the time it was released. By writing the solution of this differential equation in this form, we are assuming that at $t = 0$, the electron is at its equilibrium position,

$x = 0$, and that x is increasing. Systems that oscillate sinusoidally are called **harmonic oscillators**, and given the general nature of the arguments that led to this result, it should describe the (classical) behavior of any electron that has been displaced from its equilibrium position in some atom/molecule by a small amount, and then allowed to respond freely.

Comment: It is now easy to understand why nuclear motions have been ignored in our discussion of the interaction of electromagnetic radiation with matter. Electric fields exert forces on both the electrons and the nuclei of atoms, and accelerations of positive charges are just as effective in generating electromagnetic radiation as accelerations of negative charges. In addition, the restoring force engendered when the electronic configuration of an atom is perturbed acts on both its nucleus and its electrons; it makes them move toward each other. Furthermore, if the magnitude of the restoring force experienced by a single electron is $|dU/dx|(= kx)$, assuming the electrons in an atom act independently, the restoring force experienced by the nucleus will by $Z|dU/dx|$, where Z is the atomic number. Hence, if the acceleration of the electrons is (F/m_{elec}), the acceleration experienced by the nucleus will be $Z(F/m_{nuc})$. However, because protons and neutrons are about 1,835 times more massive than electrons, by comparison with electrons, even high atomic number nuclei are barely accelerated at all. Because scattered energies are proportional to accelerations squared, the amount of electromagnetic radiation scattered by the nucleus of an atom is tiny compared to the amount scatterd by its electrons.

The harmonic oscillator model must be modified before it can be used to describe the response of an electron bound to an atom or molecule that is continuously exposed to electromagnetic radiation. In addition to being acted on by a restoring force, that electron will experience two additional forces: (1) the force the external electric field exerts on it, and (2) a frictionlike force that arises because the radiation it emits is constantly dissipating its mechanical energy. Thus, at steady states the differential equation that describes the motion of such an electron will be:

$$m(d^2x/dt^2) = -kx - q_e E_o \exp(-i\omega t) - \eta(dx/dt). \qquad (3.8)$$

In this equation, m is the mass of the electron; q_e is the magnitude of the charge on the electron; E_o is the amplitude of the electric field of the electromagnetic radiation, which is oriented along the x-axis of the local coordinate system; ω is the angular frequency of that radiation; and η is the frictional coefficient. The time-dependent part of the electric field is written using an exponential function instead of a sine or cosine function because, as we shall shortly see, the electron in such a system need not oscillate in phase with the external electric field, and the out-of-phase component of its motion, if any, is best described using complex numbers.

As appendix 3.2 shows, the polarizability of an atom that behaves this way will have the following dependence on ω:

$$\alpha(\omega) = +\left(q_e^2/m\varepsilon_o\right)\left[\left(\omega_o^2 - \omega^2\right) + i(\eta\omega/m)]/[(\omega_o^2 - \omega^2\right)^2 + (\eta\omega/m)^2\right], \quad (3.9)$$

where ω_o is again $(k/m)^{1/2}$. [NB: The sign of the imaginary part of α would be negative if the time-dependent part of the incident electric field were written $\exp(i\omega t)$ instead of $\exp(-i\omega t)$.] Thus:

$$a(\omega) = +(q_e^2/m\varepsilon_o)(\omega_o^2 - \omega^2)/[(\omega_o^2 - \omega^2)^2 + (\eta\omega/m)^2],$$

and:

$$b(\omega) = +(q_e^2/m\varepsilon_o)(\eta\omega/m)/[(\omega_o^2 - \omega^2)^2 + (\eta\omega/m)^2].$$

3.7 WHEN THE IMAGINARY PART OF α IS LARGE, ENERGY IS ABSORBED

As equation (3.9) shows, the imaginary part of α is appreciable only when ω is close to ω_0, the natural frequency of the oscillating atom, a condition called **resonance** (see comment), and it is easy to show that the value of b is proportional to the amount of energy being lost from the incident radiation beam as a result of electronic motions (see appendix 3.3). At the resonant frequency, the work done on our model atom by the radiation beam is $[\pi(q_eE_o)^2/\eta\omega_o]$ per cycle of the radiation, and if the beam does work on the atom, its energy (intensity) must fall. We know what that atom does with the energy it gains from the beam; in steady state, it scatters it.

It is easy to understand intuitively why energy transfer occurs efficiently only when ω_o and ω are about equal. Under those circumstances, if there were no friction, the excursions of electrons from their equilibrium positions would become very large because the polarizability would be very large, as you can easily show yourself by examining how α would behave if the frictional coefficient were 0 (see equation (3.9)). The tendency of the motions of electrons to become infinite at resonant frequencies, that is, for the mechanical energy of the atom/molecule to become indefinitely large, is countered by a proportionate increase in the amount of energy radiated by its electrons because the bigger the excursions of those electrons from their equilibrium positions, the larger the accelerations required to make those excursions happen in the time allotted.

Comment: In the derivation of equation (3.8) it is assumed that electronic displacements are so small that only the second derivative term in the Taylor series expansion of energy as a function of displacement needs to be taken into account. As ω approaches ω_0, displacements increase in size, and as they do so, the higher-order terms in the Taylor series expansion will become more important, and the model for atomic behavior embodied in equation (3.8) will become increasingly inaccurate. Thus, even if quantum mechanical effects could be totally ignored, which they cannot, equations (3.8) and (3.9) would not accurately describe what happens close to resonance.

3.8 THE REFRACTIVE INDEX OF SUBSTANCES FOR X-RAYS IS LESS THAN 1.0

This highly simplified model for the response of atoms to external electric fields predicts that atomic/molecular polarizabilities should have several of the qualitative properties we already know they do. For example, it predicts that atomic polarizabilities ought to be complex, and that both their real and imaginary parts should depend very strongly on frequency. However, it also predicts something we had not anticipated, namely that at high frequencies, $\alpha(\omega)$ will be negative.

When visible light interacts with a transparent substance such as glass or water, the frequency of the light is (usually) well below that of any of the frequencies at which electronic transitions in these substances take place, which tend to be in the ultraviolet range. Thus, our model for polarizabilities predicts that at low frequencies atomic polarizabilities will be:

$$\alpha = +\left(q_e^2 / m_e \varepsilon_0 \omega_o^2\right),$$

and refractive indices will be greater than 1.0, indicating the visible light moves less rapidly through these substances than it does through a vacuum. For X-rays, the situation is different. In that case, the frequency of the radiation is well *above* the frequencies associated with any of the many electronic transitions possible for substances. In that case,

$$\alpha = -\left(q_e^2 / m_e \varepsilon_0 \omega^2\right),$$

and indices of refraction should be less than 1.0.

It is easy to understand why these two limiting expressions differ in sign. As we already know, the electrons associated with atoms scatter X-rays the way unbound electrons do, and it is the accelerations of charges that makes them emit radiation. In this case, accelerations are proportional to $E_o\exp[-i\omega(t - r/c)]$, the inducing field, because $\mathbf{F} = m\mathbf{a} = q\mathbf{E}$. By contrast, if the electron responsible for scattering is bound to an atom and the frequency of the radiation is small compared to resonant frequency of the atom, it is its displacement from its equilibrium position—$x(t)$ in the equations in appendix (3.1)—that is proportional to $E_o\exp[-i\omega(t - r/c)]$ (see section 3.6). Therefore, for the bound electron, accelerations are proportional to $d^2(E_o\exp[-i\omega(t - r/c)])/dt^2$, which is $-\omega^2 E_o\exp[-i\omega(t - r/c)]$. Consistent with this analysis, if $-\left(q_e{}^2/m\varepsilon_o\omega^2\right)$, the (nominal) polarizability of a free electron, is substituted into equation (1.4), we find that the scatter produced by a free electron ought to be: $\mathbf{E}_s = (1/4\pi\varepsilon_o mr)(q_e/c)^2 \sin(\varphi)E_o\exp[-i\omega(t - r/c)]\mathbf{1}_\varphi$, which is Thomson's expression for the scattering of a free electron.

Comment: The preceding limiting case equations make sense qualitatively. The resonant frequency of an electron bound to an atom, ω_o, is a measure of how fast it can respond to changes in an external electric field. If the frequency of that field is much lower than ω_o, the electron will respond to its changes so quickly that the force *exerted on it by the nucleus is always (effectively) equal to and opposite to the force exerted on it by the exernal field. Under these circumstances, we would expect that polarizability would be independent of the frequency of the external electromagnetic radiation, as the preceding equations predict. On the other hand, if the frequency of the external field is much higher than* ω_0*, the electron will be incapable of tracking it accurately. Indeed, at the high-frequency limit, as the previous equations predict, the polarizability of the atom to which the electron belongs should fall to 0 because the electron is unable to accelerate fast enough to start moving in one direction in response to the external field before the sign of that field changes and starts pushing it in the opposite direction. It will thus respond to such fields as though their field strengths were 0.*

It is not hard to demonstrate experimentally that the refractive indices of solid materials for X-rays are indeed less than 1.0. You may recall that when a beam of light traveling through a piece of glass encounters a flat glass–air interface at a glancing angle that is low enough, total internal reflection ensues. Instead of exiting from the glass, the beam is reflected by the interface back into the glass, which has a higher refractive index than air, as though the glass–air interface were a mirror. The phenomenon is called **total internal reflection**, and the roof prisms in binoculars exploit it. The angle below which total internal reflection occurs (the **critical angle**) is determined by the difference in n between glass and air, and the bigger that difference is, the bigger the critical angle. When X-rays traveling through air (or a vacuum) encounter a flat liquid or solid surface at a glancing angle, the same thing happens; the beam is reflected back into the medium of higher refractive index, which in this case happens to be—surprise—air.

The reflection of X-rays by air–solid interfaces, which might be called *total external reflection,* has practical uses. For example, if a beam of X-rays is totally reflected by a concave glass surface of the appropriate shape, it will be focused, and focusing elements that exploit this phenomenon are often included in X-ray instrumentation.

3.9 THE WAVELENGTH DEPENDENCES OF THE PROCESSES THAT CONTROL LIGHT AND X-RAY POLARIZABILITIES ARE DIFFERENT

When the frequency dependence of polarizabilities is analyzed using quantum mechanics, as it must be if accurate results are to be obtained, it turns out that the intrinsic angular frequencies of atoms (i.e., their ω_0s) are related to the energy changes that occur when the electrons in atoms and molecules change orbitals as follows: $h\omega/2\pi = \Delta E_{\text{orbital}}$. Furthermore, quantum mechanics predicts that when an atom is exposed to electromagnetic radiation having a frequency close to the ω_0 appropriate for some transition, its electronic configuration will oscillate sinusoidally between the ground state and the corresponding excited state. This is consistent with the classical model, which also calls for large amplitude oscillatory motions close to resonance.

In the optical case, the changes in electronic configuration of concern correspond to the promotion of outer shell electrons into empty, excited-state orbitals, and by comparison to those energy changes, the differences in energy between adjacent excited-state orbitals are large. By contrast, the relevant changes in electronic configuration for X-rays are the promotions of inner shell electrons (e.g., K-shell or L-shell electrons) into empty excited-state orbitals, or even into the vacuum, in other words to outright ionization. In that case, the energy differences between adjacent, empty, exited-state orbitals are very small compared to the amount of energy it takes to promote a K-shell electron into even the lowest lying empty orbital. Thus, in the optical case, each transition has a specific energy associated with it; it will not happen if the energy of an incident photon is too big any more than it will happen if its energy is too small. On the other hand, threshold effects are observed in the X-ray case. If the energy of the incident photon is too small, no transition will be seen, but photons having energies larger than the minimum required to extract an electron from an inner shell orbital are effective too. The "extra" energy of an above-threshold photon can be "used" either to promote the inner shell electron into a higher energy excited-state orbital or to make it leave the atom entirely, possibly with kinetic energy to spare. In this connection, it is worth noting that the X-ray photons used for structure determinations have energies around 10^4 eV, and that the differences in energy of the outer orbitals of atoms are generally < 1 eV. These differences in energy scales lead to marked differences between the way polarizability changes with wavelength as an optical transition is crossed and the way it changes when an X-ray transition is crossed.

3.10 ON THE FREQUENCY DEPENDENCE OF ATOMIC SCATTERING FACTORS FOR X-RAYS

As you already know, scattering factors for X-rays are defined as the ratio of the scattered field generated by the interaction of radiation with some atom or molecule to the scattered field that would have been observed under the same circumstances if the scattering entity had been a single, free electron. Using equation (1.4) to describe the scatter produced by a single electron associated with an atom, Thomson's equation for the scattering produced by a free electron (equation (1.6)), and our simple model of the atom, we can predict that the scattering factor of an electron bound to an atom will be:

$$\begin{aligned} \mathrm{f}(\omega) &= -\left(m/q_e^2\right)\varepsilon_o\alpha(\omega)\omega^2 \\ &= \left[\omega^2\left(\omega_o^2-\omega^2\right)+i(\eta\omega^3/m)\right]/\left[\left(\omega_o^2-\omega^2\right)^2+(\eta\omega/m)^2\right]. \end{aligned} \quad \textbf{(3.10)}$$

Equation (3.10) implies that for a whole atom, it might be reasonable to use an expression resembling the following for $\mathrm{f}(|\mathbf{S}|)$:

$$\left[\omega^2\left(\omega_o^2-\omega^2\right)+i(\eta\omega^3/m)\right]/\left[\left(\omega_o^2-\omega^2\right)^2+(\eta\omega/m)^2\right] \times \int \rho(\mathbf{r})\exp(-2\pi i\mathbf{r}\cdot\mathbf{S})dV_r,$$

where $\rho(\mathbf{r})$ is the electron density distribution of the atom.

Crystallographers divide atomic scattering factors into three parts:

$$f = f_o + \Delta f' + i\Delta f'',$$

where f_o is the scattering that would be observed if all the electrons in the atom were free electrons, in other words, $f_o = \int \rho(\mathbf{r}) \exp(-2\pi i\mathbf{r} \cdot \mathbf{S})dV$. The wavelength-dependent component of the real part of f is captured in $\Delta f'$, whereas $\Delta f''$ is its entire imaginary part. (f_o, $\Delta f'$ and $\Delta f''$ are all measured in number of electrons.) It is easy to show that if atoms responded to X-rays the way our model predicts that:

$$\Delta f' = \left[\omega_o^2 \left(\omega^2 - \omega_o^2\right) + (\eta\omega/m)^2\right]\left[\left(\omega_o^2 - \omega^2\right)^2 + (\eta\omega/m)^2\right]^{-1} \times \int \rho(\mathbf{r}) \exp(-2\pi i\mathbf{r} \cdot \mathbf{S})dV_r,$$

and:

$$\Delta f'' = -\left[(\eta\omega^3/m)\right]\left[\left(\omega_o^2 - \omega^2\right)^2 + (\eta\omega/m)^2\right]^{-1} \int \rho(\mathbf{r}) \exp(-2\pi i\mathbf{r} \cdot \mathbf{S})dV_r.$$

3.11 REAL X-RAY ABSORPTION AND DISPERSION SPECTRA DO NOT LOOK THE WAY CLASSICAL THEORY PREDICTS

One way to understand the implications of equation (3.10) is to use it to compute the real and imaginary parts of some particular scattering factor as a function of frequency. Before that can be done, η/m must either be estimated theoretically or evaluated experimentally. We might start by using the classical expression for η/m, $\eta/m = (2/3)\left(q_e{}^2/4\pi\varepsilon_o mc^3\right)\omega^2$, which is valid if radiation emission is the sole mechanism for dissipating the mechanical energy of the system of interest.

Taking the resonant frequency of a single electron system to be 2×10^{19} rad s^{-1}, which is the angular frequency of X-rays having wavelengths of 1 Å, the frequency dependence of the real and imaginary parts of the scattering factor can be computed using equation (3.9), with the results shown in figure 3.2. Thus, classical theory predicts that both the real and imaginary parts of atomic scattering factors will vary dramatically with wavelength over a very narrow frequency range close to the relevant resonant frequencies, in other words, at frequencies/wavelengths that differ from the resonance by less than $\sim 0.01\%$. Moreover, it predicts that absorption peaks will have Lorentzian shapes, which optical absorption peaks generally do.

Comment: Plots of the real part of α against frequency are called ***dispersion*** *curves because of their connection to the refractive indices of substances.*

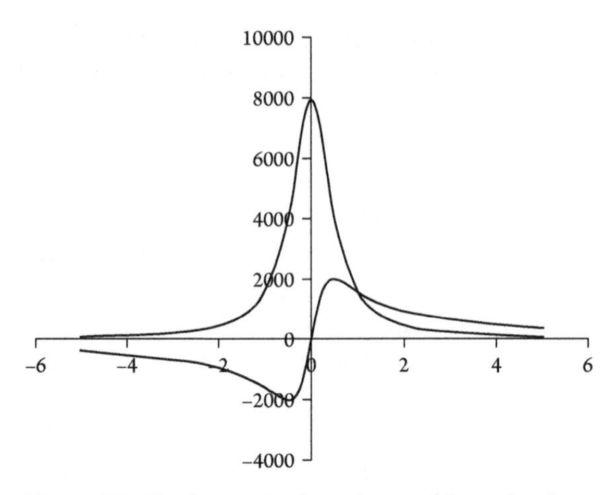

Figure 3.2 The frequency dependence of the real and imaginary parts of the scattering factor of a single, bound electron that has a resonant frequency of 2×10^{19} rad s^{-1}. The vertical scale is the amplitude of the scattering factor in units of electrons, and the horizontal scale is the difference (in radians per second) between the resonant frequency and the frequency at the point of observation in units of 10^{15} rad s^{-1}. The real part is the low amplitude, sigmoid curve. The imaginary part, which is proportional to the extinction coefficient, is the symmetric peak.

Although the classical model for scattering factors is useful conceptually, it has serious shortcomings, as can be appreciated by comparing figure 3.2 with figure 3.3, which shows the way the far more realistic theory of Cromer and Liberman predicts $\Delta f'$ and $\Delta f''$ should vary with wavelength for a selenium atom close to the absorption edge it has at about the same frequency. As we anticipated, real X-ray absorption spectra are a succession of steplike discontinuities similar in form to the single step shown in figure 3.3 (top) not a series of Loretzian peaks centered at different frequencies. Furthermore, if real X-ray absorption spectra do not have the shape our classical model predicts, it can hardly come as a surprise that the dispersive components of their scattering factors (i.e., $\Delta f'$), do not vary with wavelength the way classical theory predicts either (see figure 3.3, bottom).

Although both qualitatively and quantitatively the spectra in figure 3.3 are far closer to reality than those in figure 3.2, even they do not fully capture what is seen when the wavelength dependencies of $\Delta f'$ and $\Delta f''$ are measured for selenium--containing molecules in the same energy/wavelength range. The chemical context of an atom can have large effects on both kinds of spectra. For example, it may alter the energy at which an absorption edge occurs by a small amount, and it can be counted on to alter the shapes of edges. Instead of being simple steps, many edges include narrow peaks called *white lines* at wavelengths slightly shorter than the edge wavelength that may have amplitudes two or three times larger than the plateau

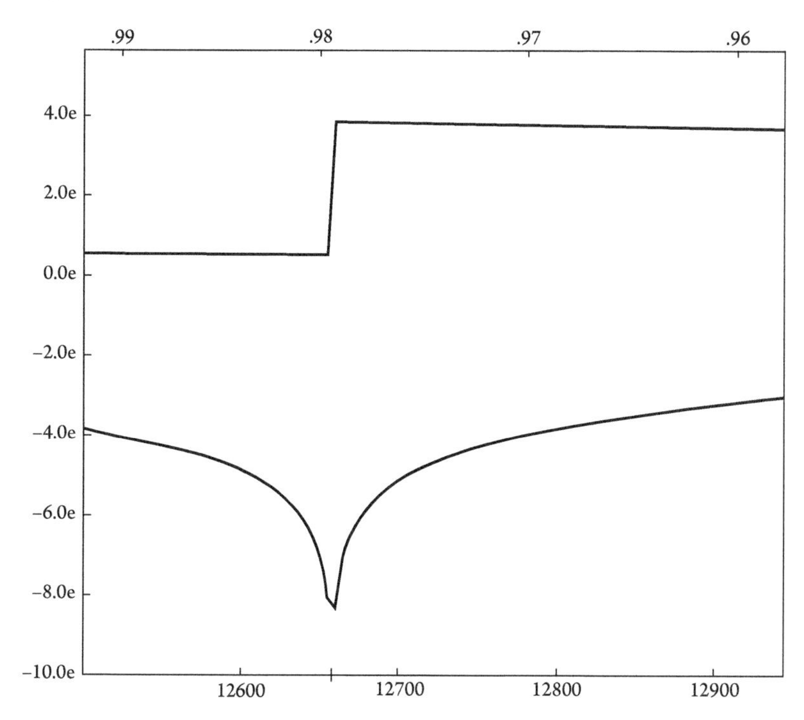

Figure 3.3 The wavelength dependence on $\Delta f''$ (top) and $\Delta f'$ (bottom) for a free selenium atom near 1 Å. The units of the ordinate are the number of electrons. The lower abscissa gives photon energies in electron volts, whereas the upper abscissa reports photon wavelengths in angstroms. (Figure produced using "X-ray Anomalous Scattering", a program developed by Ethan A. Merritt [1995] that uses the Cromer/Liberman approach for computing f'. Available at: http://www.bmsc.washington.edu:scatter/.)

value of the edge. In addition, at energies higher beyond that of white lines, $\Delta f''$ commonly oscillates in an irregular way before settling down to its asymptotic value.

Comment: EXAFS (extended X-ray absorption fine structure) spectroscopy is the experimental field denoted to the extraction of information about the chemical environment of metal atoms from the fine structure of the X-ray absorption edges observed for them in molecules of different kinds, including, for example, metalloproteins.

3.12 THE IMAGINARY COMPONENT OF F CAN BE DETERMINED BY MEASURING MASS ABSORPTION COEFFICIENTS

The absorption of any type of radiation by matter obeys the Lambert-Beer law, as mentioned earlier, and thus the ratio of the intensity of the radiation after it has passed through a sample of some material that has a thickness of t, I, to the intensity of the incident radiation, I_o, obeys the following relationship:

$$(I/I_o) = \exp(-\mu t),$$

where μ is the linear absorption coefficient of that particular material, which has dimensions of meters^{-1}. For X-rays, μ depends primarily on the atomic composition of the material in question, which is to say the numbers of atoms of each type present in some material, per unit volume. If the substance is a pure element, we know from equation (3.1) that μ must equal $[(\omega/c)N]\text{Im}(\alpha)$, where $\text{Im}(\alpha)$ is the imaginary part of the polarizability, and N is the number of atoms per unit volume. Furthermore, comparison of equations (3.9) and (3.10) reveals that:

$$\text{Im}(\alpha) = \left(q_e^2/m_e\varepsilon_o\right)(\Delta f''/\omega^2),$$

and hence:

$$\mu(\omega) = \left(Nq_e^2/m_e\varepsilon_o c\right)(\Delta f''(\omega)/\omega).$$

Thus, by measuring the mass absorption coefficient of some materal as a function of X-ray frequency, we can obtain experimental values for $\Delta f''(\omega)$ for the different elements it contains. (Remember, the absorption edges of one element are most unlikely to fall at the same values of ω as the absorption edges of another.) It turns out that $\Delta f''(\omega)$ and $\Delta f'(\omega)$ are related to each other by a mathematical operation called the **Kronig-Kramers transformation,** which we will not discuss here. Thus, once $\Delta f''(\omega)$ has been measured experimentally, $\Delta f'(\omega)$ can be computed.

3.13 SCATTERING CAN BE DESCRIBED USING SCATTERING LENGTHS AND CROSS SECTIONS

All expressions for the scattering of X-rays by electrons contain the same proportionality constant, $q_e^2/4\pi\varepsilon_o mc^2$, and when that constant is analyzed dimensionally, it turns out to be a length. Its magnitude is 2.828×10^{-15} m, and it is called the **scattering length** of the electron. Equation (1.6), which can be written:

$$E_s = -\left(q_e^2/4\pi\varepsilon_o m_e c^2\right)(\sin\varphi/R)E_o \exp[-i\omega(t - R/c)],$$

is thus the scattering length of the electron divided by R, which is also a length, times the product of the incident field and a polarization factor. Thus, we find that at 1 m, ignoring the $\sin(\varphi)$ term, E_s is weaker than the field that induced it by a factor of 3.5×10^{14}!

$E_s{}^2$ is proportional to energy, as we already know, and the fraction of the energy incident on a single free electron that gets scattered by it is $(E_s/E_o)^2$, integrated over the surface of a sphere of unit radius surrounding that electron. When that integration is done, we find that the fraction in question is $(8\pi/3)\left(q_e{}^2/4\pi\varepsilon_o mc^2\right)^2$, a quantity that has the dimensions of area (see question 4). Areas like this are called **cross sections.** By tradition, the unit of cross section is the **barn,** as in the playground taunt, "You couldn't hit the broad side of a barn door" ($1\,\mathrm{b} = 1 \times 10^{-28}\,\mathrm{m}^2$). The cross section of the electron for X-rays is 0.67 b.

At small scattering angles, the scattering lengths atoms are $Z \times (2.828 \times 10^{-15}\,\mathrm{m})$, but they fall off with $|\mathbf{S}|$, as we already know. Molecules also have cross sections, but crystallographers usually ignore them because they do not find them particularly useful. However, molecular cross sections are interesting to electron microscopists, and so we will discuss them later (chapter 12.2). Suffice it to say that the cross sections of atoms and molecules depend strongly on X-ray wavelength. In addition, molecular cross sections depend on the orientation of the beam relative to the molecule responsible for the scattering being observed, the structure of that molecule, and for condensed phase samples, even their supermolecular organization.

3.14 NEUTRON SCATTERING CAN BE USED TO STUDY MOLECULAR STRUCTURE

The scattering of neutrons by atoms and molecules is always described using cross sections and scattering lengths. Moving neutrons have de Broglie wavelengths ($\lambda = h/\mathrm{momentum}$), and a neutron moving at about 4,000 m/s has a de Broglie wavelength of about 1 Å. Because the kinetic energy of such a neutron corresponds to the average kinetic energy of an atom or molecule in a gas having temperature about 600 K, neutrons having wavelengths in the angstrom range are called **thermal neutrons**. Neutron beams of this sort can be obtained both from nuclear reactors and from accelerator-based devices called **spallation sources**. They can be used to investigate the structure and properties of matter on angstrom-length scales just the way beams of X-rays are.

What does the spatial distribution of scattered neutron radiation look like when a beam of monochromatic neutrons encounters a single atom? As we already know, when photons (i.e., electromagnetic radiation), are scattered by atoms, the distribution of scattered radiation obeys the following expression (see equation (1.10)):

$$\mathrm{I_d}(\mathbf{S}) =)1/2\mu_o c)\left(q_o^2 E_o \sin\varphi/4\pi\varepsilon_o m_e c^2 R\right)^2 \left[\mathrm{f}(|\mathbf{S}|)^2\right].$$

What has to be done to this equation to convert it into an expression appropriate for neutrons? In the first place, there is no $\sin(\varphi)$ factor for neutron scattering experiments. In the second place, $\left(E_o{}^2/2\mu_o c\right)$ is the energy flux in the incident radiation beam, and because energy flux is the same as a quantum/photon flux, that part of the expression can be replaced by $\mathrm{I_i}$, the flux of the incident neutron beam, measured in neutrons meter^{-2} second^{-1}. In the third place, we know that the scattering length of an atom for X-rays is $\left(q_e{}^2/4\pi\varepsilon_o m_e c^2\right)\mathrm{f}(|\mathbf{S}|)$. However, neutrons are neutral particles that are much more massive than electrons, and they are scattered primarily by atomic nuclei, the physical sizes of which are many orders of magnitude smaller than 1 Å. Thus, to a thermal neutron, an atom looks like a delta function, and the neutron equivalent of $\left(q_e{}^2/4\pi\varepsilon_o m_e c^2\right)\mathrm{f}(|\mathbf{S}|)$, that is, the scattering length of an atom for neutrons, is a constant independent of $|\mathbf{S}|$, which is usually designated by the letter

b, the value of which varies from atom to atom. Thus, for neutrons, the scattering produced by a single atom is given by the following equation:

$$I_d(\mathbf{S}) = I_i(b/R)^2.$$

It follows that the neutron cross section of an atom is $4\pi b^2$.

The neutron scattering lengths of atoms tend to be smaller than their X-ray scattering lengths, but of the same order of magnitude. In the energy range of interest to biologists, they are only weakly dependent on neutron wavelength. However, unlike X-ray scattering lengths, they do not depend on either atomic number or atomic mass in any simple way because the interactions of neutrons with nuclei are far more complicated than the simple, billiard-ball collisions we might anticipate. Consequently, the neutron scattering lengths of the different isotopes of an element can differ significantly.

The reason neutron scattering lengths vary with Z in an unintuitive way is that during the scattering process, neutrons transiently merge with atomic nuclei to form compound nuclei that have the same atomic number they did before, but a mass number larger by 1. Nothing comparable happens when photons or electrons are scattered by atoms, and its consequences can only be described using quantum mechanics. Furthermore, because neutrons have a spin of 1/2, the spin of that transient nucleus will either be $(I + 1/2)$ or $(I - 1/2)$, where I is the spin of the nucleus of the atom in question, depending on whether the two particles encounter each other with their spins parallel or anti parallel, which are the only two possibilities. It makes a difference because the scattering length associated with the spins-parallel interaction of some atom, b_+, is usually not the same as the scattering length of its spins-antiparallel interaction, b_-. The two quantities can even differ in sign! (Welcome to the wonderful world of nuclear physics!)

Spin orientation effects lead to a kind of scattering incoherence that has no electromagnetic or electronic equivalent. Under ordinary conditions, the nuclear spins in samples are randomly oriented, as are the orientations of the spins of the neutrons in the beam being used to study their structures. Under these conditions, the number of different spin states available to the two kinds of transient nuclei are $[2(I + 1/2) + 1]$ for the spins-parallel interaction, and $[2(I - 1/2) + 1]$ for the spins-anti parallel interaction. Because all of these states are equally probable, the probability that the outcome of a single encounter will be a b_+ event is $[(I + 1)/(2I + 1)]$, and the probability that a b_- event will occur is $[I/(2I + 1)]$.

The total cross section we measure for an atom, σ, is the cross section it displays, averaged over many neutron–atom encounters. Thus, the total cross section of a single atom is:

$$\sigma = [4\pi/(2I + 1)][(I + 1)b_+^2 + Ib_-^2].$$

Similarly, its (coherent) scattering length, b, which is again the average over many encounters, is:

$$b = [(I + 1)b_+ + Ib_-]/(2I + 1).$$

Note, however, that if I is not equal to 0, which it often is not, and b_+ and b_- differ, which they always do, then σ will be *larger* than $4\pi b^2$, which may seem peculiar. The difference between the two quantities—the portion of the total cross section, σ, that is not accounted for by $4\pi b^2$—is called **incoherent scatter** because that portion of the total neutron scatter of an atom does not contribute to interference phenomena. It gives rise instead to a background scatter that has no dependence of **S**, except the dependence it acquires through temperature factor effects (see chapter 5). Note also that if a nucleus has a spin of 0, there will be only one interaction that needs to be taken into account, b_+, and σ will equal $4\pi b^2$. Nuclei that have nuclear spins of 0 do not scatter neutrons incoherently.

The nucleus responsible for the lion's share of the incoherent scattering observed when the structures of biological samples are studied by neutron scattering/diffraction is ^{1}H, which is a spin 1/2 nucleus. It turns out that for ^{1}H, b_+ is $\sim 1.04 \times 10^{-14}$ m and b_- is $\sim -4.7 \times 10^{-14}$. When these numbers are plugged into the formulae just given, we discover that for ^{1}H, b is about -0.4×10^{-14} m, which implies a coherent cross section of about 2 b, but its total cross section is about 80 b. Thus, almost all the neutrons scattered by ^{1}H are scattered incoherently! More than that, hydrogen atoms will correspond to negative features in the neutron equivalent of an electron density distribution (i.e., a scattering length density disribution), whereas (almost) all other atoms will contribute positive features.

Because the nuclei of the different isotopes of the same element differ in spin, they differ not only in (average) scattering length but also in incoherent cross section. Of particular interest to biologists in this regard is the distinction between ^{1}H, the dominant isotope of hydrogen, and its most abundant variant ^{2}H, deuterium. The scattering length of deuterium is $+0.667 \times 10^{-14}$ m, whereas that of ^{1}H is -0.4×10^{-14} m, as we already know. In addition, the incoherent cross section of ^{2}H is about 2 b, whereas the incoherent cross section of ^{1}H is almost 80 b. Thus, the structure factor of a biomolecule for neutron scattering can be altered dramatically by replacing the ^{1}H isotopes it normally contains with ^{2}H isotopes. This effect has been exploited in many of the experiments that have been done to explore biological structure with neutrons. That same isotopic substitution will also reduce the contribution that incoherent scatter makes to the background on top of which coherent scattering profiles must be measured. Thus, it should come as no surprise that heavy water (D_2O) is often the solvent of choice for biological neutron scattering and diffraction experiments.

Two points should be made in closing. First, unless special efforts are made, the atoms of each atomic number found in biological macromolecules will be the mixtures of isotopes normally found in nature. Because those isotopes differ in scattering length, isotopic heterogeneity will also contribute to the total of incoherent scattering cross sections of these molecules (see problem 5). (The contribution is usually small.) Second, thermal neutrons can be absorbed by atomic nuclei, and absorption probabilities depend very strongly on neutron energy. In the neighborhood of the energy at which the probability of the absorption of thermal neutrons by some nucleus reaches its maximum value, absorption probabilities and scattering lengths vary with energy in much the same way classical theory indicates they should for X-rays (see figure 3.2).

3.15 ELECTRONS ARE STRONGLY SCATTERED BY ATOMS AND MOLECULES

From the standpoint of most structural biologists, electron scattering is much more important than neutron scattering, and in the hearts and minds of many biologists, it competes with X-ray scattering on an equal footing. It is easy to generate well-collimated beams of highly monochromatic, short wavelength electrons. Furthermore, as we shall see later (chapter 11), devices can be built that focus these beams the way glass lenses focus light. They can be used to build microscopes that use electrons as their working radiation. These microscopes have proven extremely useful for studying the structures of biological materials.

Electrons are scattered by atoms as a result of their electrostatic interactions with atomic nuclei and the electrons that surround them. The elastic scattering lengths of atoms for electrons, $f_e(|\mathbf{S}|)$, are given by the following equation:

$$f_e(|\mathbf{S}|) = \left(m_e q_e^2/2\pi\varepsilon_o h^2\right)(Z - f_x(|\mathbf{S}|))/|\mathbf{S}|^2,$$

where Z is the atomic number and $f_x(|\mathbf{S}|)$ is the X-ray scattering factor of that same atom. (Note that the contribution made by nuclear scatter is opposed by that made by electrons, consistent with the difference in sign of their charges.) For very small scattering angles, where the equation just provided gives an ill-defined result:

$$f_e(|\mathbf{S}|) = \left(m_e q_e^2/2\pi\varepsilon_o h^2\right) Z\langle r^2\rangle,$$

where $\langle r^2\rangle$ is the mean radius squared of the atom. (NB: In contrast to the X-ray scattering factors discussed earlier, which are all dimensionless numbers, the quantities designate by "f" in descriptions of electron scattering/diffraction effects are scattering lengths.) Because the velocities of the electrons used for electron microscopy are a substantial fraction of the speed of light, relativistic effects must be taken into account when computing f_e. The mass that must be used is $m_e(1 + q_e E_o/m_e c^2)$, where m_e is the rest mass of the electron, as usual, and E_o is the accelerating voltage. The velocity of the electrons in the beam is: $c[1 - (1 + q_e E_o/m_e c^2)^{-2}]^{1/2}$, and the appropriate wavelength to use for computing $|\mathbf{S}|$ is $h[2q_e m_e E_o(1 + q_e E_o/2m_e c^2)]^{-1/2}$.

Using the equations just provided for $f_e(|\mathbf{S}|)$, we discover that at low scattering angles, the elastic scattering length of carbon for electrons is 2.45×10^{-10} m, which is about four orders of magnitude *larger* than the scattering length for carbon for X-rays at low angles (1.70×10^{-14} m), and the same is true of other atoms. The large magnitudes of the scattering lengths of atoms for electrons have important practical consequences. The bad news is that if you are going to study structures using electrons and you want the first Born approximation to be valid, your sample had better be very thin—1 micron or less. X-ray samples (and neutron samples) can be much thicker than that—a millimeter or so. The good news is that the amount of material that must be prepared and expended to characterize some biological object by electron microscopy is much, much less than the amount required for any X-ray or neutron experiment.

3.16 ELECTRONS ARE SCATTERED INELASTICALLY BY ATOMS

The equations given in section 3.15 for the electron scattering lengths of atoms describe the way atoms scatter electrons elastically. Electrons are also scattered inelastically by atoms, and it is not hard to understand why. As we already know, electrons are scattered by their interactions with both the nuclei of atoms and their electrons. Classical mechanics tells us that when a rapidly moving object of low mass, such as an electron, has a collisional interaction with an object that has a high mass, such as an atomic nucleus, the transfer of kinetic energy between them will be inefficient, and both are likely to emerge with about the same kinetic energies they had before they collided. On the other hand, when two objects having the same mass collide, kinetic energy will be efficiently transferred between them. Thus, this simple-minded argument suggests that electron–atom encounters may result in inelastic scattering because some of the interactions that ensue are electron–electron interactions. From the point of view of the structural biologists, inelastic scatter is far more important in the context of electron scattering than it is in the context of light, X-ray, or neutron scattering.

The inelastic scattering of electrons has three qualitative properties that are worth noting. First, for low Z atoms, elastic and inelastic cross sections (i.e., the probabilities for inelastic and elastic scattering) are roughly equal for electrons having energies in the range of interest to structural biologists. Second, as Z increases, elastic cross sections grow significantly, but inelastic cross sections do not. Third, the range of scattering angles over which electrons are scattered inelastically by atoms is much smaller than the range for elastically scattered electrons. Most inelastically scattered electrons are found inside $2\theta = 1°$. This fact makes sense intuitively because nuclei are pointlike objects even to electrons moving at the velocities they do in electron microscopes, but the electron density distributions associated with nuclei are comparatively huge.

PROBLEMS

1. Show that when z is very large, equation (3.1) predicts that the light scattered by a transparent object of radius r_m and polarizability per unit volume of α' will be the same as that implied by the Fraunhoffer scattering equations provided in chapter 1 (e.g., equation 1.4). (Hint: Replace the exponential in equation (3.1) with the appropriate power series.)
2. The Lambert-Beer law that is used to describe the way substances absorb light is summarized by the following equation:

$$\log_{10}(I/I_o) = -\varepsilon(\lambda)ct,$$

where I is the intensity of the light emerging from the sample, I_o is the intensity of the light incident on the sample, $\varepsilon(\lambda)$ is the extinction coefficient of that material at the wavelength λ (dimension: mole^{-1} centimeter^{-1}), c is the molar concentration of the absorbing molecule in the sample, and t is the thickness of the sample in centimeters. What is the relationship between $\varepsilon(\lambda)$ and the quantity b in equation (3.6)?

3. The (static) dielectric constant of water is about 80, and its refractive index is 1.3328. The dielectric constant of liquid benzene is 2.28, and its refractive index is 1.498. Use these observations to obtain two estimates for the polarizabilities of both water and benzene. Are these pairs of estimates internally consistent? Explain the discrepancies, if any. (NB: The density of liquid benzene is 879 kg/m^3.)
4. Starting with equation (1.6), demonstrate that the cross section of a free electron for X-rays is indeed $(8\pi/3)(q_e{}^2/4\pi\varepsilon_o mc^2)^2$, as claimed in the text.
5. A sample of nickel has been obtained that is 69.7% ^{58}Ni, 26.6% ^{60}Ni, and 3.6% ^{62}Ni, all of which have nuclear spins of 0. Their scattering lengths are: 1.44×10^{-14} m, 0.28×10^{-14} m, and -0.87×10^{-14} m, respectively. What is the average cross section for the nickel atoms in this sample? Will this sample have an appreciable incoherent cross section because of its isotopic heterogeneity? If so, what will it be?
6. The density of copper is 8,960 kg m^{-3}; its atomic weight is 63.5; and its atomic number is 29. Assuming that all the electrons in a copper atom respond to X-rays that have a wavelength of 1.0 Å, and that the expression for the polarizability of an electron given in section 3.8 is valid, what should the refractive index of copper be at that wavelength, and what would the critical angle be for the total (external) reflection of those X-rays for a copper surface? (NB: The critical angle is $\sin^{-1}(n_{\text{low}}/n_{\text{high}})$, and a critical angle of 0 corresponds to normal incidence of the radiation beam on the surface of interest.)

APPENDIX 3.1 FORWARD SCATTER FROM A THIN SLAB

To evaluate the forward scatter produced when a beam of radiation encounters a thin slab of material of thickness $d\tau$, start by computing the electric field of the scattered radiation at a point downstream from the sample at $(0, 0, z)$, which lies on the beam axis (see figure 3.1). The component of the scattered field perpendicular to the z-axis produced by the scattering from the volume element at $(r,\ \alpha,\ 0)$ will be (see equation (1.4)):

$$E_s(R, 2\theta, \varphi, t) = (1/4\pi c^2)(\omega^2\alpha'/R)\sin(\varphi)E_o\exp[-i\omega(t - R/c)]\cos(2\theta)dV,$$

where R is the distance from $(r,\ \alpha,\ 0)$ to $(0, 0, z)$, α' is the polarizability per unit volume of the sample, and dV is the differential element of volume. (The cylindrical symmetry of the system guarantees that the component of the scattered field parallel to the z-axis, averaged over the whole sample, will be 0.) The distance from the sample to the point where the scattered radiation is to be observed, $(0, 0, z)$, is chosen to be more than sufficient to guarantee the validity of equation (1.5)—tens of thousands of angstroms at least, but by human standards, not very far, for example, 10 cm. Because the scattering angles to be considered are all very small, φ, the angle between the direction of polarization and the direction of scatter toward $(0, 0, z)$, will be so close to 90° for volume elements in the illuminated part of the sample that $\sin(\varphi)$ is effectively 1.0 for all of them. Thus, the total contribution made to the electric field measured at $(0, 0, z)$ by the volume elements in the ring of material that is r from the center of the beam and dr wide will be:

$$E_{sz} = (1/4\pi c^2)(E_o\omega^2\alpha') \exp(-i\omega t)[\cos(2\theta)/R] \exp(i\omega R/c) 2\pi r dr d\tau.$$

Now, $\cos(2\theta) = z/R$, and because $R^2 = z^2 + r^2$, and z is large compared to the radius of the beam, R^{-1} can be replaced by a Taylor series expansion that is truncated after the second term. Thus, $R^{-1} \approx z^{-1}(1 - r^2/2z^2)$, and $\cos(2\theta) = (1 - r^2/2z^2)$. It follows that:

$$E_{sz} = (1/4\pi c^2)(E_o\omega^2\alpha') \exp[-i\omega(t - z/c)](d\tau/z) \times \int_0^{r_{\max}} (1 - r^2/2z^2) \exp(i\omega r^2/2cz) 2\pi r dr,$$

where $r_{\max}$ is the radius of the illuminated disk of material. Because $(r_{\max}/z)$ is much less than 1.0, $(1 - r^2/2z^2)$ is 1.0 for all intents and purposes, and thus the integral that needs to be evaluated is $\int \exp(i\omega r^2/2cz)(2\pi r)dr$. When it is all over, we find that:

$$E_{sz} = (\omega/2c)(\alpha' d\tau) \exp(i\pi/2) E_o \exp[-i\omega(t - z/c)][\exp(i\omega r_{\max}^2/2cz) - 1].$$

APPENDIX 3.2 A CLASSICAL MODEL FOR THE MOTION OF ELECTRONS IN THE PRESENCE OF ELECTROMAGNETIC RADIATION

As was explained in the main body of the chapter, the motion of a single electron that is part of an atom or molecule that is being acted on by electromagnetic radiation can be modeled classically using the following differential equation:

$$md^2x/dt^2 = -kx - q_e E_o \exp(-i\omega t) - \eta dx/dt.$$

Steady-state solutions to differential equations of this sort have the following form (see any textbook on differential equations):

$$x(t) = A \exp(-i\omega t),$$

where A is a constant that is evaluated by substituting this expression for x(t) into the differential equation. The result is:

$$- A\omega^2 \exp(-i\omega t) + \omega_o^2 A \exp(-i\omega t) - i(\eta\omega/m)A \exp(-i\omega t)$$
$$= -(q_e E_o/m) \exp(-i\omega t). \quad (3.2.1)$$

Equation (3.2.1) will be true if:

$$A = -(q_e E_o/m)(\omega_o^2 - \omega^2 - i\eta\omega/m)^{-1}.$$

Thus:

$$x(t) = -(q_e E_o/m)(\omega_o^2 - \omega^2 - i\eta\omega/m)^{-1} \exp(-i\omega t).$$

Because $\mu = -q_e\mathbf{x} = \alpha\varepsilon_o\mathbf{E}_o$, we obtain:

$$\alpha = (q_e^2/m\varepsilon_o)(\omega_o^2 - \omega^2 - i\eta\omega/m)^{-1}. \quad (3.2.2)$$

If both the numerator and the denominator of α are multiplied by $({\omega_o}^2 - \omega^2 + i\eta\omega/m)$, α can be separated into its real and imaginary parts:

$$\alpha = \left(q_e^2/m\varepsilon_o\right)\left[\left(\omega_o^2 - \omega^2\right) + i\eta\omega/m\right]\left[\left(\omega_o^2 - \omega^2\right)^2 + (\eta\omega/m)^2\right]^{-1}. \quad (3.2.3)$$

This equation describes the behavior of this system under steady-state conditions, which implies that the atom or molecule has been exposed to electromagnetic radiation for so long that the magnitudes of all of the energy flows related to its interactions with the incident radiation have ceased changing with time. Assuming the displacements of electrons in multielectron atoms are independent of one another, the polarizability of an atom of atomic number Z will be Z times equation (3.2.3).

APPENDIX 3.3 ENERGY ABSORPTION AND THE IMAGINARY PART OF α

Suppose an atom is being acted on by an electric field that obeys the equation $E_o \cos(-\omega t)$. The dipole moment at any instant will be $\alpha\varepsilon_o E_o \cos(-\omega t)$, and the position of the electron will be $-(1/q_e)\alpha\varepsilon_o E_o \cos(-\omega t)$. If α is written as (Re + i Im), then:

$$x(t) = -\mathrm{Re}[(1/q_e)(\mathrm{Re} + i\,\mathrm{Im})\varepsilon_o E_o \cos(-\omega t)]$$
$$= -(1/q_e)\mathrm{Re}[(\mathrm{Re} + \exp(i\pi/2)\mathrm{Im})\varepsilon_o E_o \cos(-\omega t)]$$
$$= -(1/q_e)[\mathrm{Re}\ \cos(-\omega t) - \mathrm{Im}\ \sin(-\omega t))\varepsilon_o E_o].$$

Now work equals force times displacement, which in this case can be written as force times the product of velocity and time. Thus:

$$W = q_e E_o^2 \varepsilon_o \int \cos(-\omega t)(\omega/q_e)[\text{Re } \sin(-\omega t) + \text{Im } \cos(-\omega t)]dt$$

$$= \omega E_o^2 \varepsilon_o \int_0^{2\pi/\omega} [\text{Re} \sin(-\omega t) \cos(-\omega t) + \text{Im } \cos^2(-\omega t)]dt.$$

NB: $2\pi/\omega$ is the duration of one cycle of oscillation. The Re part of the integral is 0, and the Im part of the integral is (π/ω)Im. Thus:

$$W = \pi E_o^2 \varepsilon_o \text{ Im} = \pi (q_e E_o)^2 (\eta\omega/m) \left[\left(\omega_o^2 - \omega^2\right)^2 + (\eta\omega/m)^2 \right]^{-1}.$$

PART TWO

Crystallography

4

On the Diffraction of X-rays by Crystals

The lion's share of what is known today about the three-dimensional arrangement of atoms in biological molecules has been extracted from the X-ray diffraction patterns of the crystals that they form. Among the molecules whose structures have been determined crystallographically are most of the small molecules found in living organisms, such as amino acids and nucleotides, and tens of thousands of biological macromolecules and macromolecular complexes. There is every reason to believe that x-ray crystallography will remain the dominant technique in structural biology, into the indefinite future.

X-ray crystallography, the science of deducing structures from diffraction patterns, got its start in 1912, shortly after von Laue's discovery that crystals diffract X-rays. W. H. Bragg and W. L. Bragg, who were father and son, solved the first crystal structures, which were those of salts like sodium chloride. The first protein crystals were prepared in the 1920s, but it was not until the late 1930s that the first useful diffraction patterns were obtained from them. What those diffraction patterns revealed was the magnitude of the challenge protein crystals posed to crystallographers; in the 1930s, no one had any idea how to go about extracting structures from diffraction data sets that complicated. Thus, unlike most of the scientific developments described by the press as "breakthroughs" today, Max Perutz's discovery in the 1950s of an experimental method for solving the structures of protein crystals was the real thing. It ushered in the modern era of structural biology.

In the late 1950s and early 1960s, Perutz and Kendrew published the first protein crystal structures that displayed the positions of atoms, an advance that proved that Perutz's approach to solving protein crystal structures works in practice. Nevertheless, for the next two decades or so, it took years of effort on the part of entire teams of scientists to solve a single protein crystal structure. Happily, crystallographic technology has improved dramatically since then. Today, diffraction data sets that would have taken months to collect in the 1960s or 1970s are measured in minutes. Furthermore, the algorithms for analyzing crystallographic data, the computer programs that realize them, and the computers that run those programs have all improved immeasurably. Macromolecular crystal structure determination is on its way to becoming a routine analytical technique, like small molecule crystallography, the fruits of which are as accessible to the uninitiated as they are to the expert.

Nevertheless, structural biologists still need to understand how crystal structures are solved. For one thing, today's technology does *not* automatically solve all macromolecular crystal structures, and in fact, many of the crystal structures in the Protein Data Bank would not be there if the individuals responsible for them had not known what they were doing. Furthermore, those who do not understand how crystal structures are solved are at risk of being led astray when they make use of structures solved by others. Finally, the last word on how to obtain structures from diffraction data has yet to be written, and only those who fully understand that art can possibly move the field forward.

This chapter is the first of four on X-ray crystallography. We begin by comparing the way single molecules scatter/diffract X-rays to the way crystals of molecules diffract X-rays.

4.1 THE FOURIER TRANSFORM OF A ROW OF DELTA FUNCTIONS IS A ROW OF DELTA FUNCTIONS

A (perfect) **crystal** is a one-, two-, or three-dimensional array of molecules, or constellation of molecules, in which each molecule, or constellation of molecules, bears exactly the same geometric relationship to its nearest neighbors as every other. This cannot be true at the boundaries of crystals, of course, but because almost all the scattering from crystals big enough to see is produced by the molecules in their interiors, boundary effects can be ignored.

The simplest crystal we could conceive of is an infinitely long line of point-atoms, each separated from its nearest neighbors by the same distance. The electron density distribution of such a crystal can be represented mathematically as follows:

$$\mathrm{p}(x) = \sum_{-\infty}^{\infty} \delta(x - na),$$

where n is an integer and a is the separation between neighboring delta functions. This function is sometimes described as the **picket fence function**, and if $a = 1$, it may be called the **shah function**. Any regular array of points like this is a **lattice**, and lattices can be one-, two-, or three-dimensional.

It is easy to see what the (one-dimensional) Fourier transform of $\mathrm{p}(x)$ must be once it is rewritten as follows:

$$\mathrm{p}(x) = \delta(x) + \sum_{1}^{\infty} [\delta(x - na) + \delta(x + na)].$$

The Fourier transform of the first term is 1.0 for all s, and the Fourier transform of each term in the sum is $2\cos(2\pi nas)$ (see chapter 2). Therefore:

$$\mathrm{P}(s) = 1 + 2\sum_{1}^{\infty} \cos(2\pi nas).$$

It may not be instantly obvious that $\mathrm{P}(s)$ is a row of delta functions in reciprocal space separated by intervals of $(1/a)$, but it is easy to see that it is plausible. In the first place, at all values of s where $s = (m/a)$ and m is an integer, $\mathrm{P}(s)$ will be ∞ because the value of all the cosine terms in the sum will be 1.0. In the second place, at all locations in reciprocal space where $s \neq (m/a)$, $\mathrm{P}(s)$ will be 1 plus a sum of cosine terms, half of which are positive and half of which are negative. Thus, at those locations, $\mathrm{P}(s)$ will be of the order of 0. Hence, it is not too big a stretch to propose that:

$$\mathrm{P}(s) = 1 + 2\sum_{1}^{\infty} \cos(2\pi nas) = (1/a)\sum_{-\infty}^{\infty} \delta(s - n/a),$$

where the $(1/a)$ term is the multiplier that has to be included to get the scaling right (see section 2.6). The Fourier transform of a picket fence function is a picket fence function. The Fourier transform of a lattice in real space is a lattice in reciprocal space.

4.2 SAMPLING IN RECIPROCAL SPACE CORRESPONDS TO REPLICATION IN REAL SPACE (AND VICE VERSA)

A one-dimensional function in real space that has been multiplied by $\mathrm{p}(x)$, like a one-dimensional function in reciprocal space that has been multiplied by $\mathrm{P}(s)$, is said to be **sampled**. The reason is that the product of $\mathrm{g}(x)$, which can be any continuous function, and $\mathrm{p}(x)$ is:

$$\mathrm{g}(x)\mathrm{p}(x) = \sum_{-\infty}^{\infty} \mathrm{g}(na)\delta(x - na).$$

Thus, multiplication by $\mathrm{p}(x)$ transforms a continuous function into a discrete string of numbers that "sample" the value of the continuous function at evenly spaced intervals. The same thing happens when functions in reciprocal space are multiplied by $\mathrm{P}(s)$.

If $\mathrm{g}(x)$ and $\mathrm{G}(s)$ are Fourier mates, what will $\mathrm{h}(x)$ be if $\mathrm{h}(x) = \mathrm{FT}^{-1}(\mathrm{P}(s)\mathrm{G}(s))$? Because the inverse Fourier transform of the product of two functions is the convolution of their inverse Fourier transforms:

$$h(x) = FT^{-1}\,(P(s)G(s)) = p(x) * g(x),$$

$$h(x) = \int_{-\infty}^{\infty} \sum_{-\infty}^{\infty} \delta(x - na - x')g(x')dx' = \sum_{-\infty}^{\infty} g(x - na).$$

Thus, $h(x)$ is the sum of an infinitely long string of $g(x)$ functions each with its own origin, and there is a new origin every $\pm a$ along x, starting at $x = 0$. Going in the other direction, the Fourier transform of a sum of $g(x)$ functions whose origins occur every a in real space is the product of $FT(g(x))$, the transform of a single such function, and a series of delta functions in reciprocal space that start at $s = 0$ and occur every $(1/a)$.

If $g(x)$ is nonzero over a range of values in x that exceeds a, it is likely to be a challenge to recognize that $h(x)$ is in fact a row of replicas of $g(x)$ because in the regions where adjacent replicas overlap, they will add. However, if the range over which $g(x)$ is nonzero is less than a, the relationship between $h(x)$ and $g(x)$ will be obvious because adjacent replicas of $g(x)$ will not overlap.

Examining the function $FT^{-1}(P(s)G(s))$ a little more carefully, we find that:

$$\begin{aligned} FT^{-1}\,(P(s)G(s)) &= (1/a)\sum \int \delta(s - n/a)G(s)\exp(2\pi isx)ds \\ &= (1/a)\sum G(n/a)\exp\left[2\pi ix(n/a)\right]. \end{aligned}$$

Thus, sampling has the effect of converting Fourier integrals into Fourier sums. In section 2.2, it was pointed out that Fourier series naturally repeat and that Fourier integrals do not. We see this phenomenon in action here. The Fourier transform of a continuous function in real space is a continuous function in reciprocal space, whereas the Fourier transform of a lattice of continuous functions in real space is a Fourier series in reciprocal space.

4.3 CRYSTALS CAN BE DESCRIBED AS CONVOLUTIONS OF MOLECULES WITH LATTICES

Since the 19th century, the locations of atoms in crystals have been described using a language that separates the arrangement of atoms in the individual molecules or groups of molecules of which a crystal is composed from the way those molecules are packed in crystals. The key to this method for describing crystal geometry is the concept of the three-dimensional lattice, and given what we already know about one-dimensional lattices, the properties of three-dimensional lattices, should be no surprise.

A three-dimensional lattice is a three-dimensional array of points of infinite extent in which every point has neighbors in exactly the same relative locations as every other point. Thus, the neighborhoods of all the points in a three-dimensional lattice are identical. The reason the lattice concept is so useful to crystallographers is that the electron density distributions of an entire crystal can be described

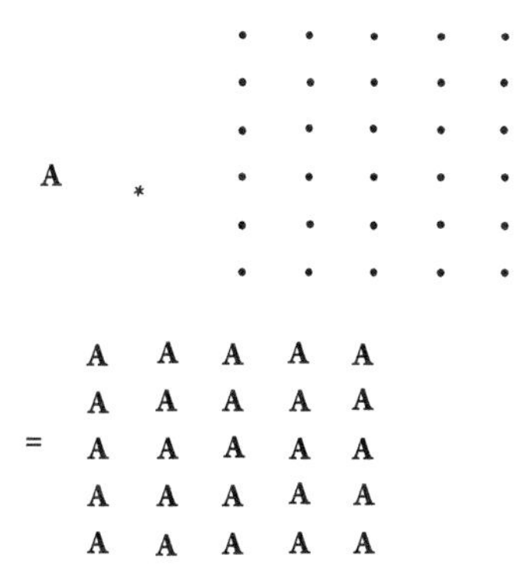

Figure 4.1 A two-dimensional crystals of "A"s represented as the convolution of a single letter A with a two-dimensional lattice of points.

mathematically as the convolution of a three-dimensional lattice with the electron density distribution of a single molecule (or single group of molecules). Figure 4.1 is a picture of a two-dimensional lattice that illustrates this idea.

It follows that the Fourier transform of the electron density distribution of a crystal will be the product of the Fourier transform of $\rho(\mathbf{r})$, the electron density distribution of the molecule or groups of molecules of which the crystal is made, and the Fourier transform of its lattice. In section 4.8, we will show that the Fourier transform of a three-dimensional lattice in real space is a three-dimensional lattice in reciprocal space, which is consistent with what we learned about the transforms of one-dimensional lattices in section 4.1. Thus, the Fourier transform of the electron density distribution of a crystal is a sampled version of $F(\mathbf{S})$, the Fourier transform of the structure of the molecule or groups of molecules found in that crystal.

The sampling that characterizes the Fourier transforms of crystals has important practical consequences. If you place an X-ray detection device such as a piece of photographic film downstream from a crystal that is being irradiated with a beam of monochromatic X-rays, the diffraction pattern recorded on that detector will *not* be a continuous function. It will instead be an array of discrete spots, each corresponding to a place in reciprocal space where the transform of the crystal lattice is not 0. For historical reasons that will become clear shortly, those spots are called **reflections**.

4.4 LATTICES "AMPLIFY" FOURIER TRANSFORMS

Using one-dimensional lattices, we can easily arrive at an important generality that holds for lattices of all dimensions. The Fourier transform of $h(x)$ (see section 4.2) can be written as follows:

$$\mathrm{FT}(\mathrm{h}(x)) = \sum \int \mathrm{g}(x - na) \exp(-2\pi ixs) dx,$$

which the shift theorem tells us is equivalent to:

$$\begin{aligned} \mathrm{FT}(\mathrm{h}(x)) &= \sum \exp(-2\pi inas) \int \mathrm{g}(x) \exp(-2\pi ixs) dx \\ &= \sum \exp(-2\pi inas) \mathrm{G}(s), \end{aligned}$$

but we already know that this sum is nonzero only when $s = m/a$, where m is some integer, and, thus, $\mathrm{FT}(\mathrm{h}(x))$ is a sampled version of $\mathrm{G}(s)$. Now, suppose the lattice of $\mathrm{g}(x)$ functions under consideration is not infinitely long, but rather is n repeats long, where n is an integer big enough so that the sin x/x functions that in the Fourier transforms of finite lattices replace the δ functions in the transform of the infinite lattices (see chapter 2) are still very narrow compared to a, then at (m/a), $\mathrm{FT}(\mathrm{h}(x)) = n\mathrm{G}(m/a)$, and the intensity of the transform at that point will be $n^2\mathrm{G}(m/a)\mathrm{G}^*(m/a)$.

In addition to amplifying scattering signals, crystals give the structural biologist something else that is very valuable. Not only is the scattering produced by single molecules so weak that under most circumstances it can hardly be detected, it is impossible to control the orientation of single molecules with the precision required if their structures are to be determined by Fourier inversion of the scattering data they provide. A crystal of molecules, on the other hand, is something big enough to see, big enough to work with, and big enough so that it can be held firmly in a beam of X-rays in whatever orientation an experimentalist may desire.

As we continue our discussion of crystals, it is important to realize that even though the crystallographer intent on determining the structure of some biological macromolecule must pay attention to the geometry of the crystals it forms, lattice geometries (usually) have little or no biological significance per se. The only property of crystals that really matters is the access they provide to $\mathrm{F}(\mathbf{S})$, the Fourier transform of the electron density distribution of the molecules they contain.

4.5 THE NYQUIST THEOREM TELLS YOU HOW OFTEN TO SAMPLE FUNCTIONS WHEN COMPUTING FOURIER TRANSFORMS

Sometimes the scattering of a specimen *can* be measured experimentally at all values of $\mathbf{S}$ (e.g., see section 8.1). When this is so, which it never is for crystallographers, an important practical question arises. How frequently in $\mathbf{S}$ must $\mathrm{I}(\mathbf{S})$ be measured if a faithful rendering of $\rho(\mathbf{r})$ is to emerge when the inverse Fourier transform of the correspondingly sampled version of $\mathrm{F}(\mathbf{S})$ is computed? Because that Fourier inversion is necessarily going to be carried out using a digital computer, $\mathrm{FT}^{-1}(\mathrm{F}(\mathbf{S}))$ is going to be evaluated using an expression that amounts to $\sum \mathrm{F}(\mathbf{S}_i) \exp[2\pi i(\mathbf{r} \cdot \mathbf{S}_i)]\Delta\mathbf{S}$, rather than its continuous equivalent, $\int \mathrm{F}(\mathbf{S}) \exp[2\pi i(\mathbf{r} \cdot \mathbf{S})] dV_s$, no matter what. Thus, the need to decide how many $\mathrm{F}(\mathbf{S}_i)$ to include in a Fourier sum cannot be escaped.

It is not hard to figure out how often $F(\mathbf{S})$ must be sampled. In the first place, $F(\mathbf{S})$ must be sampled on a regular lattice in reciprocal space. Otherwise, the transform of the array of points on which $F(\mathbf{S})$ is sampled will not be a lattice in real space. Because an inverse Fourier transform computed using $F(\mathbf{S})$ data sampled on a lattice in reciprocal space is a Fourier sum, the rendering of $\rho(\mathbf{r})$ that emerges will repeat in real space on a real space lattice, the shape and size of which is reciprocally related to the sampling lattice in reciprocal space. Thus, to ensure that adjacent images of the molecule do not overlap, the sampling interval in each direction in reciprocal space should not be less than the reciprocal of the maximum linear dimension of the molecule (or constellation of molecules) in the corresponding direction.

The idea that the transform of a real space function must be sampled no less frequently than the reciprocal of that function's maximum extent to ensure that the sampled transform adequately represents the function's continuous transform is called **Nyquist's theorem**. It should be noted that the sampling of molecular transforms seen in crystal diffraction patterns is certain to satisfy the Nyquist criterion because the space occupied by one molecule cannot be invaded by its neighbors.

Comment: Because adjacent molecules touch each other in crystals, it might be thought that the sampling of molecular transforms in crystal diffraction patterns reduces the number of values of $F(\mathbf{S})$ *that can be accessed experimentally to the minimum that can be tolerated if the electron density distributions of the molecules in crystals are to be recovered from the data. That would be so if molecular electron density distributions could have any functional form whatever, but this is not the case. Molecules are made of atoms, which are quasi-spherical objects, having radii that are never less than* $\sim$ *1.0 Å. Furthermore, the electron density inside crystals is everywhere* ≥ 0*. Thus, the set of electron density distributions possible for molecules is much smaller than the set of all conceivable three-dimensional distribution functions. This fact has useful consequences that will be discussed in chapter 7.*

The Nyquist theorem also has implications for those computing the Fourier transforms of real-space functions. Judgment must be exercised when computing the transforms of functions in real space because in principle, $F(\mathbf{S})$ can extend to $|\mathbf{S}| = \infty$, even if $\rho(\mathbf{r})$ is of finite extent in real space. Thus, if $F(\mathbf{S})$ does have significant amplitude at very large $|\mathbf{S}|$, as a practical matter, it may be impossible to compute $F(\mathbf{S})$ accurately because no matter how finely $\rho(\mathbf{r})$ is sampled, adjacent replicas of $F(\mathbf{S})$ will overlap significantly in the rendering of reciprocal space that emerges. Fortunately, if for no other reason than the fact that atoms have finite sizes, the Fourier transforms of the electron density distributions of real molecules tend to diminish in amplitude with increasing $|\mathbf{S}|$. If it is known that beyond some value of $|\mathbf{S}|$, $|\mathbf{S}|_{max}$, the transform of $\rho(\mathbf{r})$ is negligibly weak, then the Nyquist theorem says that a decently accurate version of $F(\mathbf{S})$ will emerge if $\rho(\mathbf{r})$ is sampled no less frequently than $1/(2|\mathbf{S}|_{max})$. (The factor of 2 in this expression is necessary because $|\mathbf{S}|_{max}$ is the *radius* of the region of interest in reciprocal space, not its *diameter.*)

4.6 LATTICES DIVIDE SPACE INTO UNIT CELLS

Even though the geometries of the lattices of macromolecular crystals seldom have biological significance, it is important to understand how they are described. Figure 4.2 is a picture of a two-dimensional lattice, and through one of its points a large number of lines have been drawn that radiate out from it in many different directions. Of the infinite number of lines that could be drawn through that point, two will intersect a greater number of lattice points, per unit length, than all the others. In figure 4.2, those two lines are labeled **a** and **b**, and they define the natural coordinate axes to use for describing the locations of objects in that lattice. We will call the vectors that join the lattice point used as the origin in this exercise to its nearest neighbors along those two lines **a** and **b** (see figure 4.3).

If lines parallel to **a** and **b** are drawn through every point in a lattice, the plane of the lattice will be divided into parallelograms that are all the same size and shape, that together entirely fill the plane, and that have a lattice point at each corner. These parallelograms are called **unit cells** (figure 4.4). The same exercise can be carried out

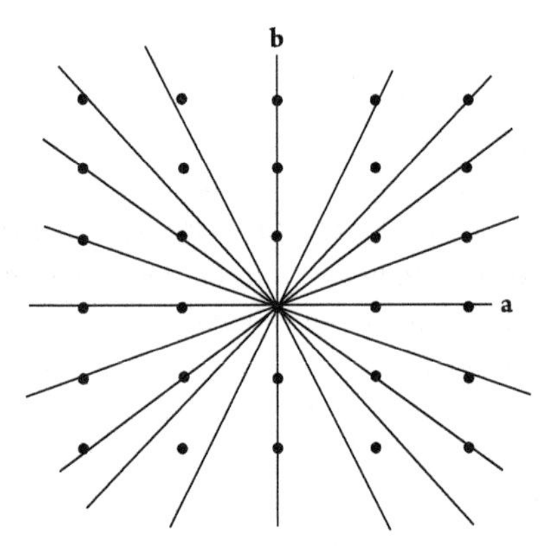

Figure 4.2 Determining the axes of a two-dimensional lattice.

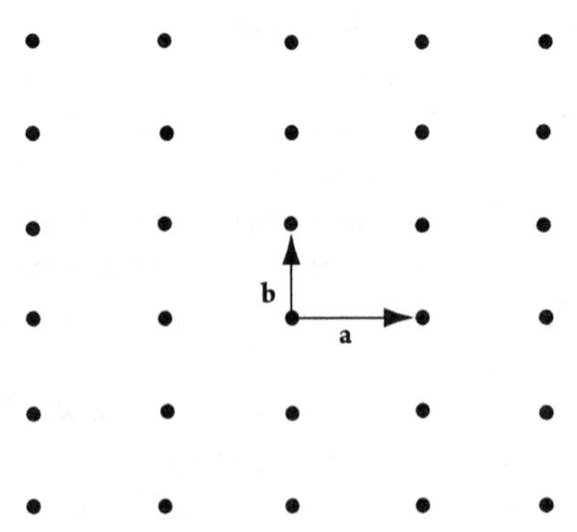

Figure 4.3 Definition of the unit cell vectors of a two-dimensional lattice.

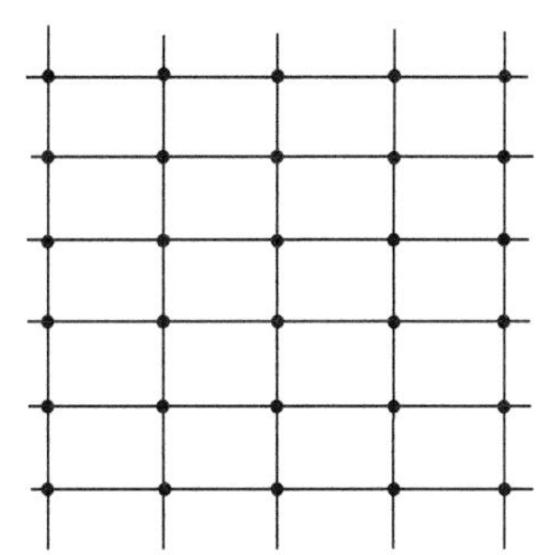

Figure 4.4 A two-dimensional lattice divided up into its unit cells.

on a three-dimensional lattice, and when it is, three preferred directions—**a**, **b**, and **c**—will emerge, and the unit cells will be parallelepipeds, the sides of which have lengths $|\mathbf{a}|$, $|\mathbf{b}|$, and $|\mathbf{c}|$. The angles those vectors make with each other are called α, β, and γ, where α is the angle between **b** and **c**, β is the angle between **a** and **c**, and γ is the angle between **a** and **b**.

A unit cell is said to be **primitive** if it has lattice points only at its corners, of which there will be eight if the lattice is three-dimensional. Furthermore, no matter what the arrangement of points in a lattice, it is always possible to define a primitive unit cell for it. However, for reasons that will be explained shortly, crystallographers often describe lattices using unit cells that are not primitive, and when they do, the axis system employed will not be the same as the one used to define the primitive unit cells of that same lattice. Furthermore, the unit cells defined by nonprimitive axis systems will always be larger than the primitive unit cells of the same lattice and, necessarily, will contain and/or contact more than eight lattice points.

Nonprimitive unit cells are used to describe crystal lattices when crystals possess rotational symmetry, but before continuing with this discussion, it is important to be clear that symmetry is a *local* property in crystals, not a global one. If a crystal is translated by a distance $|\mathbf{a}|$ along the **a** axis of its unit cell, for example, even though the crystal as a whole will have moved, its local appearance, which is to say the appearance it presents to someone whose field of view is restricted to, say, 20 unit cells in the middle of the crystal, will not have changed at all. Thus, (locally) all crystals have the translational symmetry that is described by their lattices. Viewed locally, they may also possess additional symmetry that is not translational, such as rotational symmetry. For example, the local appearance of some crystal may not change when it is rotated by 180° about an axis that passes through its lattice points in some special direction. If the rotational symmetry a crystal possesses is not obvious in the primitive unit cells of its lattice, a larger, nonprimitive unit cell that does clearly display that symmetry will often be used to describe its lattice instead.

Figure 4.5 illustrates why nonprimitive unit cells can make sense. In this example, the lattices of the two, two-dimensional crystals depicted in that figure happen to be

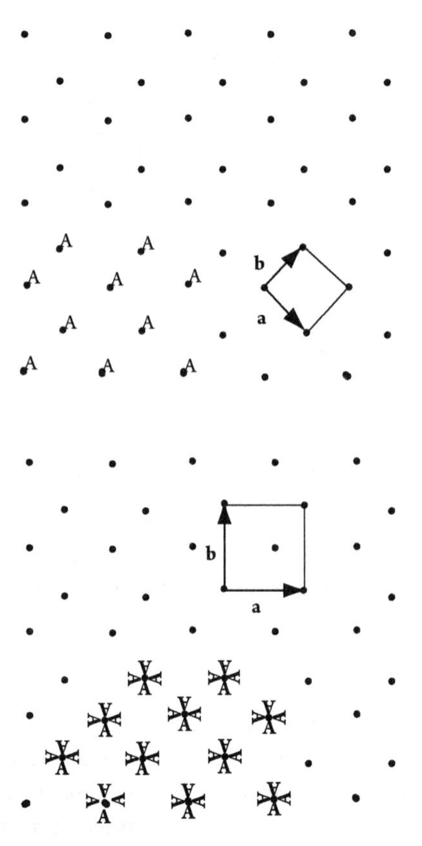

Figure 4.5 Two crystals that have identical lattices, but different rotational symmetries. Both lattices are shown with only a small number of the "molecules" of which they are composed in place so that the underlying lattices will be evident.

identical, but the contents of their unit cells are not. In the top "crystal", each lattice point is associated with a single "molecule", which in this case looks remarkably similar to the letter "A". In the bottom "crystal", the molecule associated with each lattice point is a cluster of four "A"s arranged in rosette that has four-fold rotational symmetry. The top crystal has no rotational symmetry, and thus the rotational symmetry of its primitive unit cells is the same as the crystal as a whole. On the other hand, the bottom crystal has four-fold rotational symmetry, which is to say that the local appearance of that crystal would not change if it were rotated by 90°, 180°, or 270° around an axis normal to the plane that passes through any one of its lattice points.

The bottom crystal can be described using the same (primitive) unit cell as the top crystal, but that unit cell would not contain any complete rosettes, which, if all you had to look at was a single unit cell, would make it hard to discern that a crystal made up of such unit cells possesses four-fold rotational symmetry. However, if the unit cell axes for the bottom crystal are chosen as shown, the larger unit cell they define will display the four-fold symmetry characteristic of the parent crystal.

Comment: As the example shown in figure 4.5 illustrates, in order for a crystal to display some particular kind of rotational symmetry, both its lattice and the contents of its unit cells must have geometries that are compatible with it.
Comment: Beyond the translation operations that characterize the lattice of a crystal, structural biologists need to concern themselves with only two other kinds of symmetry operations: rotations, and the combined translation–rotation operations called "screw axis rotations". The rotations possible in crystals are rotation by 180°, 120°, 90°, and 60°, which are often described as two-, three-, four-, and six-fold rotations, respectively, and we note that if 90° rotation about some axis is a symmetry operation for some crystal, for example, so too will 180° and 270° rotation about that same axis. An n_m-fold screw axis rotation corresponds to translation along a unit cell axis by an amount equal to m/n times its length followed by rotation by $360°/n$. (NB: m must be an integer greater than 0 but less than n. Two-, 3-, 4-, and 6-fold screw axes appear in the crystals formed by biological macromolecules.)

All possible three-dimensional lattices were identified in the 19th century. They are called the **Bravais lattices**, and there are 14 of them (see appendix 4.1). Based on differences in rotational symmetry, the 14 can be divided into 7 systems. Within each system, more than one kind of unit cell may be possible, which is the reason there are 14 Bravais lattices, instead of 7. The molecules that fill the unit cells of a crystal are often packed together in a symmetric manner, and when the different symmetries of the contents of unit cells that are compatible with each of the Bravais lattices are taken into consideration, 230 three-dimensional space groups emerge. The space group to which a crystal belongs is determined by the totality of the translational operations that relate its lattice points to each other, and all the other symmetry operations its structure accommodates. Crystals of biological macromolecules can belong to only 64 of the 230 possible space groups because biological macromolecules contain chiral centers, and 166 space groups include symmetry operations that invert chiral centers. The 230 space groups are described in the *International Tables for Crystallography*.

4.7 THE MINIMAL ELEMENT OF STRUCTURE IN ANY UNIT CELL IS ITS ASYMMETRIC UNIT

A crystal can be thought of as constructed of identical "bricks", or unit cells, each of which encompasses structures that together constitute at least one molecule. (If you look carefully at the top crystal in figure 4.5, you will realize that its unit cell does not enclose a single, complete "A". What it contains instead are unconnected fragments of "A"s that come together to make complete "A"s only when unit cells are assembled to make the crystal. (NB: The boundaries of the unit cells of a crystal need not partition the molecules it contains in a chemically sensible way!) However, as we already pointed out, crystals of biological molecules commonly possess more than just translational symmetry, and when they do, it may be possible conceptually, to regenerate the structure of an entire crystal once the structure of an entity smaller than the contents of an entire unit cell has been worked out. This minimal element of structure is the **asymmetric unit** of a crystal. The structure of the contents of a single unit cell of a crystal can be recovered by operating on the structure of its asymmetric unit using all the symmetry operations of that crystal (except for its translational symmetry operations), and once this has been done, the structure of the crystal as a whole can be regenerated by translation.

It sometimes happens in biological crystals that the molecular assembly in the unit cell has a higher symmetry than that displayed by the crystal of which it is a part, or has internal symmetry elements that do not align with any of the symmetry elements of the crystal as a whole. These situations are described by saying that the molecule in the crystal possesses **noncrystallographic symmetry**. Noncrystallographic symmetry can be taken advantage of when solving crystal structures, as we shall see.

4.8 THE TRANSFORM OF A THREE-DIMENSIONAL LATTICE IS NONZERO ONLY AT POINTS IN RECIPROCAL SPACE THAT OBEY THE VON LAUE EQUATIONS

A two-dimensional lattice in a three-dimensional real space can be described mathematically as the convolution of a line of points separated by a constant distance a and oriented along a vector $\mathbf{a}$ and a noncollinear line of points that are separated by intervals of b and oriented along a vector $\mathbf{b}$. If that two-dimensional lattice is then convoluted with a third row of equally spaced points that are separated by c and aligned with $\mathbf{c}$, a vector that does not lie in the $(\mathbf{a}, \mathbf{b})$ plane, a three-dimensional lattice will emerge. Because this is so, once the three-dimensional Fourier transform of an equally spaced, linear array of points has been worked out, the (three-dimensional) Fourier transform of a three-dimensional lattice can be expressed as the product of the Fourier transforms of three such linear arrays by virtue of the convolution theorem.

The first step in the derivation of the three-dimensional Fourier transform of a line of regularly spaced points in real space is to write down a three-dimensional function in real space, $\mathrm{L}(\mathbf{a})$, that represents a linear string of three-dimensional Dirac delta functions:

$$\mathrm{L}(\mathbf{a}) = \sum_{-\infty}^{\infty} \delta(\mathbf{r} - n\mathbf{a}).$$

In this equation, $\mathbf{r}$ is a vector from the delta function in that array that has been chosen as its origin to any other point in real space, and $\mathbf{a}$ is one of the three vectors that defines the unit cell of the lattice of interest. It extends from one lattice point to that lattice point's next neighbor along $\mathbf{a}$. By extending the arguments made in section 4.1, we can show that:

$$\mathrm{FT}(\mathrm{L}(\mathbf{a})) = \int \sum \delta(\mathbf{r} - n\mathbf{a}) \exp(-2\pi i \mathbf{r} \cdot \mathbf{S})$$

$$= \sum_{-\infty}^{\infty} \exp(-2\pi i \mathbf{a} \cdot \mathbf{S}) = (\infty/a) \sum \delta(\mathbf{a} \cdot \mathbf{S} - h).$$

(NB: a is the length of **a**.) Thus, the transform in question will be nonzero only at places in reciprocal space where $\mathbf{a} \cdot \mathbf{S}$ is an integer, and everywhere else it will be 0. We will henceforth call that integer h.

FT(L(**a**)) is a family of parallel planes in reciprocal space, and the distance between nearest neighbors planes in the direction normal to those planes, which is the **a** direction, is $(1/a)$. The vector **S** can point in any direction in reciprocal space, and what the condition $\mathbf{a} \cdot \mathbf{S} = h$ implies is that the transform of the corresponding row of equally spaced points in real space will be 0 unless the far end of **S** lies in one of the planes in that family.

The Fourier transform of the second row of points, the ones aligned along **b**, will also be a set of parallel, equally spaced planes in reciprocal space. Their separation in the direction of **b** will be $(1/b)$, and the equation that describes their locations in reciprocal space is nonzero only when $\mathbf{b} \cdot \mathbf{S}$ equals an integer we will call k. The product of these two transforms:

$$(\infty/a) \sum \delta(\mathbf{a} \cdot \mathbf{S} - h)(\infty/b) \sum \delta(\mathbf{b} \cdot \mathbf{S} - k)$$

is the Fourier transform of the two-dimensional lattice. It is nonzero only where **b** planes and **a** planes (which are not parallel) intersect in reciprocal space, and the intersection of two planes is a line. Thus, the three-dimensional Fourier transform of a two-dimensional lattice is a family of parallel lines perpendicular to the (**a**, **b**) plane in real space. Neighboring lines are separated by integer multiples of $(1/a)$ in the **a** direction and integer multiples of $(1/b)$ in the **b** direction, and along all such lines, both $\mathbf{a} \cdot \mathbf{S}$ and $\mathbf{b} \cdot \mathbf{S}$ are integers.

The Fourier transform of the **c** family of points is yet another set of parallel planes, and for these planes, $\mathbf{c} \cdot \mathbf{S}$ must be an integer, which we will call l. Because the intersections of planes with lines are points, the Fourier transform of a three-dimensional lattice in real space is a three-dimensional lattice in reciprocal space, as promised, and each point in that lattice must simultaneously satisfy three integer equations:

$$\mathbf{a} \cdot \mathbf{S} = h; \quad \mathbf{b} \cdot \mathbf{S} = k; \quad \mathbf{c} \cdot \mathbf{S} = l. \tag{4.1}$$

These three equations are called **von Laue's equations**, and using them, you can predict the locations of the points in reciprocal space where the Fourier transform of a crystal can have a value different from 0. Appendix 4.2 shows more formally how the lengths and orientations of the vectors describing unit cells in reciprocal space lattices, $\mathbf{a}^*$, $\mathbf{b}^*$, and $\mathbf{c}^*$, are related to the unit cell vectors of the corresponding real-space lattice, **a**, **b**, and **c**. The approach outlined there is general enough to cope with real-space lattices, the unit cell vectors of which are not mutually perpendicular, which is often the case.

4.9 THE FOURIER TRANSFORMS OF CRYSTALS ARE USUALLY WRITTEN USING UNIT CELL VECTORS AS THE COORDINATE SYSTEM

Once the reciprocal lattice of a crystal has been worked out (appendix 4.2), it is easy to write down an expression for **S** that describes where the Fourier transform of a crystal can be nonzero. At those locations:

$$\mathbf{S}\,(h,k,l) = h\,\mathbf{a}^* + k\,\mathbf{b}^* + l\,\mathbf{c}^*,$$

where h, k, and l are von Laue integers, and $\mathbf{a}^*$, $\mathbf{b}^*$, and $\mathbf{c}^*$ are the unit cell vectors of the reciprocal lattice of the crystal. The (h, k, l) integers of each $\mathbf{S}(h, k, l)$ identify a specific orientation of some crystal and a specific scattering angle, and if that crystal is placed in an X-ray beam of specified wavelength in that orientation, diffraction will (usually) be seen at the corresponding scattering angle. It follows that each reflection in the diffraction pattern of a crystal can be assigned a unique three-integer identifier (h, k, l), where the three integers are the von Laue integers of that reflection. A crystal diffraction pattern is said to have been **indexed** when the (h, k, l) values of all of its reflections have been determined.

If $\mathrm{F}(\mathbf{S})$ is the Fourier transform of the contents of the unit cell of some crystal, then:

$$\mathrm{F}(\mathbf{S}) = \sum_i \mathrm{f}_i \exp(-2\pi i \mathbf{r}_i \cdot \mathbf{S}).$$

The vector $\mathbf{r}_i$, which specifies the position of the ith atom in the unit cell, can be written as the sum of the components of $\mathbf{r}_i$ in the **a**, **b**, and **c** directions as follows: $\mathbf{r}_i = x_i\mathbf{a} + y_i\mathbf{b} + z_i\mathbf{c}$, where x, y, and z are measured not in angstroms, for example, but instead in fractional unit cell coordinates. Thus, if the position of the ith atom is such that it is 50 Å away from the lattice point chosen as the origin of the unit cell coordinate system in the **a** direction and $|\mathbf{a}| = 100$ Å, its fraction coordinate, x_i, will be 0.5. Adhering to this convention, and using the expression for **S** as a sum of components we find:

$$\mathrm{F}(h,k,l) = \sum_i \mathrm{f}_i \exp\left[-2\pi i(x_i\mathbf{a} + y_i\mathbf{b} + z_i\mathbf{c}) \cdot (h\mathbf{a}^* + k\mathbf{b}^* + l\mathbf{c}^*)\right].$$

If **a**, **b**, and **c** are parallel to $\mathbf{a}^*$, $\mathbf{b}^*$, and $\mathbf{c}^*$, respectively, and perpendicular to each other, it will be true that:

$$\mathrm{F}(h,k,l) = \sum_i \mathrm{f}_i \exp\left[-2\pi i(hx_i\mathbf{a} \cdot \mathbf{a}^* + ky_i\mathbf{b} \cdot \mathbf{b}^* + lz_i\mathbf{c} \cdot \mathbf{c}^*)\right],$$

and because $|\mathbf{a}| = 1/|\mathbf{a}^*|$, $|\mathbf{b}| = 1/|\mathbf{b}^*|$, and $|\mathbf{c}| = 1/|\mathbf{c}^*|$:

$$\mathrm{F}(h,k,l) = \sum_i \mathrm{f}_i \exp\left[-2\pi i(hx_i + ky_i + lz_i)\right]. \qquad (4.2)$$

Equation (4.2) is true even when the unit cell vectors of a crystal are not mutually orthogonal. To prove that assertion, we start by pointing out that because $\mathbf{a}^* = \mathbf{b} \times \mathbf{c}/V$ (see appendix 4.2), $\mathbf{a} \cdot \mathbf{a}^* = (\mathbf{a} \cdot \mathbf{b} \times \mathbf{c})/V$. However, $(\mathbf{a} \cdot \mathbf{b} \times \mathbf{c}) = V$, and therefore $\mathbf{a} \cdot \mathbf{a}^*$ is 1.0. Using the same logic, $\mathbf{b} \cdot \mathbf{b}^* = \mathbf{c} \cdot \mathbf{c}^* = 1.0$. It follows that the step from the equation immediately before equation (4.2) to equation (4.2) is valid. The only issues that remain have to do with the values of terms such as $(x_i\mathbf{a}) \cdot (k\mathbf{b}^* + l\mathbf{c}^*)$, which the sequence of preceding equations implies must be 0, and it is easy to show that they are. For example, consider $\mathbf{a} \cdot \mathbf{b}^*$. It equals $\mathbf{a} \cdot (\mathbf{c} \times \mathbf{a}/V)$, but by definition, $\mathbf{b}^*$, the vector defined by the term inside the parentheses, is perpendicular to both $\mathbf{c}$ and $\mathbf{a}$, and the dot product of a vector with any vector that is perpendicular to it is 0. Thus, equation (4.2) is universally true.

The inverse Fourier transform of $\mathrm{F}(h, k, l)$ is $\rho(x, y, z)$, where x, y, and z are fractional coordinates. The relationship between the two is:

$$\rho(x,y,z) = (1/V)\sum_h \sum_k \sum_l \mathrm{F}(h,k,l) \exp\left[2\pi i(hx + ky + lz)\right],$$

where the sums are over all h, all k, and all l, and V is the volume of the unit cell. [NB: $\mathrm{F}(h, k, l)$ is a complex number.]

Comment: The (1/V) component on the right side of this equation is the three-dimensional equivalent of the (1/d) component on the right side of equation (2.3). Remember that X-ray structure factors (i.e., Fs), are numbers of electrons, and all exponential terms are dimensionless. Thus, if the expression on the right-hand side of this equation is to have the dimensions of number of electrons per unit volume (i.e., electron density), the expression inside the triple sum has to be divided by a volume.

4.10 BRAGG'S LAW PROVIDES A SECOND WAY TO DESCRIBE CRYSTALLINE DIFFRACTION PATTERNS

W. L. Bragg did not think about crystalline diffraction the way von Laue did. Bragg knew that crystals were lattices of atoms/molecules, and that you can draw families of parallel, equidistant planes through lattices that together intersect all the points in those lattices. He thought of these planes, which in a real crystal are populated with molecules, as reflecting X-rays the way mirrors reflect light. You will recall that when light is reflected by mirrors, the angle of incidence equals the angle of reflection. However, Bragg proposed that unlike the reflection of light from mirrors, which can occur over a wide range of angles, X-rays of a given wavelength will be

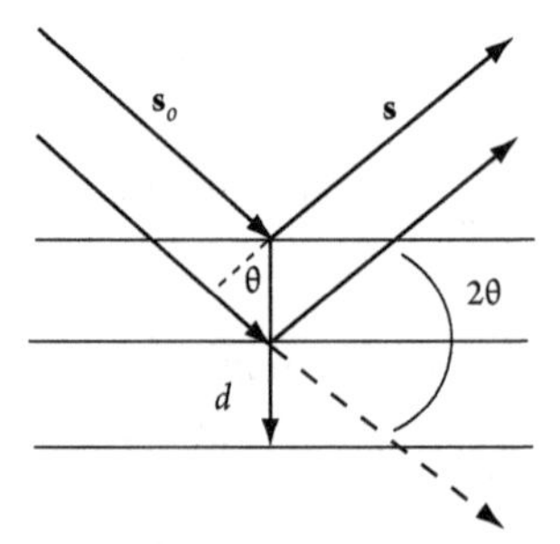

Figure 4.6 The geometry of X-ray diffraction, as envisioned by Bragg.

reflected effectively by parallel planes of atoms/molecules in a crystal only at special angles, namely at those angles of incidence for which the radiation reflected by one plane in a crystal interferes constructively with the radiation reflected by all the other planes in the crystal that are parallel to it. Figure 4.6 illustrates the geometry Bragg had in mind.

The horizontal lines in figure 4.6 represent parallel planes of atoms/molecules in some crystal, and, as usual, $\mathbf{s}_o$ and $\mathbf{s}$ are unit vectors in the direction of the incident X-ray beam and the diffracted X-ray beam, respectively, whereas $\mathbf{d}$ is the vector normal to the family of planes that connects one plane to its next neighbor. Assuming that X-rays reflect off planes of atoms/molecules in a crystal the way light reflects off a mirror, the angle of incidence will be equal to the angle of reflection, and the angle between $\mathbf{d}$ and $\mathbf{s}_o$ will be θ, which is one-half the scattering angle, 2θ, and the angle between $\mathbf{s}$ and $\mathbf{d}$ will be $[(\pi/2) - \theta]$. It follows that the distance X-rays reflected from the second plane must travel from the source to the detector will be longer than the distance X-rays reflected from the top plane travel by an amount equal to $(2d \sin \theta)$. If the X-rays that reflect from the two planes are to interfere constructively at the detector, this difference in distance must be an integer multiple of the wavelength of the radiation. If it is not, the radiation reflected off the different planes in that family of planes will cancel at the detector due to mutual interference. Thus, diffraction will be observed only if:

$$n/d = 2 \sin \theta/\lambda, \tag{4.3}$$

where d is the distance between nearest neighbor planes in the set of parallel planes responsible for some reflection, θ is one-half of the scattering angle, which is the angle between $\mathbf{s}$ and $\mathbf{s}_o$, and n is an integer, which is commonly taken to be 1, although it need not be. Equation (4.3) is **Bragg's law**. The geometry of diffraction/reflection implies that, $\mathbf{S}$, the scattering vector associated with any reflection is normal to the family of crystal planes responsible for it, and because $|\mathbf{S}| = 2 \sin \theta/\lambda$, Bragg's law tells us that $|\mathbf{S}| = (1/d)$.

4.11 VON LAUE'S INTEGERS ARE MILLER INDICES

Because von Laue's equations and Bragg's law are alternative descriptions of the same reality, they must be consistent with each other, and they are. However, before we can understand their relationship, we must take a short detour through some 19th-century science. In the 18th and 19th centuries, the geometries of crystals were investigated by chemists and mineralogists who were interested in substances like quartz and diamond that are naturally crystalline. They discovered that even though the overall shape, or **habit**, of the crystals of many substances can vary a lot, the relative orientations of the faces of the crystals formed by a given material are much less variable. In fact, the angles describing those orientations are so characteristic that they can be used to identify the substance of which a crystal is composed.

On the basis of observations of this sort, 19th-century crystallographers concluded that crystals must be three-dimensional arrays of atoms/molecules organized on lattices, and that the exterior faces of crystals must be parallel to the lattice planes that are densely populated with atoms/molecules. Because the geometry of a crystal's lattice determines the directions of the normals to those planes, the angles between the normals of the faces of a crystal big enough to see must be the same as the angles between the normals of the corresponding planes in the crystal's lattice. (When crystals are cleaved, they tend to shear along directions perpendicular to the normals of their densely populated planes rather than shearing across such planes, because stabilizing intermolecular interactions tend to be more abundant within such planes than they are between them.) Starting with these ideas, and with the information they had about the orientations of crystal faces, 19th-century crystallographers were able to work out the orientations and relative magnitudes of the unit cell vectors of many crystals.

Ever since the 19th century, the orientations of planes that form the faces of crystals have been specified using sets of integers called **Miller indices**. Imagine a set of parallel planes each separated from its nearest neighbor by the same distance such that, collectively, the set of planes intersects every point in some lattice. The Miller indices that describe the orientation of that family of planes are a triplet of integers, (h, k, l), where h is the number of times that **a** is intersected by members of that family of planes, k is the number of times **b** is intersected, and l is the number of times **c** is intersected. Figure 4.7 shows how Miller indices work on a two dimensional lattice. Two families of lines are shown, the $(1, 1)$ family and the $(2, 1)$ family. Note that the family of planes normal to **a** are the $(1, 0)$ planes, whereas the planes normal to **b** are the $(0, 1)$ planes.

Bragg's Law tells us that when X-rays are reflected by the (h, k, l) family of planes in a crystal, **S**, the scattering vector, is normal to those planes. By definition, $\mathbf{a}/h$ is a vector connecting a point in one member of this family of planes to a point in its nearest neighbor. Thus, its component in the direction of a unit vector normal to those planes must equal the distance between planes. Because $|\mathbf{S}| = 2 \sin \theta/\lambda$, $[\mathbf{S}/(2 \sin \theta/\lambda)]$ is the normal vector of unit length for that family of planes, and, thus, $[\mathbf{S}/(2 \sin \theta/\lambda)] \cdot (\mathbf{a}/h) = d$. Furthermore, Bragg's law tells us

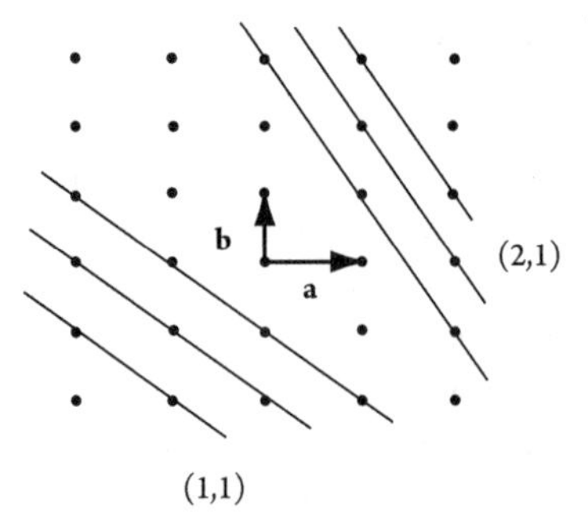

Figure 4.7 Miller planes and Miller indices on a two-dimensional lattice.

that $2\ \sin\ \theta/\lambda = (1/d)$, and, thus, if $[\mathbf{S}/(2\ \sin\ \theta/\lambda)] \cdot (\mathbf{a}/h) = d$, then $\mathbf{a} \cdot \mathbf{S} = h$. Similar reasoning applied to $\mathbf{b}/k$ and $\mathbf{c}/l$ yields the other two von Laue equations. Thus, Bragg's law and von Laue's equations are equivalent, and von Laue's integers are Miller indices.

Comment: In the era in which crystal lattices and Miller indices were worked out, no one knew how to measure the sizes of atoms and molecules, and in fact, until circa 1900, there were still reputable scientists who doubted the existence of atoms, let alone the existence of lattices of atoms. It is awe-inspiring that the early crystallographers got as far as they did in understanding the microscopic organization of crystals, and if that is not enough to fill you with wonder, you should contemplate the fact that the 230 space groups were also discovered in the 19th century by members of that same fraternity.

Some of the reflections observed in the diffraction patterns of crystals have indices that do not make sense as Miller indices. For example, consider the set of all planes in the lattice of some crystal that are parallel to the plane in which the unit cell vectors **b** and **c** lie. If the unit cell is primitive, that family of planes will intersect all lattice points, and its Miller indices will be $(1, 0, 0)$. Unless forbidden by systematic absences, a phenomenon we will discuss later, the diffraction pattern of that crystal will have a $(1,\ 0,\ 0)$ reflection, but, equally consistent with both von Laue's equations and Bragg's law, it is also all but certain to include reflections that have indices $(2, 0, 0)$, $(3,\ 0,\ 0)$, $\ldots\ (n,\ 0,\ 0)$. It is not hard to work out the orientation of the $(2, 0, 0)$ planes of a crystal; they are parallel to the $(1, 0, 0)$ planes. However, it would make no sense to describe the faces of a crystal as corresponding to its $(2, 0, 0)$ planes because every other member of the $(2, 0, 0)$ family intersects no lattice points whatever.

Reflections having Miller indices that can be written $(nh,\ nk,\ nl)$ are called **higher-order reflections**, where n is the order. When the higher-order versions of some reflections have diffracted intensity associated with them all it means is that it takes several Fourier terms to describe the variation in electron density in the corresponding crystal in the direction normal to the parent $(h,\ k,\ l)$ family of Miller planes, which should not be surprising. How many crystals can there be that have electron density distributions so simple that the variation in electron density in any given direction inside the crystal is a sinusoidal oscillation having a single spatial frequency?

4.12 EWALD'S CONSTRUCTION PROVIDES A SIMPLE TOOL FOR UNDERSTANDING CRYSTAL DIFFRACTION

P. P. Ewald, a theoretical physicist who was deeply involved with X-ray crystallography in its early days, devised a graphical method for analyzing the geometry of crystal diffraction that is as useful today as it was when he invented it almost a century ago. Figure 4.8 is a picture of Ewald's construction. A sphere, which is called the **sphere of reflection** for reasons that will soon emerge, or sometimes the **Ewald sphere**, is drawn the center of which is (conceptually) the position occupied by the crystal of interest, and which has a radius of $(1/\lambda)$ in the scale of the drawing. A line is drawn through the center of that sphere to represent the direction of the incident beam of X-rays. In figure 4.8, that line is labeled $\mathbf{s}_0/\lambda$ because $\mathbf{s}_0$ is the vector of unit length that points in the direction of the incident beam. At the point where the beam vector intersects the surface of the sphere beyond the crystal, the reciprocal lattice of the crystal is drawn to the scale of the drawing with its origin—its $(0, 0, 0)$ point—at that intersection. The orientation of the reciprocal lattice is determined by the orientation of the crystal, and it changes when the crystal is rotated. In order for the $(h,\ k,\ l)$ reflection to be observable, the reciprocal lattice must be oriented so that its $(h,\ k,\ l)$ point intersects the sphere of reflection. In figure 4.8, a single lattice point, the $(0, 2, 0)$ point, meets that criterion. It is easy to show that both Bragg's law and von Laue's equations are satisfied for any reflection that intersects the sphere. Thus, Ewald's geometric construction tells you how to orient a crystal in the beam so that the intensity associated with a specific reflection can be measured, and it also tells you what the scattering angle will be for that reflection when the crystal is so oriented.

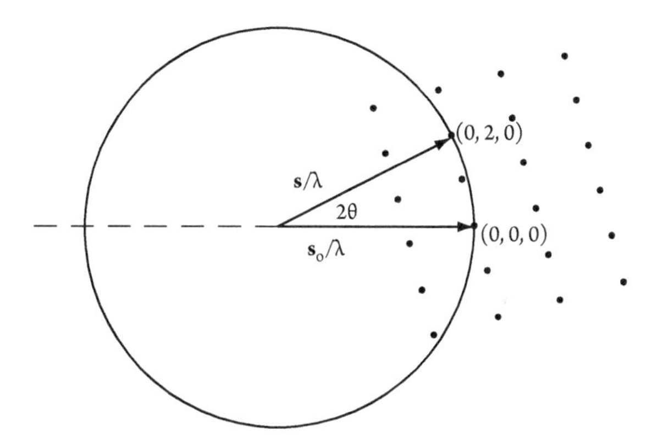

Figure 4.8 Ewald's construction.

Three facts about crystal diffraction are obvious once you grasp the principles behind Ewald's construction. First, if the unit cell lengths of some crystal are small (i.e., it is a crystal of small molecules), the density of lattice points in reciprocal space will be low, as it is in figure 4.8, and at any given orientation of that crystal in the X-ray beam, the number of lattice points intersecting the sphere of reflection will be small. It could be 0. On the other hand, if the crystal of interest contains macromolecules, the number of reflections intersecting the sphere will be large in any given orientation because of the large size of the unit cell. Second, crystals *must* be rotated in X-ray beams if the intensities of all their reflections are to be measured. Third, no matter how you orient a crystal in the beam, you will never be able to observe its diffraction pattern beyond $|\mathbf{S}| = 2/\lambda$, because that is all the sphere of reflection affords. The only way to make the sphere of reflection larger, so that a larger volume of the reciprocal space of some crystal can be examined experimentally, is to decrease the wavelength of the X-rays used to explore it.

PROBLEMS

1. A diffraction pattern is being collected from a crystal that has an orthorhombic unit cell ($a = 123$ Å; $b = 87$ Å; $c = 37$ Å) using X-rays that have a wavelength of 1 Å.
   a. At what scattering angle, 2θ, will the $(2, 5, 10)$ reflection be observed?
   b. If you were to produce an indexed list of all the reflections the diffraction pattern of this crystal might be measurable under these conditions, what are the largest values of h, k, and l that could appear in that list?
   c. Estimate the number of reflections that will be found in the diffraction pattern of this crystal between $|\mathbf{S}| = 0.25$ Å^{-1}, and $|\mathbf{S}| = 0.35$ Å^{-1}
2. Inversion is a mathematical operation that takes a molecule, each atom of which is located (x, y, z), and transforms it into a molecule the atoms of which are located at $(-x, -y, -z)$. For some molecules (e.g., methane), inversion is a symmetry operation, which means that the inverted and uninverted molecules are indistinguishable; in other words, they will superimpose if rotated appropriately.

   Molecules for which inversion is a symmetry operation sometimes form crystals for which inversion is also a symmetry operation. In other words, in each unit cell, for every atom at (x, y, z), there is an identical atom at $(-x, -y, -z)$. Prove that the phases for all the reflections of such crystals must be either 0 or π. [Hint: What is the total contribution made to $F(h, k, l)$ by a particular atom—the one at (x, y, z)—and its symmetry mate—the one at $(-x, -y, -z)$?]
3. Suppose diffraction data are being measured from a crystal that has unit cell dimensions of $a = 40$ A, $b = 50$ A, and $c = 60$ A, and $\alpha = \beta = \gamma = 90°$ using X-rays having a wavelength of 1.0 A. Suppose that the crystal has been mounted so that its b-axis is perpendicular to the direct beam and aligned with the spindle of the device included in the instrument being used to collect data that makes it possible to rotate crystals in the beam

under computer control. (For the purposes of answering the questions that follow, assume the crystal diffracts beyond 2 A^{-1}.)

a. Can any diffraction spots having a k index of 51 be observed with the crystal mounted this way? If so, which ones?
b. What will the Miller indices be of the $(h, 25, 0)$ reflection observed that has the smallest value of h?
[Hint: Remember Ewald.]

4. What is the Bragg spacing that corresponds to a scattering angle of 90° if the radiation being used to measure scatter is 1.54 Å.

APPENDIX 4.1 THE BRAVAIS LATTICES

Three-dimensional lattices are classified on the basis of their rotational symmetries, and when this is done, we discover that there are only seven types, or systems, of lattices possible: **triclinic, monoclinic, orthorhombic, tetragonal, trigonal/rhombohedral, hexagonal**, and **cubic**. Triclinic crystals have no rotational symmetry whatever. Each unit cell of a monoclinic crystal must have a two-fold rotation axis or a two-fold screw axis coincident with one its axes, which is often (but not always) taken to be the b axis. Each unit cell of an orthorhombic crystal must have three mutually perpendicular two-fold rotation or two-fold screw axes that are aligned with its own axes. Tetragonal crystals all have a single four-fold rotation axis or four-fold screw axis coincident with the c axis of the unit cell. Trigonal/rhombohedral crystals have a three-fold or three-fold screw axis axis coincident with **c**, and some trigonal unit cells can also be represented using a rhombohedral lattice. The unit cells in hexagonal crystals have the same shapes as the unit cells in trigonal crystals, but there is a six-fold rotation or screw axis along their c axes instead of a three-fold or three-fold screw axis. Each unit cell of a cubic crystal must have four three-fold axes orientated along its body diagonals. Lattices that belong to all of the Bravais lattice systems, except the triclinic system, may display symmetries beyond the minimum required for system membership.

Within each lattice system there may by more than one type of unit cell possible. All systems are compatible with primitive unit cells, which are designated "P", or in the case of the rhombohedral crystals, "R" unit cells. However, they may also be compatible with nonprimitive cells that have either an extra lattice point in the middle of the face of the unit cell that is coplanar with **a** and **b**, i.e. be C-centered cells, or have extra lattice points in middle of all unit cell faces, in which case they are called face (or F)-centered cells, or have an extra lattice point in the center of each unit cell. Unit cells of the last type are said to be body-centered, and are designated "I". The 14 Bravais lattices emerge when these possibilities are taken into account (see figure 4.1.1).

It should be obvious that the rotational symmetry of lattices belonging to the cubic system absolutely requires that their unit cells be cubes. Rotational symmetry imposes similar restrictions on the geometries of the unit cells belonging to all the other systems except (as usual) the triclinic system (see figure 4.1.1). That said, it is important to realize that there is nothing to keep a crystal that has cubic unit cells

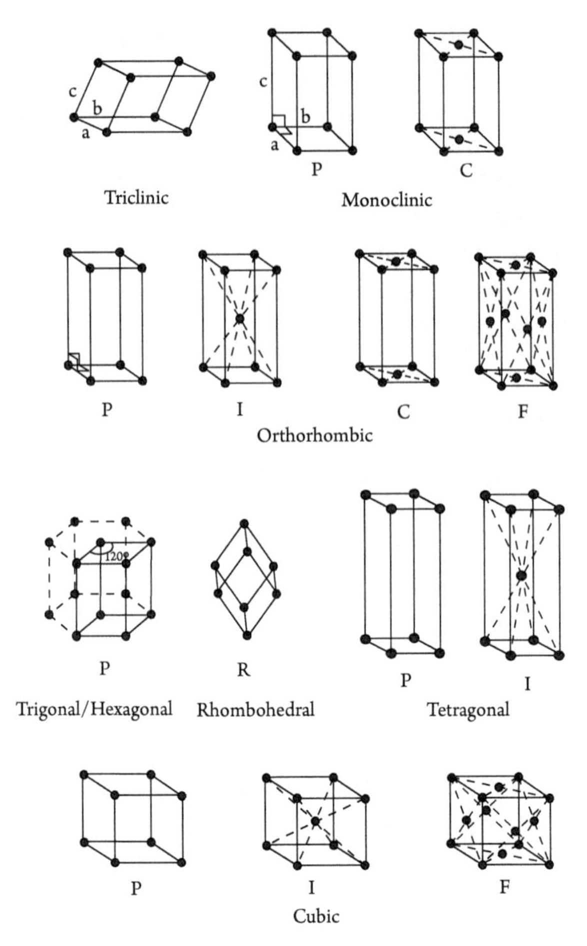

Figure 4.1.1 The 14 Bravais lattices.

from being triclinic. All that is required is that contents of its unit cells be arranged in such a way that the crystal lacks rotational symmetry. Thus, a lattice cannot belong to some lattice system unless the geometry of its unit cells is consistent with the rotational symmetry of that system, but obedience to those rules is *not* a sufficient condition for system membership.

Finally, unlike the directions and lengths of the unit cell vectors in a crystal, which are completely determined by the translational symmetry of the crystal, the placements of its unit cells in three-dimensional space is to some degree arbitrary. If the crystal is triclinic, the origin of the unit cell can be placed wherever the crystallographer chooses, relative to the molecules found therein. However, if the crystal is monoclinic, for example, all lattice points must lie on two-fold axes, but the position of the origin of the unit cell along the two-fold axis can be chosen as the crystallographer decides. A few moments thought will reveal that for orthorhombic crystals, the locations of lattice point locations are completely determined by lattice symmetry.

APPENDIX 4.2 ON THE RELATIONSHIP BETWEEN UNIT CELLS IN REAL SPACE AND UNIT CELLS IN RECIPROCAL SPACE

The Fourier transform of a three-dimensional lattice in real space is a lattice in reciprocal space. Thus, the reciprocal lattice of a crystal must divide its reciprocal space up into unit cells the same way its real-space lattice divides its real space up into unit cells. Call the three vectors that define the unit cells in some crystal's reciprocal space, $\mathbf{a}^*$, $\mathbf{b}^*$, and $\mathbf{c}^*$. What is the relationship between **a**, **b**, and **c**, and $\mathbf{a}^*$, $\mathbf{b}^*$, and $\mathbf{c}^*$?

In the first place, the volume of a unit cell in real space, V, equals $\mathbf{a} \cdot \mathbf{b} \times \mathbf{c}$. This is an elementary result from vector algebra, and we anticipate (correctly) that the volume of unit cells in the reciprocal space of the crystal will be $(1/V)$. In the second place, when it comes to working out unit cell dimensions in reciprocal space, the issues that need to be addressed are the relative orientations of planes of atoms/molecules in the crystal, not the orientations of the corresponding rows of points in real space, the Fourier transforms of which are used to work out the geometry of the reciprocal lattice. What we know for certain is that the vectors **b** and **c** define a plane in real space, that the vector given by the cross product of **b** and **c**, $\mathbf{b} \times \mathbf{c}$, is normal to that plane, and that there are a very large number of planes parallel to that plane in the crystal. The direction of the cross product of **b** and **c** will be roughly the same as that of **a**, but unlike **a**, it will always be the direction along which the distance between adjacent (**b**, **c**) planes in real space is the smallest it can be, and thus the direction of that cross-product vector is the one it makes sense to take as the direction of $\mathbf{a}^*$, and the reciprocal of the interplane distance in that direction should be the length of $\mathbf{a}^*$. This thought process leads to the following definitions for $\mathbf{a}^*$, $\mathbf{b}^*$, and $\mathbf{c}^*$:

$$\mathbf{a}^* = \mathbf{b} \times \mathbf{c}/V; \ \mathbf{b}^* = \mathbf{c} \times \mathbf{a}/V; \ \mathbf{c}^* = \mathbf{a} \times \mathbf{b}/V. \tag{4.2.1}$$

Note that if **a**, **b**, and **c** are mutually perpendicular then $V = abc$, and $\mathbf{a}^*$ will be parallel to **a** and have a length $(1/a)$, $\mathbf{b}^*$ will be parallel to **b** and have a length $(1/b)$, and $\mathbf{c}^*$ will be parallel to **c** and have a length $(1/c)$, which all makes sense. Furthermore, it is easy to show that $\mathbf{a}^* \cdot \mathbf{b}^* \times \mathbf{c}^* = 1/V$.

5

On the Appearance of Crystalline Diffraction Patterns

In chapter 4, we learned that the diffraction patterns of crystals are sampled, and because the geometry of that sampling is determined by the crystal lattice, the lengths and orientations of the unit cell vectors of a crystal can be deduced from the locations of the reflections in its diffraction pattern. However, unit cell size and shape are not the only information about a crystal that can be gleaned directly from its diffraction pattern. Here, we explain what that other information is and how it is accessed. We begin with a brief description of how diffraction data are collected from macromolecular crystals.

5.1 DIFFRACTION DATA ARE COLLECTED FROM MACROMOLECULAR CRYSTALS USING THE OSCILLATION METHOD

Over the last 20 years, a single approach to data collection has emerged as the industry standard in macromolecular crystallography. It uses instruments like the one shown schematically in figure 5.1, which are called, generically, **diffractometers**, or more precisely **oscillation diffractometers**. Crystals are mounted on a **goniometer**, which is a mechanical device that makes it possible to rotate them in the X-ray beam by accurately known amounts under computer control. Downstream of the goniometer there is an X-ray detector that may or may not be mounted on an arm so that it can be moved in an arc, the axis of which passes through the location where the crystal is mounted on the goniometer, again under computer control.

The X-ray detectors used today are area, or position-sensitive, detectors that measure not only the total amount of X-ray energy reaching them, but also the locations on their surfaces where that energy is impinging, and they are read out

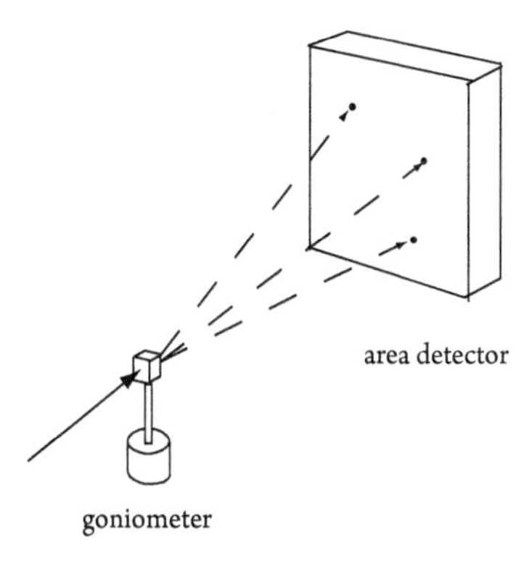

Figure 5.1 A simple diffractometer for macromolecular crystallography.

electronically. Displayed graphically, the output from such a detector looks like the image that would have resulted if a piece of photographic film had been exposed to the diffracted X-rays in the same location. For any given orientation of a crystal in the X-ray beam, the detector will register all the reflections in the diffraction pattern of that crystal that intersect the Ewald sphere inside the region delimited by the edges of its sensitive area. Finally, a beam-stop is placed between the crystal and the detector to prevent the undiffracted part of the incident beam from reaching the detector (not shown in figure 5.1). That component of the beam is invariably so intense that it would permanently damage the detector if allowed to reach it.

A typical data collection run starts with the crystal mounted on the diffractometer and the X-ray beam off. The beam is turned on, and the crystal made to rotate back and forth around some predetermined axis over a degree of arc or so (i.e., to **oscillate**) for a period of time that is long enough so that the average reflection will deliver enough energy to the detector to be measured accurately. The beam is then turned off; the image that has formed on the face of the detector is transferred into the memory of a computer; and the detector reinitialized. The crystal is then rotated by a degree or so about some axis, and the process is repeated.

A single detector image is called a **frame**, of which figure 5.2 is a typical example. The oscillations of the crystal that occur as each frame is exposed cause a subset of its reflections to move back and forth across the sphere of reflection. Thus, the data recorded on a single frame correspond to a spherical wedge in the crystal's reciprocal space. The bigger the angular range of the oscillation, the larger the volume of reciprocal space included in each wedge, and the smaller the number of frames it takes to record all of reciprocal space. However, the larger these wedges, the more likely it is that different reflections in a crystal's diffraction pattern will accidentally intersect the sphere of reflection at the same location and so register at the same position on the surface of the detector. Oscillation angles of 1° are often satisfactory for macromolecular crystals, but when unit cells are large, smaller oscillation angles may have to be used to reduce overlap.

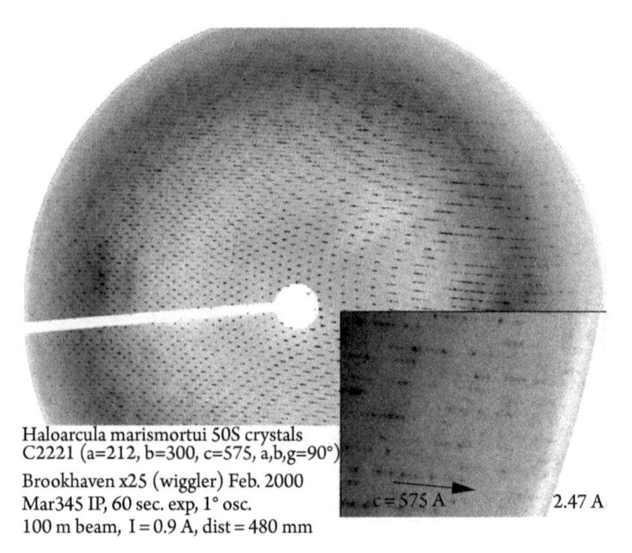

Figure 5.2 A typical frame of oscillation data. The unit cell dimensions and other relevant information are provided on the figure. (Figure provided by Professors Poul Nissen (Univ. Aarhus), Nenad Ban (ETH), and Thomas Steitz (Yale Univ.))

The wavelengths of the X-rays used for crystallographic data collection range from about 1.6 to 0.9 Å. Crystals are likely to be 10^{-3} to 10^{-4} m in linear dimensions. The distance from the crystal to the detector is typically 0.3 to 0.5 m. The diameter of the X-ray beam is usually adjusted so that it will illuminate no more than the entire volume of the crystal because X-ray photons that do not pass through the crystal contribute nothing to diffracted intensities, but they can contribute to background by being scattered either by gas molecules in the flight path of the direct beam, or, possibly, by components of the experimental apparatus itself.

Over the years, several different technologies have been used for position-sensitive X-ray detection, including, originally, photographic film. Charge-coupled devices (CCDs) dominate today, but may not in the future. The typical CCD X-ray area detector unit has a sensitive area that is $\sim 0.05 \times \sim 0.05$ m in size, and the detectors of diffractometers are often square arrays of several such units, which increases the volume of reciprocal space that can be recorded at each setting of the instrument. Like the CCDs found in home cameras, the images produced by each of these devices are composed of several million pixels.

Data about the relative intensities of reflections are extracted from frames using computer programs that rely on user judgment to some degree. The input includes not only a set of frames, but information about their relative orientations, the oscillation angle, the wavelength of the X-rays used, the crystal-to-detector distance, the location where the beam would have hit the detector if it had been allowed to do so, and the size and location of the sensitive area of the detector. The user identifies several reflections on a single frame and determines the area each reflection occupies on that frame. (It is usually significantly larger than a single pixel.) Once this has been done, modern data reduction programs can (usually) determine the geometry and orientation of the reciprocal space lattice of the crystal, and then automatically find and index all the reflections recorded on all the other frames in the data set. If data collection has been done correctly, these programs will assign to each reflection an intensity proportional to the total energy that it delivered to the detector, which in turn will be proportional to I(**S**).

5.2 MEASURED INTENSITIES MUST BE CORRECTED FOR SYSTEMATIC ERROR

The reason we measure the intensities of the reflections in a crystal's diffraction pattern is to harvest the information they contain about I(**S**), the square of the amplitude of the crystal's Fourier transform. However, the correspondence between the measured intensities produced by data reduction programs and I(**S**) is invariably imperfect because of experimental error, some of it systematic, and the rest statistical (e.g., counting error or the equivalent). The most obvious source of systematic error is detector background, which can be estimated by measuring the signal registered in areas of the detector where reflections did not impinge. Clearly background must be subtracted from all measured intensities.

We already know how to deal with polarization, which is another important source of systematic error (see section 1.7). The data correction procedure described in section 1.7 is appropriate if the X-ray beam used for data collection was unpolarized, which it will have been if it was produced by a conventional laboratory X-ray source. The X-ray beams generated by synchrotron light sources, on the other hand, are highly polarized, and, thus, a different correction factor is required for synchrotron data sets.

Two other kinds of systematic error need to be considered. First, all else being equal, the farther a reflection is in reciprocal space from the axis around which a crystal is oscillated, the shorter the length of time that reflection will spend intersecting the sphere of reflection, as the frame of data in which it appears is collected and the weaker the signal it will produce on the detector. The adjustments in reflection intensities that must be made to correct for this effect are called **Lorentz corrections**. Diffraction data must always be corrected for Lorentz effects. Second, the data may also have to be corrected for absorption. About 60% of the photons in a beam of X-rays having a wavelength of 1.54 Å, the wavelength of the CuK_{α} radiation commonly produced by laboratory X-ray generators, are absorbed when the beam passes through 1 mm of water, and the crystals one studies can be that large. Absorption reduces the measured strengths of reflections, and if a crystal has an asymmetric shape, the magnitude of that reduction will vary from one reflection to the next, depending on crystal orientation, and, thus, will lead to systematic error.

Prior to the advent of synchrotron beam lines, diffraction data sets were routinely corrected for absorption, but it is seldom done today for two reasons. First, the wavelengths of the X-rays used for data collection today are usually around 1 Å, which is considerably shorter than 1.54 Å that was the norm in presynchrotron days, and absorption is proportional to λ^3. Second, the crystals commonly used today are commonly much smaller than those used in the past. If the crystal under study is 0.1 mm thick instead of 1 mm thick, and 1 Å X-rays are used instead of 1.54 Å X-rays, the variation in absorption with crystal orientation may be less than a few percent, no matter how irregular the shape of the crystal, and that is hardly worth worrying about.

The programs used for the initial reduction of diffraction data generate files that list the indices of each reflection observed, its measured intensity, corrected for systematic error, and an estimate of its standard deviation. If a reflection was measured more than once as data were accumulated, which is often the case, its average intensity will be reported, and the standard deviation assigned to its intensity will be estimated on the basis of the variation between individual observations. Inevitably, but accidentally, some of the reflections in a crystal's diffraction pattern will not have been fully recorded on any single frame in the data set because the ranges of the angles explored were such that they did not pass all the way through the sphere of reflection on any one of them. The intensities for partially recorded reflections are pieced together by combining data found on more than one frame.

5.3 RADIATION DAMAGE KILLS CRYSTALS

Absorption has an effect on X-ray data collection that cannot be corrected for after the fact, and that can never be ignored. X-rays are a form of ionizing radiation. Thus, when crystals absorb X-rays, ions and radicals are produced that react with the molecules in those crystals. These radiation-induced reactions disorder crystals by destroying the molecules of which they are made. The longer a crystal is exposed to radiation, the weaker its diffraction pattern becomes, and weakening invariably begins at large $|\mathbf{S}|$, and progresses inward toward small $|\mathbf{S}|$. The dose-dependent deterioration of diffraction patterns is referred to **radiation damage**, and it limits the amount of data that can be obtained from any one crystal. For reasons that are poorly understood, the amount of radiation it takes to "kill" the diffraction pattern of the crystals of one kind of macromolecule can be quite different from the amount it takes to degrade the diffraction pattern of the crystals of another. Experimentation is the only way to find out how much radiation a given type of crystal can tolerate. If the crystals of some macromolecule are very sensitive to radiation damage, it may be necessary to merge partial data sets collected from several different crystals to obtain a complete data set of acceptable quality.

It is worth noting that, as we pointed out earlier, the amounts of energy delivered to a detector by the reflections from some crystal are proportional to λ^3. Thus, absorption and diffraction intensities change in parallel with X-ray wavelength. Long wavelengths lead to strong diffraction, but also to strong absorption. There is no free lunch.

5.4 DIFFRACTION PATTERNS TEND TO BE CENTROSYMMETRIC

X-ray diffraction patterns commonly display a symmetry that has important consequences for almost everything that is done with diffraction data. We know already that the intensities in an experimental X-ray diffraction pattern that has been corrected for its systematic errors are proportional to $F(\mathbf{S})F^*(\mathbf{S})$, where $F(\mathbf{S})$ is the Fourier transform of $\rho(\mathbf{r})$, the electron density distribution of the unit cell, or, perhaps, just a single, isolated molecule. In chapter 3, we discovered that atomic scattering factors can be complex numbers, and, thus, $F(\mathbf{S})$ should be written as follows:

$$F(\mathbf{S}) = \int \left[\rho(\mathbf{r}) + i\rho'(\mathbf{r})\right] \exp\left(-2\pi i \mathbf{r}\cdot\mathbf{S}\right) dV_r,$$

where $\rho(\mathbf{r})$ is the electron density distribution that corresponds to the real parts of the atomic scattering factors, and $\rho'(\mathbf{r})$ is the electron density distribution that corresponds to the imaginary parts of those scattering factors. Hence:

$$\begin{aligned} I(\mathbf{S}) &= F(\mathbf{S})F^*(\mathbf{S}) \\ &= \iint \left[\rho(\mathbf{r})\rho\left(\mathbf{r}'\right) + \rho'(\mathbf{r})\rho'(\mathbf{r}')\right] \exp\left[-2\pi i(\mathbf{r}-\mathbf{r}')\cdot\mathbf{S}\right] dV_r dV_{r'} \\ &\quad + i\iint \left[\rho'(\mathbf{r})\rho(\mathbf{r}') - \rho(\mathbf{r})\rho'(\mathbf{r}')\right] \exp\left[-2\pi i(\mathbf{r}-\mathbf{r}')\cdot\mathbf{S}\right] dV_r dV_r. \end{aligned}$$

Now for $(\mathbf{r}-\mathbf{r}')$ term in the first double integral, there is an equally weighted $(\mathbf{r}'-\mathbf{r})$ term, and as a result, all the sine terms in the first double integral cancel out leaving only cosine terms. In contrast, when $\mathbf{r}$ and $\mathbf{r}'$ are interchanged, the ρ term in the second integral changes sign, and all the cosine terms in the second integral cancel, leaving only sine terms. Thus:

$$\begin{aligned} I(\mathbf{S}) &= \iint \left[\rho(\mathbf{r})\rho(\mathbf{r}') + \rho'(\mathbf{r})\rho'(\mathbf{r}')\right] \cos\left[2\pi(\mathbf{r}-\mathbf{r}')\cdot\mathbf{S}\right] dV_r dV_{r'} \\ &\quad + \iint \left[\rho'(\mathbf{r}')\rho(\mathbf{r}) - \rho(\mathbf{r}')\rho'(\mathbf{r})\right] \sin\left[2\pi(\mathbf{r}-\mathbf{r}')\cdot\mathbf{S}\right] dV_r dV_{r'}. \end{aligned} \tag{5.1}$$

If $\rho'(\mathbf{r})$ is 0, or so small compared to $\rho(\mathbf{r})$ that its contribution to the diffraction pattern cannot be measured, which amounts to the same thing, $I(\mathbf{S})$ will equal $I(-\mathbf{S})$ because $\cos(x) = \cos(-x)$. Thus, if the electron density distribution of a molecule is real, or nearly so, its diffraction pattern will be centrosymmetric. A diffraction pattern that is centrosymmetric is said to obey **Friedel's law**.

At the wavelengths of the X-rays commonly used for macromolecular crystallography, $\Delta f''$ for H, C, N, and O is less than 0.5% of $(f_o + \Delta f')$, and for P and S, the two high atomic number atoms abundant in biomolecules, $\Delta f''$ is still only ~3% of $(f_o + \Delta f')$. Thus, for these molecules, the cosine term in equation (5.1) utterly dominates, and only if care is taken can the deviations of their diffraction patterns from Friedel's law be measured reliably.

Friedel's law has two obvious implications. First, if Friedel's law holds, only one hemisphere's worth of $I(\mathbf{S})$ data need be collected for a molecule because once half of the sphere has been measured, the other half can be calculated using the relationship $I(\mathbf{S}) = I(-\mathbf{S})$. Second, if the diffraction pattern of a crystal of a molecule that contains chiral centers obeys Freidel's law, it will be possible to determine the relative hands of all of its chiral centers from its diffraction pattern, but *not* their absolute hands. [Remember, if the *i*th atom of a chiral molecules is located at $\mathbf{r}$, the corresponding atom in its enantiomer will be found at $-\mathbf{r}$, and thus $\rho(\mathbf{r}) = \rho_{enant}(-\mathbf{r})$.] For biochemists, this is seldom a problem because the absolute hands of most of the chiral molecules they work with were determined long ago. What happens in practice is that as the crystal structures of a chiral biomolecule, for example, a protein, is solved using diffraction data that obeys Freidel's law, an arbitrary choice gets made at some point that determines the absolute hand of the electron density maps that emerges. (If Murphy's law did not apply, the absolute hands of these initial maps would be right half the time.) However, if the absolute hands of the chiral centers in such a molecule are already known, it is easy to correct the initial map if it has the wrong hand. All we have to do is replot that electron density distribution map so that the value for the electron density at $\mathbf{r}$ in the new map equals to $\rho(-\mathbf{r})$ in the original map.

5.5 ANOMALOUS DIFFRACTION CAN PROVIDE USEFUL INFORMATION ABOUT THE CHEMICAL IDENTITIES OF ATOMS IN ELECTRON DENSITY MAPS

If anomalous scattering makes an appreciable contribution to the diffraction pattern of a crystal, its diffraction pattern will not obey Friedel's law because the sine component of the intensities of reflections (equation (5.1)) changes sign when $\mathbf{S}$ is replaced by $-\mathbf{S}$. For crystals of this sort, a data set corresponds to a full sphere of data, which means more data must be collected, and in principle, when it comes time to compute electron density maps from the data, both an $(f_o + f')$ map, and an f'' map ought to be computed separately, which is something macromolecular crystallographers seldom do.

One reason you might want to know what $\rho'(\mathbf{r})$ looks like is that the contribution a given kind of atom makes to it is proportional to $\Delta f''$, which varies strongly with atomic number. Thus, for example, a $\rho'(\mathbf{r})$ map of a crystalline protein could reveal the locations of its sulfur atoms, of which there is likely to be only a few, and if the protein includes high-Z atoms, like the iron atoms found in heme proteins, their positions will also be evident in such maps.

5.6 ANOMALOUS DIFFRACTION EFFECTS CAN BE USED TO DETERMINE THE ABSOLUTE HAND OF CHIRAL MOLECULES

Not only can anomalous data reveal the locations of the high Z atoms a crystal may contain, they can also be used to determine the absolute hand of its chiral centers. The diffraction patterns of pairs of enantiomeric molecules containing atoms that anomalously scatter X-rays are related to each other. If the structure of one of the enantiomers is $[\rho(\mathbf{r}) + i\rho'(\mathbf{r})]$, then $[\rho(\mathbf{r}) + i\rho'(\mathbf{r})]_{enant} = [\rho(-\mathbf{r}) + i\rho'(-\mathbf{r})]$. Hence, if equation (5.1) describes the scatter produced by one enantiomer of some structure, then the scatter given by the other enantiomer will be:

$$\begin{aligned} I_{enant}(\mathbf{S}) = &\int\int \left[\rho(-\mathbf{r})\,\rho\left(-\mathbf{r}'\right) + \rho'(-\mathbf{r})\,\rho'\left(-\mathbf{r}'\right)\right] \\ &\cos\left[-2\pi\left(\mathbf{r}-\mathbf{r}'\right)\cdot\mathbf{S}\right] dV_r dV_{r'} \\ &+\int\int \left[\rho'\left(-\mathbf{r}'\right)\rho(-\mathbf{r}) - \rho\left(-\mathbf{r}'\right)\rho'(-\mathbf{r})\right] \\ &\sin\left[-2\pi\left(\mathbf{r}-\mathbf{r}'\right)\cdot\mathbf{S}\right] dV_r dV_{r'}. \end{aligned}$$

A few moments of reflection will reveal that $I_{enant}(\mathbf{S}) = I(-\mathbf{S})$, which means that the diffraction pattern of one isomer of the molecule will be related to that of its enantiomer by the transformation of **S** to –**S**.

The reason this matters is that the coordinate systems used to describe the locations of atoms in any crystal are chiral. Unit cell axes are usually defined so that **c** points in the same direction as $\mathbf{a} \times \mathbf{b}$, and the convention for defining unit cell vectors in reciprocal space, $\mathbf{a}^*$, $\mathbf{b}^*$ and $\mathbf{c}^*$, also usually conforms to the right-hand rule. However, it does not matter whether the coordinate system used is right-handed or left-handed because the absolute hands of chiral molecules are determined crystallographically by comparing the chirality implied by their diffraction patterns to the chirality of the coordinate system used to specify the locations of their atoms.

As we shall see in chapter 6, the structures of molecules are easily determined using only the cosine component of $I(\mathbf{S}) = (1/2)[I(\mathbf{S}) + I(-\mathbf{S})]$, which tends to dominate $I(\mathbf{S})$ in any case. Because the cosine part of $I(\mathbf{S})$ obeys Freidel's law, that component of the diffraction data produced by a crystal composed of molecules that contain chiral centers will be equally compatible with the structures of *both* of its enantiomers. As we already noted, the structure of a molecule containing chiral centers of unknown absolute hand obtained from analysis of the cosine part of the data its crystals produce will have an absolute hand to be sure, but it is as likely to be wrong as it is to be right. Starting with the model for the structure of the unit cell that we can obtain from that initial electron density map, we can compute the sine part of $I(\mathbf{S})$ (i.e., its anomalous part) for *both* enantiomers of the molecule using equation (2.9):

$$\begin{aligned} I_{sin}(h,k,l) = &\sum\sum\left(f_i''f_j - f_if_j''\right) \\ &\sin\left(2\pi\left[\left(x_i - x_j\right)h + (y_i - y_j)k + (z_i - z_j)l\right]\right). \end{aligned}$$

When those two sets of computed intensities are compared to the measured anomalous data $(= (1/2)[I(\mathbf{S}) - I(-\mathbf{S})])$, it will be found that one of them is more consistent with the measured data than the other, and, on that basis, a decision can be made about the absolute hand of the molecule.

5.7 CRYSTAL SYMMETRY RESULTS IN RECIPROCAL SPACE SYMMETRY

In addition to providing the crystallographer with unit cell parameters and reflection intensities, modern data reduction programs also identify the space group to which a crystal belongs. Sometimes the program's output will indicate that data are consistent with just one space group. Other times, the program will indicate that the data are consistent with a small number of different, but related space groups. Crystallographers need to understand how those programs determine space groups because they do not always get them right. Space group information is used at many stages in any crystallographic structure determination, so if a crystal's space group has been misidentified in the beginning, trouble will surely follow.

As we already know, the space group to which a crystal belongs is determined by its rotational symmetry, and given the close relationship that exists between lattice geometry in real space and lattice geometry in reciprocal space, it should not come as a surprise that the diffraction patterns of crystals are dramatically affected by their rotational symmetries. We will illustrate the procedure used to work out the reciprocal space consequences of real-space symmetry by looking at the effects that two-fold rotational symmetry and two-fold screw axis symmetry have on diffraction patterns.

Consider a crystal that has a single two-fold axis running through each of its lattice points parallel with the c axis of its unit cell. Each atom in its asymmetric unit can be thought of as making two contributions to the transform of the unit cell: its own contribtion, and that of its symmetry mate. In this instance, the contribution of the ith atom, $\mathrm{F}_i(h,\ k,\ l)$, will be:

$$\mathrm{F}_i(h,k,l) = \mathrm{f}_i\left(\exp\left[-2\pi i(hx_i + ky_i + lz_i)\right] + \exp\left[-2\pi i\left(-hx_i - ky_i + lz_i\right)\right]\right)$$

because rotation around a two-fold axis coincident with $\mathbf{c}$ takes $(x,\ y,\ z)$ to $(-x,\ -y,\ z)$. The transform of the entire unit cell will be a sum of such terms, where the sum runs over all atoms in the asymmetric unit. The preceding equation can be rewritten as follows:

$$\mathrm{F}_i(h,k,l) = 2\mathrm{f}_i \exp\left(-2\pi i l z_i\right)\cos\left[2\pi\left(hx_i + ky_i\right)\right],$$

and once this is done, it becomes obvious that $|\mathrm{F}(h,\ k,\ l)| = |\mathrm{F}(-h,\ -k,\ l)|$, which means that the diffraction patterns of crystals that have two folds along $\mathbf{c}$ have two-fold axes along $\mathbf{c}^*$, and furthermore, that $|\mathrm{F}(-h,\ k,\ l)| = |\mathrm{F}(h,\ -k,\ l)|$. These relationships are true no matter whether the f_i are real or complex. With some additional effort, it can also be shown that if the f_i are real, which is often at least approximately the case, $|\mathrm{F}(h,\ k,\ l)| = |\mathrm{F}(-h,\ -k,\ -l)|$ (Friedel's law), and $|\mathrm{F}(h,\ k,\ l)| = |\mathrm{F}(h,\ k,\ -l)|$.

Two-fold screw axes have a more interesting effect on diffraction patterns, and again we will assume that our crystal has a single two-fold screw axis parallel to $\mathbf{c}$. Here too, every atom in the asymmetric unit will have a single, symmetry-related mate, and we write:

$$\begin{aligned}\mathrm{F}_i(h,k,l) = \mathrm{f}_i\,&\left[\exp\left[-2\pi i(hx_i + ky_i + lz_i)\right]\right.\\&\left.+\exp\left(-2\pi i\left[-hx_i - ky_i + l(z_i + 1/2)\right]\right)\right]\end{aligned}$$

(Remember that in this case the operation that superimposes one atom on its symmetry mate is rotation by 180° and translation along **c** by 1/2.) If the lz_i part of this expression is factored out in a judicious way, the previous equation can be written:

$$F_i(h,k,l) = f_i \exp\left[-2\pi i l\left(z_i + 1/4\right)\right] \left(\exp\left[-2\pi i\left(hx_i + ky_i - l/4\right)\right] + \exp\left[-2\pi i\left(-hx_i - ky_i + l/4\right)\right]\right).$$

This expression has an interesting implication. If $h = k = 0$, the term in bold parentheses is $2\cos(-\pi l/2)$, and, thus, whenever $h = k = 0$ and l is odd, the contribution made by each pair of symmetry-related atoms to the transform of the unit cell will be 0; hence, the transform of the entire unit cell will be 0 also. Thus, if a crystal has a two-fold screw axis along **c**, every other one of its (0, 0, l) reflections will have an intensity of 0. When its diffraction pattern is displayed graphically, it will be obvious there are **systematic absences** in it: in other words, that there is a patterned set of lattice points in reciprocal space where the rest of the diffraction pattern suggests that intensity ought to be seen, but is not.

Two-fold screw axis symmetry has other effects on the appearance of diffraction data sets. When the real and imaginary parts of $F_i(h, k, l)$ are considered separately, and $|F(h, k, l)|$ worked out, it becomes clear that $|F(h, k, l)| = |F(-h, -k, l)|$, and that $|F(-h, k, l)| = |F(h, -k, l)|$, just as is the case for two-fold axes. In addition if the f's are all real, $|F(h, k, l)| = |F(-h, -k, -l)| = |F(h, k, -l)|$. In short, except for systematic absences along $\mathbf{c}^*$, two-fold screw axes have the same effect as two-fold axes.

Data reduction programs assign space groups to crystals by searching for evidence for the existence of rotation axes, screw axes, and such in their diffraction patterns using symmetry rules of the sort just described. That information is then combined with the information the program has extracted about the lengths and angles of unit cell vectors to identify the space group to which the crystal belongs. Nonprimitive lattices are considered provided: (1) that the shape of the lattice permits, and (2) that when a nonprimitive unit cell is used to describe the lattice, symmetries become apparent in diffraction patterns that are not evident when the same lattice is described using a primitive unit cell. In short, there is a lot of crystallographic common sense embedded in data reduction programs, and they save crystallographers lots of time. However, it is important to realize that they may not arrive at the right answer. This can happen when the background in the frames being processed is high or is distributed in unusual ways, or when data sets are incomplete. There are also a number of more profound crystal pathologies that can lead to confusion that we will not discuss here (e.g., twinning).

Comment: In the crystallographic literature, negative values for Miller indices are often written as an integer with a bar over it, instead of as an integer with a negative sign in front of it, for example, $(\bar{h}, \bar{k}, \bar{l})$ *instead of* $(-h, -k, -l)$*. We will always use negative signs here.*

5.8 REAL CRYSTALS ARE NOT PERFECTLY ORDERED

So far, we have assumed that all crystals are perfectly ordered; that is, that the positions of corresponding atoms within each unit cell are exactly the same from one unit cell to the next. Like many such assumptions, this one is only a (useful) first approximation to the truth, and it is now time to consider the effects that deviations from perfect order have on the diffraction patterns of crystals. Three kinds of disorder will be considered: (1) thermal disorder, (2) mosaic disorder, and (3) polymorphism. Thermal disorder, as its name implies, is the disordering caused by the small, thermally driven excursion of atoms from their average locations that constantly occur in all molecules, but any other phenomenon that gives rise to small, random displacements of atoms from their average positions will have the same effect crystallographically. Mosaic disorder arises when the lattice in one part of a macroscopic crystal is not perfectly aligned with the lattice in another part. We will use the term *polymorphism* to describe what happens when there are a small number of conformations possible in some region of a large molecule that has been crystallized, and the differences between them are large compared to the deviations in atomic positions caused by ordinary thermal motions.

5.9 DISORDER WEAKENS BRAGG REFLECTIONS

From the point of view of the working crystallographer, what matters most about disorder is what it does to the intensities of reflections, and so that is where our discussion of disorder will begin. As we already know (see equation (1.10)), the intensity of the (h, k, l) reflection in the diffraction pattern of a perfect crystal is proportional to:

$$\iint \rho(\mathbf{r})\rho(\mathbf{r}') \exp\left[-2\pi i(\mathbf{r} - \mathbf{r}') \cdot \mathbf{S}(h,k,l)\right] dV_r dV_{r'}.$$

What happens if the atoms in a crystal are constantly in motion due to thermal effects? To make this question easy to answer, we will assume that the thermal motions of the atoms in that crystal have three properties.

1. The time-averaged positions of corresponding atoms are the same in all unit cells; that is, the crystal would be perfect if there were no thermal motions. It follows from this condition that the time-averaged displacement of every atom in the crystal from its average position is 0.
2. The distributions of the displacements experienced by corresponding atoms are the same from one unit cell to the next.
3. The motions of corresponding atoms in the different unit cells of the crystal are uncorrelated. If this is so and condition 2 holds, then at any instant in time, averaged over all unit cells, the displacements of corresponding atoms from their average positions will be 0.

If the scattering factors of the atoms in some crystal are all real, at any instant in time, the intensity of its Bragg reflections will be proportional to:

$$\sum_n \sum_m \iint \left[\rho_a(\mathbf{r}) + \delta_n(\mathbf{r})\right]\left[\rho_a(\mathbf{r}') + \delta_m(\mathbf{r}')\right]$$
$$\exp\left[-2\pi i(\mathbf{r} - \mathbf{r}') \cdot \mathbf{S}(h,k,l)\right] dV_r dV_{r'}.$$

Both sums are over all unit cells, $\mathbf{r}$ is measured relative to the origin of unit cell n or unit cell m, as the case may be, and the integrations are over the volumes of the two unit cells. $\rho_a(\mathbf{r})$ is the electron density at $\mathbf{r}$, averaged over all the unit cells in the crystal; $\delta_n(\mathbf{r})$ is the deviation from that average electron density at $\mathbf{r}$ in unit cell n; and $\delta_m(\mathbf{r})$ is the corresponding quantity for unit cell m. Multiplying the two density terms together, we find that $I(h, k, l)$ is the sum of four components, three of which can be written as follows:

$$\textstyle\sum_n \int \left[\rho_a(\mathbf{r}) \text{ or } \delta_n(\mathbf{r})\right] \exp\left[-2\pi i\mathbf{r} \cdot \mathbf{S}(h, k, l)\right] dV_r \int \left[\sum_m \delta_m(\mathbf{r}')\right]$$
$$\exp\left[2\pi i\mathbf{r}' \cdot \mathbf{S}(h, k, l)\right] dV_{r'}.$$

However, because $\sum_m \delta_m(\mathbf{r}') = 0$, all these terms are 0, and hence:

$$I(h,k,l) \propto \iint \rho_a(\mathbf{r})\rho_a(\mathbf{r}') \exp\left[-2\pi i(\mathbf{r} - \mathbf{r}') \cdot \mathbf{S}(h,k,l)\right] dV_r dV_{r'}. \qquad (5.2)$$

Thus, the intensities of the Bragg reflections produced by an imperfectly ordered crystal report on the *average* electron density distribution in its unit cells. That said, it is vital to recognize that this average electron density distribution is *not* the same as the electron density distribution that would have been seen if the coordinates of corresponding of atoms were exactly the same in all cells, and all the atoms in the crystal were motionless. The average distribution is that idealized electron density distribution blurred/smeared by atomic motions, as we will now demonstrate.

The effects that atomic motions have on diffraction data are best understood starting with an expression for the time-averaged value of $F(h, k, l)$ for entire crystals that are based on atomic positions and scattering factors rather than on electron density distributions. The position of the ith atom in the nth unit cell of the crystal can be written $[\mathbf{r}_i + \boldsymbol{\Delta}_{in}(t)]$, where $\mathbf{r}_i$ is the average position of that atom in the unit cell, where the average is over all unit cells and a long period of time, and $\boldsymbol{\Delta}_{in}(t)$ is the instantaneous displacement of that atom from its average position in unit cell n. Because this is so:

$$\langle F(h,k,l)\rangle = \left\langle \sum_n \sum_i f_i \exp\left[-2\pi i(\mathbf{r}_i + \boldsymbol{\Delta}_{in}(t)) \cdot \mathbf{S}(h,k,l)\right] \right\rangle,$$

where the summations are over all unit cells and all atoms in the unit cell, and the notation $\langle x\rangle$ means the value of x, averaged over all unit cells and over the time it takes to measure reflections, which is normally much, much longer than the characteristic time of thermal vibrations. Because only the part of the equation that has to do with $\boldsymbol{\Delta}_{in}(t)$ varies with time and differs between unit cells, this equation can be rewritten as follows:

$$\langle \mathrm{F}(h,\ k,\ l)\rangle = \sum_i \mathrm{f}_i \exp\left[-2\pi i \mathbf{r}_i \cdot \mathbf{S}(h,\ k,\ l)\right]\left\langle \sum_n \exp\left[-2\pi i \boldsymbol{\Delta}_{in}(t) \cdot \mathbf{S}(h,k,l)\right]\right\rangle.$$

Thus:

$$\langle \mathrm{I}(h,\ k,\ l)\rangle \propto \sum_i \sum_j \mathrm{f}_i \mathrm{f}_j \exp\left[-2\pi i(\mathbf{r}_i - \mathbf{r}_j) \cdot \mathbf{S}(h,\ k,\ l)\right] \times$$
$$\langle \exp\left[-2\pi i \boldsymbol{\Delta}_i(t) \cdot \mathbf{S}(h,\ k,\ l)\right]\rangle \langle \exp\left[2\pi i \boldsymbol{\Delta}_j(t) \cdot \mathbf{S}(h,\ k,\ l)\right]\rangle \quad \mathbf{(5.3)}$$

(The n and m subscripts are dropped from the $\boldsymbol{\Delta}$ terms in this equation because, by hypothesis, the distributions of displacements of corresponding atoms are the same in all unit cells, and so time averages are the same as space averages.) We should not forget that almost all measured diffraction data sets provide information about this averaged value of $\mathrm{I}(h,\ k,\ l)$, that is, the value of I described by equation (5.3), not its instantaneous value. Appendix 5.1 shows how the averages in equation (5.3) are evaluated, and when they are, we find that:

$$\langle \mathrm{I}\ (h,\ k,\ l) \propto \sum_i \sum_j \mathrm{f}_i \exp\left[-(B_i/4)|\mathbf{S}|^2\right]$$
$$\times\ \mathrm{f}_j \exp\left[-\ (B_j/4)\ |\mathbf{S}|^2\right] \exp\left[-2\pi i(\ \mathbf{r}_i - \mathbf{r}_j) \cdot \mathbf{S}(h,\ k,\ l)\right] \quad \mathbf{(5.4)}$$

The B_i that appears in equation (5.4) is called a **temperature factor**, or **Debye-Waller factor**, or **B-factor**, and $B_i = [(8\pi^2/3){<}\Delta_i{}^2{>}]$, where ${<}\Delta_i{}^2{>}$ is the time-averaged value of the square of the displacement of the ith atom from its average position. (A statistician would call it the variance of the atom's position.) B-factors have the dimensions of length squared.

Comment: Implicit in equation (5.4) is the notion that the thermal motions of atoms in crystals are isotropic; in other words, that the magnitude of their displacements are the same in all directions, which in fact is almost never so. The anisotropies of atomic motions in crystals can be accounted for by replacing the rotationally averaged atomic B-factor functions in equation (5.3) with similar functions that instead of being spherical, are triaxial ellipsoids, and that vary in axis length and orientations from one atom to the next.

B-factors have important consequences both in reciprocal space and in real space. To appreciate their influence in reciprocal space, imagine you were collecting data from a macromolecular crystal, the individual atoms of which all had B-factors around 20 Å^2, which is on the small side for macromolecular crystals. As equation (5.4) shows, at $|\mathbf{S}| = 0.5\text{Å}^{-1}$, the intensities of reflections observed would be only 8% (!) of what they would have been if B had been 0. Thus, thermal disorder reduces the intensities of Bragg reflections, and the magnitude of that reduction increases rapidly with scattering angle. Obviously, the larger the thermal disorder in a crystal, the lower the value of $|\mathbf{S}|$ at which reflections become so weak they cannot be measured effectively, $|\mathbf{S}|_{max}$. Because the rendering of $\rho_a(\mathbf{r})$ that emerges from a data set that ends at $|\mathbf{S}|_{max}$ cannot include features that have length scales much less than $1/|\mathbf{S}|_{max}$, crystal disorder plays a major role in determining the size of the smallest features that can be discerned in experimental electron density maps. (This issue is discussed in greater depth in section 7.3.) Clearly, large values of B are bad.

The consequences of thermal fluctuations in real space are equally easy to understand. If there were no thermal fluctuations, in the electron density distributions derived from crystal diffraction patterns, each atom would be a spherically symmetric object, the electron density distribution of which is given by the square of its (rotationally averaged) electronic wave function, as we already know. As equation (5.4) shows, if the positions of atoms are varying due to thermal fluctuations, the form factors for each atom, $f_i(|\mathbf{S}|)$, are effectively replaced by the quantity $[f_i(|\mathbf{S}|)\exp(-B_i(|\mathbf{S}|/2)^2)]$, and in real space, this will alter the image of each atom from $FT^{-1}[f_i(|\mathbf{S}|)]$ to $(FT^{-1}(f_i(|\mathbf{S}|)){*}FT^{-1}[\exp(-B_i(|\mathbf{S}|/2)^2)])$.

It turns out that the Fourier transform of a Gaussian is a Gaussian, and when the math is done, we discover that:

$$FT^{-1}\left(\exp\left[-(B_i/4)|\mathbf{S}|^2\right]\right) = 8\,(\pi/B_i)^{3/2}\exp\left(-4\pi^2|\mathbf{r}^2|/B_i\right),$$

where $\mathbf{r}$ is the vector from the origin to some point in real space. Because atomic wave functions are Gaussian-like, the convolution of the electron density distribution of an atom with a thermal Gaussian will be a spherically symmetric function that is also roughly a Gaussian because the convolution of a Gaussian with a Gaussian is a Gaussian. If $B = 20$ Å^2, for example, then the value of $\mathbf{r}$ where the inverse transform of the B-factor Gaussian falls to 0.37 of its value at $\mathbf{r} = 0$ will be 0.71 Å, which is of the order of magnitude of the radius of an atom. Thus, the temperature factor contributions to the apparent sizes of atoms in electron density maps can be considerable, and the bigger the temperature factor of an atom, the bigger it will appear to be in an electron density map. This B-factor–dependent broadening of the images of individual atoms is the source of the blurring/smearing of averaged electron density distributions referred to earlier. In addition to making the images of atoms larger than they "should be", temperature factor smearing also reduces the value of $\rho(\mathbf{r})$ at the centers of atoms.

5.10 DISORDER MAKES CRYSTALS SCATTER IN DIRECTIONS THAT ARE NOT ALLOWED BY VON LAUE'S EQUATIONS

As appendix 5.1 shows, the Bragg reflections produced by a disordered crystal do not account for all of the radiation it scatters. The difference between the total scatter a crystal produces and its Bragg scatter, which we will call I_{dif} for reasons that will shortly become evident, is given by the following equation (see equation (5.1.3)):

$$\langle \mathbf{I}(\mathbf{S}, t) \rangle_{\text{dif}} = \sum_n \sum_m \exp\left[-2\pi i(\mathbf{u}_n - \mathbf{u}_m) \cdot \mathbf{S}\right]$$
$$\sum_i \sum_j \mathrm{f}_i \mathrm{f}_j \, \exp[-2\pi i(\mathbf{r}_i - \mathbf{r}_j) \cdot \mathbf{S}]$$
$$\times \exp[-2\pi^2(\langle[\boldsymbol{\Delta}_{in}(t) \cdot \mathbf{S}]^2\rangle + \langle[\boldsymbol{\Delta}_{jm}(t) \cdot \mathbf{S}^2\rangle)]$$
$$[\exp(4\pi^2\langle[\boldsymbol{\Delta}_{in}(t) \cdot \mathbf{S}][\boldsymbol{\Delta}_{jm}(t) \cdot \mathbf{S}]\rangle) - 1]. \quad (5.5)$$

The implications of equation (5.5) become clear when it is used to determine the consequences of a simple model for crystal disorder, which is associated with the names of Debye and Waller in the minds of most crystallographers, but, in fact, was first explored by Einstein. In a crystal that behaves as this model requires, the thermal motions of all atoms are random, isotropic, and completely uncorrelated with each other; in other words, the motions of every atom are totally unrelated to the motions of all its neighbors. If this is so, when $i = j$ and $n = m$, the average value of $< [\boldsymbol{\Delta}_{in}(t) \cdot \mathbf{S}][\boldsymbol{\Delta}_{jm}(t)] \cdot \mathbf{S}) >$ will be $< [\boldsymbol{\Delta}_{in}(t) \cdot \mathbf{S}]^2 >$, which is a positive number, but when $i \neq j$ and/or $n \neq m$, its average value will be 0 because the $\boldsymbol{\Delta}_{in}$ term in the product is just as likely to be negative when the $\boldsymbol{\Delta}_{jm}$ term is positive as it is to be positive and vice versa. Thus, when $i \neq j$ or $n \neq m$, the value of the last square brackets in equation (5.5) will be 0, which means that all such pairs of atoms make no contribution whatever to the average diffuse scatter. When those terms are eliminated from equation (5.5), we obtain:

$$\langle \mathrm{I}(\mathbf{S}) \rangle \propto \sum_i \mathrm{f}_i^2[1 - \exp(-4\pi^2\langle[\boldsymbol{\Delta}_i(t) \cdot \mathbf{S}]^2\rangle)]$$
$$\propto \sum_i \mathrm{f}_i^2(1 - \exp[-(B_i/2)/|\mathbf{S}|^2]). \quad (5.6)$$

Thus, the diffraction pattern of such a crystal will include a component that has a magnitude of 0 when the scattering angle is 0, but increases with scattering angle up to an asymptotic value of $\sum_i \mathrm{f}_i^2$. It will be amorphous—it will have no preferred directions—and, thus, be visible between Bragg reflections, which explains why this kind of scattering is called **(thermal) diffuse scattering**. The diffraction patterns of many macromolecular crystals have a background that qualitatively, at least, resembles what Einstein's model for thermal disorder predicts. The frame of data shown in figure 5.2 is typical in this regard. In it, the diffuse background becomes conspicuous at $|\mathbf{S}|^{-1} \sim 3$ Å.

5.11 THERMAL DIFFUSE SCATTER NEED NOT BE ISOTROPIC

Einstein's model for thermal disorder is grossly oversimplified. The thermal motions of atoms in crystals are bound to correlate with those of their neighbors not least because of the chemical bonds that join them. In addition, anisotropic motions are likely to be the rule rather than the exception because atoms are bound to collide with their neighbors as they move around, and their surrounds are seldom spherically symmetric. Thus, the diffuse scatter seen between the reflections in macromolecular diffraction patterns need not be as amorphous as the Einstein model predicts; at any given value of $|\mathbf{S}|$, it can vary a lot with direction. In fact, if the diffuse scatter seen in the diffraction pattern obtained from a crystal is not amorphous, it is proof positive that the thermal motions of atoms in that crystal are correlated in some way. Unfortunately, as interesting as it might be to know how those motions correlate in some crystal, there is no simple way to extract that information from diffuse scattering data. The best that can be done is to find a model for those motions that explains the observed diffuse scattering pattern reasonably well, but there is no guarantee that any such model is the only one that could account for the data, let alone the best of all such models.

To obtain some insight into what correlated disorder can do to the thermal scatter of a crystal, consider a one-dimensional crystal that consists of a large number of point-atoms, the positions of which obey the following, time-dependent equation:

$$x_n = na + \varepsilon \sin(\omega t - 2\pi na/\lambda).$$

In this equation, x_n is the position of the nth atom in the crystal, the average position of which is na, and $\varepsilon \sin(\omega t - 2\pi na/\lambda)$ represents a small, time-dependent, sinusoidal displacement of that atom from its average position that has an amplitude ε, a frequency of ω, and a wavelength of λ. Thus, the displacements this equation describes correspond to a compression/decompression (i.e., a sound) wave of wavelength λ passing through the entire lattice, and the motions of its atoms are perfectly correlated, rather than being totally uncorrelated, as Einstein assumed. The Fourier transform of this array, $\mathrm{FT}(s, t)$, is:

$$\mathrm{FT}(s, t) = \sum \exp(-2\pi isna) \exp[-2\pi i\varepsilon \sin(\omega t - 2\pi na/\lambda)],$$

and:

$$\begin{aligned}\mathrm{I}(s, t) = {} & \sum\sum \exp[-2\pi is(n - m)a] \\ & \exp\big(-2\pi i\varepsilon[\sin(\omega t - 2\pi na/\lambda) - \sin(\omega t - 2\pi ma/\lambda)]\big).\end{aligned}$$

Provided ε is small compared to a, the second exponential term in the expression for $\mathrm{I}(s, t)$ can be expanded as a power series, and the time-averaged value of $\mathrm{I}(s, t)$ evaluated without too much pain (see appendix 5.2). The equation that emerges is quite interesting:

$$\begin{aligned}\langle \mathrm{I}(s, t)\rangle = {} & \sum\sum \exp[-2\pi is(n - m)a][1 - 2(\pi\varepsilon s)^2] \\ & + (\pi\varepsilon s)^2 \sum\sum \exp[-2\pi i(s + 1/\lambda)(n - m)a] \\ & + (\pi\varepsilon s)^2 \sum\sum \exp[-2\pi i(s - 1/\lambda)(n - m)a].\end{aligned}$$

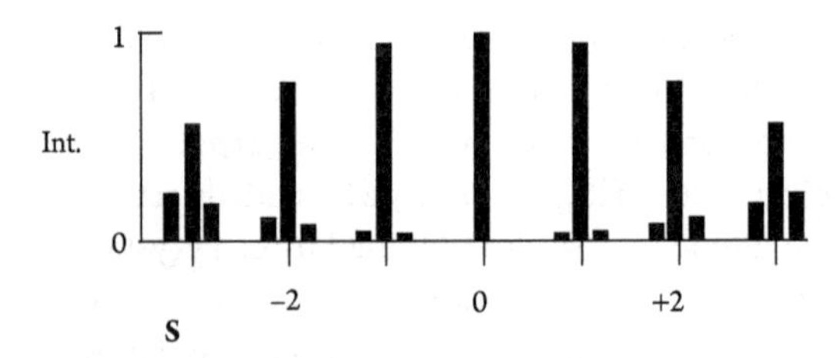

Figure 5.3 Fourier transform of a one-dimensional lattice undergoing a sinusoidal, time-dependent modulation of the positions of its lattice points. The average spacing between real space points is 1.0. The amplitude of their positional modulation is 0.05, and the wavelength of the modulation is 5.0.

The first term in this equation is a set of delta functions that occur every $(1/a)$ in reciprocal space, which is exactly what would be seen if all the atoms in the one-dimensional lattice were stationary and located at their average positions. However, the strength of this Bragg-like signal decreases with s. The expression $[1 - 2(\pi s\varepsilon)^2]$ corresponds to the first two terms of the power series expansion of $\exp[-2(\pi s\varepsilon)^2]$. Thus, if we were to replace $[1 - 2(\pi s\varepsilon)^2]$ with that function and write it as $\exp[-(B/2)s^2]$, B would equal $8\pi^2\varepsilon^2$, which should look familiar. The second and third terms in this expression represent the scatter due to atomic motions. If those motions were random, this part of the scatter would be completely amorphous (equation (5.6)), but here, because the motions are highly correlated, it is perfectly Bragg-like. It is nonzero only when:

$$s = (m/a) \pm (1/\lambda),$$

and the intensities of these "satellite" Bragg peaks increase in proportion to s^2. Figure 5.3 shows what the small s portion of this function looks like.

Comment: Satellite peaks can sometimes be observed flanking the Bragg reflections in the diffraction patterns of the crystals of small molecules, the amplitudes of which increase with scattering angle, just as figure 5.3 illustrates. But often there are so many of them that individual peaks cannot be resolved. Nevertheless, their presence is indicative of the existence of preferred modes of lattice vibration in such crystals.

Two general conclusions can be drawn from this demonstration. First, it confirms that correlations between the thermal motions of the atoms in a crystal make the contributions of those motions to the overall scattering of the crystal depend on both the direction and magnitude of **S**. Furthermore, the tighter those correlations, and the larger the groups of atoms involved, the more strikingly structured the diffuse scattering patterns of crystals will be. Second, whether the thermal motions of atoms are correlated or not, the contributions atomic motions make to the total scatter of crystals increase as $|\mathbf{S}|^2$.

5.12 AVERAGE B-FACTORS CAN BE DETERMINED DIRECTLY FROM DIFFRACTION DATA

In the 1940s, A.J.C. Wilson introduced the crystallographic community to a concept that has been used subsequently in many different contexts. Wilson pointed out that if the unit cell of a crystal contains a lot of atoms, even though their positions are not random, the quantities $\exp[-2\pi i(hx_i + ky_i + lz_i)]$ that are calculated for each atom when computing $F(h, k, l)$ will resemble a list of random numbers, especially when $|h|$ and/or $|k|$ and/or $|l|$ are large. Thus, if the contribution each atom in the unit cell makes to a particular value of $F(h, k, l)$ is plotted on an Argand diagram, the trajectory traced out as each atom's contribution is added to the next will look like a random walk (figure 5.4). $F(h, k, l)$ is the vector from the origin of that trajectory to the far end of the last vector added to it. In other words, it is the sum of all those vectors.

This observation suggests that diffraction data sets might have statistical properties resembling those of two-dimensional random walks, the statistics of which are well understood (see appendix 5.3). If this is so, the probability that a reflection in the diffraction data set collected from a crystal that contains N atoms, where N is large, will have an amplitude between $|F|$ and $(|F| + d|F|)$ should be, approximately:

$$I(|F|)d|F| = \left(2|F|/\sum_i f_i^2\right) \exp\left(-|F|^2/\sum_i f_i^2\right), \qquad (5.7)$$

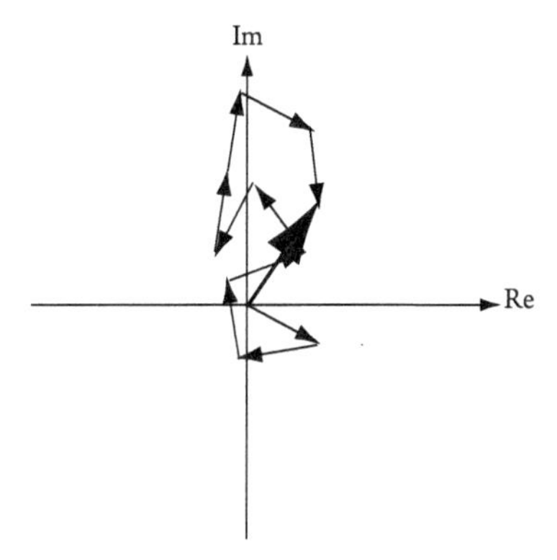

Figure 5.4 $F(h, k, l)$ (heavy vector) represented on an Argand diagram as the vector sum of the contributions of individual atoms (light vectors). In this instance, for simplicity's sake, the molecules responsible for the reflection depicted contain only 10 atoms, and they all have identical scattering factors.

where f_i is the scattering factor of the *i*th atom in the molecule, |F| is the amplitude of the reflection, and the sums run over all atoms. Using this expression, it is easy to show that the average value of I $(= |F|^2)$ is $\sum f_i^2$, a result we will often make use of later.

Using this last result, we can extract an estimate for the average value of *B* for a crystal directly from the diffraction data it yields. The first step is to divide the data into concentric shells of radius $|\mathbf{S}|$, and thickness $\Delta|\mathbf{S}|$, where $\Delta|\mathbf{S}|$ is big enough so that there will be a reasonably large number of reflections (e.g., hundreds or more) in each shell, but small enough so that the number of shells is reasonably large (~20). Provided $|\mathbf{S}|$ is large enough, the average value of the intensity within each shell should obey the equation:

$$\langle \mathrm{I}\,(|\mathbf{S}|)\rangle_{\mathrm{shell}} = N\langle f_i^2\,(|\mathbf{S}|)\rangle \exp\left[-\,(B_{\mathrm{ave}}/2)\,|\mathbf{S}|^2\right],$$

where *N* is the number of atoms in the unit cell, and the scattering factor term is the average value of f_i^2 for all the atoms in the unit cell, and B_{ave} is their average B-factor. Thus, plots of the log of the average intensity versus $|\mathbf{S}|^2$, which are called **Wilson plots**, should have the following (linear) form:

$$\ln(\langle \mathrm{I}(|\mathbf{S}|)\rangle_{\mathrm{shell}}) = \ln(N\langle \mathrm{f}_i^2(|\mathbf{S}|)\rangle) - (B_{\mathrm{ave}}/2)|\mathbf{S}|^2.$$

At $|\mathbf{S}| = 0$, the value of such a plot will be $\ln(NZ^2{}_{\mathrm{ave}})$, and if atomic scattering factors did not depend on $|\mathbf{S}|$, the slope of such a plot would be half the temperature factor. However, atomic scattering factors decrease as $|\mathbf{S}|$ increases, as we already know, but happily the natural log of that falloff is roughly linear with $|\mathbf{S}|^2$ for all atoms, which means that Wilson plots should still be linear. If the average contribution of $\mathrm{f}^2(|\mathbf{S}|)$ to that limiting slope is taken into account, an estimate for the average value of *B* for the crystal should emerge from such an analysis.

When data from macromolecular crystals are analyzed in this manner, it is generally found that within spherical shells of diffraction data P(|F|) does indeed obey equation (5.7), even when $|\mathbf{S}|$ is quite small. However, at small values of $|\mathbf{S}|$, Wilson plots are not linear in $|\mathbf{S}|^2$ because the positions of atoms in macromolecules correlate on intermediate length scales. Macromolecular Wilson plots usually become linear only when $|\mathbf{S}|$ exceeds ~0.3 Å^{-1}. (NB: The data sets used to solve macromolecular crystal structures often do not extend that far into reciprocal space.)

5.13 MOST CRYSTALS ARE MOSAIC

The crystal lattices we have considered theoretically are infinite in all three dimensions, and their reciprocal lattices are similarly infinite arrays of three-dimensional Dirac delta functions. No real crystal is infinite, and in principle, we already know what finite extent does to the transforms of lattices. The Dirac delta function found at every reciprocal lattice point in the transform of an infinite lattice is replaced by the Fourier transform of the overall shape of the crystal. If the unit length vectors that specify the edges of some crystal that has a cubic shape are $\mathbf{l}_1$, $\mathbf{l}_2$, and $\mathbf{l}_3$, and its length in all threee directions is 0.1 mm, the transform of its shape will be:

$$\prod_{1}^{3} \sin(5 \times 10^{-5}\, \pi \mathbf{S} \cdot \mathbf{l}_i)/\pi\, \mathbf{S} \cdot \mathbf{l}_i.$$

If diffraction data were collected from this crystal using 1-Å wavelength X-rays, we would expect the intensity of its reflections to die off incredibly rapidly as the difference in **S** between the center of any one of its reflections and the place where that scatter is measured increases. Beyond 1 arc second in the (= ~0.0001°) from the center of each reflection, the intensity would be effectively 0. However, the angular sizes of the reflections from macromolecular crystals are commonly tenths of a degree to even a few degrees in width.

Why are the Bragg reflections of real crystals so wide? There are a number of experimental factors one might want to consider, such as the range of wavelengths in the "monochromatic" beam used to collect the data, and the transverse coherence length of the radiation used. However, the wavelength spreads used for data collection are usually quite small ($\Delta\lambda/\lambda = 10^{-3}$ to 10^{-4}, or better), and transverse coherence lengths are usually better than 1,000 Å. Thus, even though no diffractometer will image a point like reflection as a perfect point—they have instrument transfer functions too—the contributions that beam characteristics and instrument transfer functions make to the measured sizes of reflections are unlikely to exceed 0.05°. Where does the rest of the width of reflections come from?

The standard explanation given when reflections are larger than they should be is that the crystal responsible for them is **mosaic**, and the effect that mosaicity has on the size of reflections is called **mosaic spread**. A mosaic crystal can be thought of as composed of many smaller crystals, called **mosaic blocks**, each of which is as well ordered as thermal effects allow. The reflections obtained from a mosaic crystal are larger than they otherwise would be because its mosaic blocks are slightly misaligned with each other so that their diffraction patterns do not add coherently. Mosaic spread has no effect on integrated intensities.

5.14 A SINGLE CRYSTAL STRUCTURE CAN REVEAL THE ALTERNATIVE CONFORMATIONS OF A MACROMOLECULE THAT IS POLYMORPHIC

Unlike the crystals of small molecules, which (usually) contain only a single kind of molecule, macromolecular crystals invariably include not only the protein(s) and/or nucleic acid(s) and/or polysaccharide(s) and/or lipid aggregates of interest, but also large amounts of solvent. They are best thought of as highly ordered gels. It is perfectly possible for the conformations of the macromolecular components of such a crystal to vary from one unit cell to the next by amounts that are large compared to the thermal variation discussed earlier, provided that variation does not alter the intermolecular interactions that stabilize the crystal. An external loop in a protein, for example, might have two conformations that differ so little in free energy that both arrangements of that loop are reasonably probable under the conditions prevailing in the crystal. What will be seen in the corresponding region of the electron density maps of that crystal when its structure is solved?

The answer to the question just posed should be obvious from section 5.9, where it is pointed out that the intensities of the Bragg reflections in the diffraction pattern of a crystal are determined by the average structure of its unit cells. Thus, if the macromolecules in some crystals exist in two conformations:

$$\mathrm{I}(h,\ k,\ l) \propto \left(\int [w_1 \rho_1(\mathbf{r}) + w_2 \rho_2(\mathbf{r})] \exp(-2\pi i \mathbf{r} \cdot \mathbf{S}) dV_r \right)^2,$$

where $\rho_1(\mathbf{r})$ is the electron density distribution of the first version of the structure, w_1 is the fraction of the unit cells in which that structure is found, and $\rho_2(\mathbf{r})$ and w_2 are the corresponding quantities for the second conformation. (Note: $w_1 + w_2 = 1$.) Provided that in at least some regions $\rho_1(\mathbf{r})$ and $\rho_2(\mathbf{r})$ imply differences in atomic locations that are significant compared to $1/|\mathbf{S}|_{\mathrm{max}}$, and provided also that the difference between w_1 and w_2 are not so large that one of the two conformations completely dominates the electron density, it may be possible to determine the conformations of both of the structures present from an averaged map of this sort. If so, we will also obtain estimates for both w_1 and w_2. These quantities are referred to as **occupancies**. In the parts of the structure of such a macromolecule where there is no conformational variation, a single structure will be seen, and the occupancies for all the atoms in those regions will be 1.0. In variable regions, atomic occupancies will be less than 1.0. (Clearly, the more variants of the conformation of the flexible regions of a macromolecule found in its crystals, the harder it will be to obtain structures for each variant from an averaged electron density map.)

Like every other kind of disorder, this kind of disorder, which is a form of a kind of disorder called **substitution disorder**, produces diffuse scatter, and in this case, it is certain to be highly structured because the positions of groups of atoms in the crystal are varying in a correlated manner. (That structured signal will be superimposed on whatever other diffuse scatter the crystal may be generating.) This component of the diffuse scatter will be a weighted sum of the squares of the transforms of the two versions of the structure that exist in the variable region(s) of the molecule in question.

PROBLEMS

1. a. Show that if the diffraction data obtained from some crystal do not obey Friedel's law, that the cosine component can be obtained from the measured data by dividing the sum of $\mathrm{I}(h,\ k,\ l)$ and $\mathrm{I}(-h,\ -k,\ -l)$ by 2, and that the sine component can be obtained by dividing the difference between $\mathrm{I}(h,\ k,\ l)$ and $\mathrm{I}(-h,\ -k,\ -l)$ by 2.
   b. If the square root of the cosine component of each reflection were taken to obtain an estimate of $|\mathrm{F}(h,\ k,\ l)|$, would that estimate correspond to what would have been measured if the electron density distribution of the crystal had been $\rho(\mathbf{r})$ instead of $\rho(\mathbf{r}) + i\rho'(\mathbf{r})$?

2. Consider two crystals: one of which has a primitive unit cell, and the second of which has the same unit cell, but the cell is c-centered instead of primitive. The term *c-centered* means that in addition to the usual lattice point at every corner, each unit cell has a lattice point located in the middle of the face of the unit cell that is opposite its *c*-axis, in other words, the plane in which both the **a** and **b** vectors of the unit cell are found. Thus, in addition to every atom associated with the lattice point chosen as the origin of each unit cell having coordinates $(x,\ y,\ z)$, there will be a second that has coordinates $(x + 1/2,\ y + 1/2,\ z)$. Leaving aside any differences there may be between the intensities of corresponding reflections, how will the two diffraction patterns differ qualitatively?
3. a. Some small molecule crystals possess mirror symmetry. If the planes of mirror symmetry in some crystals were taken as the *c*-face of the unit cell, i.e. the face in which the **a** and **b** vectors lie, then for every atom at $(x,\ y,\ z)$ there would have to be another atom at $(x,\ y,\ -z)$. What are the consequences of this kind of symmetry for the diffraction patterns of crystals? Assume that anomalous scattering can be ignored.
   b. Are there any restrictions on the angles between unit cell axes that are possible for such a crystal that has this kind of mirror symmetry?
4. There are many ways that crystals can be disordered, each with its own consequences for the appearance of the diffuse scatter crystals generate. Suppose the dominant kind of disorder in some crystal is caused by small, rigid-body translations of the centers of gravity of the molecules in its unit cells that vary randomly from one unit cell to the next.
   a. Derive an expression for the diffuse scatter that should be observed when diffraction data are collected from such a crystal.
   b. Describe in words what it will look like.

APPENDIX 5.1 DEBYE-WALLER FACTORS AND DIFFUSE SCATTERING

Equation (5.3), which is derived in the body of the text, shows how the average intensities of Bragg reflections are modified by the motions of the atoms in a crystal:

$$\langle \mathrm{I}(h,k,l)\rangle \propto \sum_i \sum_j \mathrm{f}_i\mathrm{f}_j \exp[-2\pi i(\mathbf{r}_i - \mathbf{r}_j)\cdot \mathbf{S}(h,k,l)]$$
$$\times \langle \exp[-2\pi i\boldsymbol{\Delta}_i(t)\cdot \mathbf{S}(h,k,l)]\rangle \langle \exp[2\pi i\boldsymbol{\Delta}_j(t)\cdot \mathbf{S}(h,k,l)]\rangle.$$

To obtain a more manageable version of this equation, the two averages on its right-hand side must be evaluated. This is done by expanding each of them as a power series, and then averaging each term separately:

$$\langle \exp[-2\pi i\boldsymbol{\Delta}_i(t)\cdot \mathbf{S}(h,k,l)]\rangle = 1 + \langle -2\pi i\boldsymbol{\Delta}_i(t)\cdot \mathbf{S}(h,k,l)\rangle$$
$$+ (1/2)\langle [-2\pi i\boldsymbol{\Delta}_i(t)\cdot \mathbf{S}(h,k,l)]^2\rangle + \ldots$$

Because by hypothesis, $\langle \mathbf{\Delta}_i(t) \cdot \mathbf{S}(h,k,l) \rangle = 0$, all the odd terms in this expansion have average values of 0, and we are left with:

$$\langle \exp[-2\pi i \mathbf{\Delta}_i(t) \cdot \mathbf{S}(h,k,l)] \rangle = 1 - 2\pi^2 \langle [\mathbf{\Delta}_i(t) \cdot \mathbf{S}(h,k,l)]^2 \rangle + \ldots$$

The first two terms in this series correspond to the first two terms in the power series expansion for $\exp\left(-2\pi^2 \langle [\mathbf{\Delta}_i(t) \cdot \mathbf{S}(h,k,l)]^2 \rangle\right)$, and so:

$$\langle \exp[-2\pi i \mathbf{\Delta}_i(t) \cdot \mathbf{S}(h,k,l)] \rangle \approx \exp\left(-2\pi^2 \langle [\mathbf{\Delta}_i(t) \cdot \mathbf{S}(h,k,l)]^2 \rangle\right).$$

In fact, if the probability that some atom will be found at some distance, Δ, from its average position is given by a Gaussian distribution, the preceding equation exact.

If the motions of an atom are isotropic, then the component of the variance of its excursions away from its equilibrium position in any direction, such as the **S** direction, will be 1/3 of the total variance of its excursions in all directions, which means that:

$$\langle [\mathbf{\Delta}_i(t) \cdot \mathbf{S}(h,k,l)]^2 \rangle = (1/3) \langle \Delta_i(t)^2 \rangle |\mathbf{S}|^2.$$

Now $|\mathbf{S}|^2 = 4 \sin^2 \theta / \lambda^2$, and so:

$$2\pi^2 \langle [\mathbf{\Delta}_i(t) \cdot \mathbf{S}(h,k,l)]^2 \rangle = (8\pi^2/3) \langle \mathbf{\Delta}_i(t)^2 \rangle \sin^2 \theta / \lambda^2.$$

The quantity $(8\pi^2/3)\langle \mathbf{\Delta}_i(t)^2 \rangle$ is called a **temperature factor**, or **Debye-Waller factor**, or **B-factor**.

It follows that:

$$\langle \mathrm{I}(h,k,l) \rangle \propto \sum_i \sum_j \mathrm{f}_i \mathrm{f}_j \exp[-2\pi i (\mathbf{r}_i - \mathbf{r}_j) \cdot \mathbf{S}(h,k,l)]$$
$$\exp[-(B_i/4 + B_j/4)|\mathbf{S}(h,k,l)|^2]. \qquad (5.1.1)$$

To understand how the non-Bragg components of the scatter produced by disordered crystals arises, you must start with an expression that describes the scattering produced by a crystal at a particular instant in time:

$$\mathrm{I}(\mathbf{S},t)_{\text{total}} = \sum_n \sum_m \exp[-2\pi i (\mathbf{u}_n - \mathbf{u}_m) \cdot \mathbf{S}]$$
$$\sum_i \sum_j \mathrm{f}_i \mathrm{f}_j \exp[-2\pi i (\mathbf{r}_i - \mathbf{r}_j) \cdot \mathbf{S}] \times$$
$$\exp\left(-2\pi i [\mathbf{\Delta}_{in}(t) - \mathbf{\Delta}_{jm}(t)] \cdot \mathbf{S}\right).$$

Both sums in the first double sum in this equation are over all unit cells in the crystal, and $\mathbf{u}_n$ is a vector that runs from the origin of the unit cell chosen to be the first unit cell in the crystal to the origin of nth unit cell. As usual, the second pair of sums are both over all atoms in the unit cell, but the $\Delta(t)$ vectors have two subscripts because at any instant in time, the displacement of the ith atom from its average position will vary from one unit cell to the next. The intensities we actually observe are averages over time, and hence:

$$\begin{aligned}\langle I(\mathbf{S},t)\rangle_{\text{total}} = &\sum_n\sum_m \exp[-2\pi i(\mathbf{u}_n - \mathbf{u}_m)\cdot\mathbf{S}]\\ &\sum_i\sum_j f_i f_j \exp[-2\pi i(\mathbf{r}_i - \mathbf{r}_j)\cdot\mathbf{S}]\times\\ &\langle\exp(-2\pi i[\boldsymbol{\Delta}_{in}(t) - \boldsymbol{\Delta}_{jm}(t)]\cdot\mathbf{S})\rangle. \qquad (5.1.2)\end{aligned}$$

Written in the same format, the Bragg scatter from the crystal is:

$$\begin{aligned}\langle I(\mathbf{S},t)\rangle_{\text{Bragg}} = &\sum_n\sum_m \exp[-2\pi i(\mathbf{u}_n - \mathbf{u}_m)\cdot\mathbf{S}]\\ &\sum_i\sum_j f_i f_j \exp[-2\pi i(\mathbf{r}_i - \mathbf{r}_j)\cdot\mathbf{S}]\times\\ &\langle\exp[-2\pi i\boldsymbol{\Delta}_{in}(t)\cdot\mathbf{S}]\rangle\langle\exp[2\pi i\boldsymbol{\Delta}_{jm}(t)\cdot\mathbf{S}]\rangle.\end{aligned}$$

The difference between the Bragg scatter equation and equation (5.1.2) is that the former is the square of the average Fourier transform, whereas the latter is the average of the square of the Fourier transform; they are not the same. We know that the Bragg expression represents the familiar, sampled function in reciprocal space. The diffuse scatter is the difference between the time averages of $I(\mathbf{S})_{\text{total}}$ and $I(\mathbf{S})_{\text{Bragg}}$. Replacing the averages of exponentials that these equations contain with their Gaussian equivalents, we discover that:

$$\begin{aligned}\langle I(\mathbf{S},t)\rangle_{\text{diffuse}} = &\sum_n\sum_m \exp[-2\pi i(\mathbf{u}_n - \mathbf{u}_m)\cdot\mathbf{S}]\\ &\sum_i\sum_j f_i f_j \exp[-2\pi i(\mathbf{r}_i - \mathbf{r}_j)\cdot\mathbf{S}]\\ &\times\left\{\exp[-2\pi^2\langle([\boldsymbol{\Delta}_{in}(t) - \boldsymbol{\Delta}_{jm}(t)]\cdot\mathbf{S})^2\rangle]\right.\\ &\left.-\exp[-2\pi^2(\langle[\boldsymbol{\Delta}_{in}(t)\cdot\mathbf{S}]^2\rangle + \langle[\boldsymbol{\Delta}_{jm}(t)\cdot\mathbf{S}]^2\rangle)]\right\},\end{aligned}$$

which can be written:

$$\begin{aligned}\langle I(\mathbf{S})\rangle_{\text{diffuse}} = &\sum_n\sum_m \exp[-2\pi i(\mathbf{u}_n - \mathbf{u}_m)\cdot\mathbf{S}]\\ &\sum_i\sum_j f_i f_j \exp[-2\pi i(\mathbf{r}_i - \mathbf{r}_j)\cdot\mathbf{S}]\\ &\times\exp[-2\pi^2(\langle[\boldsymbol{\Delta}_{in}(t)\cdot\mathbf{S}]^2\rangle + \langle(\boldsymbol{\Delta}_{jm}(t)\cdot\mathbf{S})^2\rangle)]\\ &[\exp(4\pi^2\langle[\boldsymbol{\Delta}_{in}(t)\cdot\mathbf{S}][\boldsymbol{\Delta}_{jm}(t)\cdot\mathbf{S}]\rangle) - 1], \qquad (5.1.3)\end{aligned}$$

APPENDIX 5.2 CORRELATED MOTIONS AND DIFFUSE SCATTER IN ONE DIMENSION

We start with the equation for $I(s, t)$ for a one-dimensional lattice of atoms undergoing a small, correlated, sinusoidal oscillations that was developed in section 5.10:

$$I(s,t) = \sum\sum \exp[-2\pi is(n-m)a] \\ \exp\left(-2\pi i\varepsilon[\sin(\omega t - 2\pi na/\lambda) - \sin(\omega t - 2\pi ma/\lambda)]\right).$$

The quantity we seek to evaluate is the average value of $I(s, t)$, and it is clearly:

$$\langle I(s,t)\rangle = \sum\sum \exp[-2\pi is(n-m)a] \\ \langle\exp\left(-2\pi i\varepsilon[\sin(\omega t - 2\pi na/\lambda) - \sin(\omega t - 2\pi ma/\lambda)]\right)\rangle.$$

Replacing $\sin(\omega t - 2\pi na/\lambda) - \sin(\omega t - 2\pi ma/\lambda)$ by "$\arg(n, m)$" for convenience, the second exponential component of this equation can be written as a power series:

$$\langle\exp[-2\pi is\varepsilon \arg(n,m)]\rangle = 1 - \langle 2\pi is\varepsilon \arg(n,m)\rangle \\ - (1/2)\langle[2\pi s\varepsilon \arg(n,m)]^2\rangle + \ldots$$

However, because $\arg(n, m)$ is the difference between two time-varying sine functions, their average over any period of time that is long compared to ω^{-1} will be very small compared to the other terms in the expansion, and for that same reason, all the odd-order terms in the power series may be ignored, and hence:

$$\langle\exp[-2\pi is\varepsilon \arg(n,m)]\rangle = 1 - (1/2)\langle[2\pi s\varepsilon \arg(n,m)]^2\rangle + \ldots$$

If ε is small compared to a, only the second term of this expansion will contribute significantly to the time-average of interest. Now:

$$(1/2)\langle[2\pi s\varepsilon \arg(n,m)]^2\rangle = 2(\pi\varepsilon s)^2\langle\sin^2(\omega t - 2\pi na/\lambda) \\ - 2\sin(\omega t - 2\pi na/\lambda)\sin(\omega t - 2\pi ma/\lambda) \\ + \sin^2(\omega t - 2\pi ma/\lambda)\rangle = 2(\pi\varepsilon s)^2 \\ [1 - 2\langle\sin(\omega t - 2\pi na/\lambda)\sin(\omega t - 2\pi ma/\lambda)\rangle].$$

(NB: The average value of $\sin^2\theta$ is 1/2.) Using the standard trigonometric expression for the product of two sine functions, we discover:

$$2\langle\sin(\omega t - 2\pi na/\lambda)\sin(\omega t - 2\pi ma/\lambda)\rangle = \langle\cos[2\pi(n-m)a/\lambda]\rangle \\ - \langle\cos[2\omega t - 2\pi(n+m)a/\lambda]\rangle,$$

but because the second term is the average value of a sinusoidal oscillation, its value will be negligible compared to the first.

Gathering all the terms together, we find that:

$$\langle \mathrm{I}(s,t)\rangle = \sum\sum \exp[-2\pi is(n-m)a][1-2(\pi\varepsilon s)^2] \\ +2(\pi\varepsilon s)^2 \sum\sum \exp[-2\pi is(n-m)a]\cos[2\pi(n-m)a/\lambda].$$

Replacing the cosine part of the second double sum with its exponential equivalent, we arrive at the final result:

$$\langle \mathrm{I}(s,t)\rangle = \sum\sum \exp[-2\pi is(n-m)a][1-2(\pi\varepsilon s)^2] \\ +(\pi\varepsilon s)^2 \sum\sum \exp[-2\pi i(s+1/\lambda)(n-m)a] \\ +(\pi\varepsilon s)^2 \sum\sum \exp[-2\pi i(s-1/\lambda)(n-m)a].$$

APPENDIX 5.3 RANDOM WALKS IN TWO DIMENSIONS

Consider a one-dimensional system in which an object is placed at the origin, a coin flipped, and depending on whether it comes up heads or tails, the object moved one position to the right or one position to the left, and the exercise repeated many times. The trajectory the object traces out as this process proceeds is called a *random walk*, and it is reasonable to wonder how far the object is likely to be from where it started after the process has been repeated N times. Because the process responsible for making the object move is random, this question can only be answered in an averaged, statistical sense.

If N is the number of steps the object has taken, and m is the number of steps to the right minus the number of steps to the left, combinatorial algebra tells us that provided $m << N$, the probability that the object will be found m steps to the right of the origin after N steps will be:

$$\mathrm{P}(N,m) = N![(1/4)(N+m)!(N-m)!]^{-1}(1/2)^N.$$

The $(1/2)^N$ part of this expression is the probability of observing any particular sequence of heads and tails, and the rest of the expression is the reciprocal of the probability that a random sequence of steps will have the net result that the object has moved m steps to the right. Like all quantities that contain factorials, $\mathrm{P}(N,\ m)$ is impractical to evaluate directly, but $\ln[\mathrm{P}(N,\ m)]$ is much more tractable. The reason it can be evaluated is that a simple expression exists for evaluating $\ln(N!)$ called Sterling's approximation:

$$\ln(N!) \approx (N+1/2)\ln N - N + (1/2)\ln 2\pi.$$

In addition:

$$\ln(1 \pm m/N) = \pm m/N - m^2/2N^2 + \ldots$$

If $m << N$, which it is bound to be if N is large, this series expansion can be terminated after the second term. Using these two mathematical facts, as well as a lot of careful algebra, it not hard to show that:

$$\mathrm{P}(N, m) = (2\pi N)^{-1/2} \exp(-m^2/2N).$$

If the length of the step that is taken every time the coin is flipped is f, then the distance from the origin to the place where the object is found will be mf, a quantity we will call F. Thus:

$$\mathrm{P}(N, F) = (2\pi f^2 N)^{-1/2} \exp(-F^2/2f^2 N).$$

[The factor of f^2 in the leading scale factor is included to preserve the normalization of $\mathrm{P}(N, F)$.] Note also that the distribution is symmetric; the object is just as likely to be m steps to the left of where it started as m steps to the right.

In two dimensions:

$$\mathrm{F}(N, F_x, F_y) = (2\pi f_x^2 N)^{-1/2} \exp(-F_x^2/2f_x^2 N)(2\pi f_y^2 N)^{-1/2} \exp(-F_y^2/2f_x^2 N).$$

Now, $F^2 = F_x{}^2 + F_y{}^2$, and if the walk is random, the average values of $f_x{}^2$ and $f_y{}^2$ will both equal $f^2/2$. Thus:

$$\mathrm{P}(N, F_x, F_y) = (\pi f^2 N)^{-1} \exp(-F^2/f^2 N).$$

What really interests us, however, is not $\mathrm{P}(N, F_x, F_y)$, but rather $\mathrm{P}(N, |F|)dF$, and:

$$\mathrm{P}(N, |F|)dF = 2\pi |F| \mathrm{P}(N, F_x, F_y)dF = (2|F|/f^2 N) \exp(-|F|^2/f^2 N)dF.$$

If the step sizes are not all equal, then Nf^2 must be replaced with $\sum f_i{}^2$, and then:

$$\mathrm{P}(N, |F|)dF = \left(2|F|/\sum f_i^2\right) \exp\left(-|F|^2/\sum f_i^2\right) dF.$$

6

Solving the Phase Problem

The structure of a macromolecular crystal is solved in a two-stage process, the steps of which are: (1) determination of the electron density distribution of its asymmetric unit, and (2) chemical interpretation of that electron density distribution. The task is complete when a three-dimensional atomic model of the molecule(s) in the asymmetric unit has been generated that makes chemical sense and that explains the diffraction data. "Makes chemical sense" implies that the sequence(s) of amino acids and/or nucleotides in the model of the asymmetric unit is/are consistent with prior knowledge, that each monomer has the appropriate geometry, that adjacent, nonbonded atoms do not violate each other's van der Waals radii, etc. "Explains the diffraction data" means that the structure factors computed for the crystal using the molecular model match the measured data, in other words, that $|F(h, k, l)_{calc}| = |F(h, k, l)_{obs}|$, for all reflections. (How that match is measured, and how good a match is good enough is discussed in chapter 7.)

As we already know, electron density distributions can be computed by Fourier transformation of measured diffraction amplitudes *only* if the phases associated with each reflection are known and that information is lost during the measurement process. In this chapter, we will discuss the three approaches that are used for recovering phases. We will start by treating the two strategies that are used to measure phases experimentally, and then we will turn to a computational method for obtaining phases that is based on prior knowledge.

6.1 THE PHASES OF REFLECTIONS ARE MEASURED BY COMPARING THEM TO A STANDARD

Both of the experimental strategies for estimating the phases of reflections work in fundamentally the same way. Phases are estimated by measuring the interference that results when the radiation in individual reflections is coherently combined with the radiation in a reference X-ray beam of the same wavelength, the phase of which is known.

It is easy to understand why these strategies work. The electric field of the radiation that corresponds to some reflection in diffraction pattern is proportional to $|F(h, k, l)| \exp[i\alpha(h, k, l)]$. Suppose the intensity of that reflection can be measured both in the presence and in the absence of a reference beam of the same wavelength, the electric field of which is proportional to $|F_r| \exp(i\alpha_r)$. In the absence of the reference radiation, the intensity of the (h, k, l) reflection will be $|F(h, k, l)|^2$, whereas its intensity in the presence of the reference beam, $I_{+r}(h, k, l)$, will be:

$$\begin{aligned} I_{+r}(h, k, l) &= \left(|F(h, k, l)| \exp\left[i\alpha(h, k, l)\right] + |F_r| \exp(i\alpha_r)\right)^2 \\ &= |F(h, k, l)|^2 + |F_r|^2 + 2|F(h, k, l)||F_r| \\ &\quad \cos\left[\alpha(h, k, l) - \alpha_r\right] \end{aligned}$$

(see section 6.4). Because $|F(h, k, l)|^2$ can be measured directly, and by hypothesis, $|F_r|^2$ is known, an estimate for $\cos[\alpha(h, k, l) - \alpha_r]$ can be obtained from $I_{+r}(h, k, l)$. If α_r is also known, that estimate will be consistent with two different values for $\alpha(h, k, l)$ because $\cos(\theta) = \cos(-\theta)$. The correct choice between the two alternatives can be made if this experiment is repeated using a reference beam that has a different phase.

The experiment just outlined is easier to describe than to execute. For it to work, $F(h, k, l)$ and F_r must be coherently related, which means: (1) that the wavelengths of the radiation represented by both must be identical; (2) that the direction of polarization of the electric field vectors of the two beams must be the same at all times; and (3) that the phase relationship between the two must not vary with time. In addition, α_r must be known for each reflection.

Holograms, which are optical interferograms that have a similar character, are produced using the light obtained from a single source that has been split into two beams using a half-silvered mirror. One beam is used to illuminate the object of interest, and the light scattered by the object is then made to interfere with the light from the second beam, and the result recorded (often) photographically to produce the hologram. Even though, in theory, beam-splitting strategies could be used to phase the X-ray diffraction patterns of crystals, this is not done for a number of practical reasons, one of them being mechanical stability. Interferometric instruments based on beam splitting will not produce useful data unless their dimensions are stable on length scales that are small compared to the wavelength of the working radiation over times intervals that are long compared to data collection times. If they are not, the signals they produce will correspond to $[|F(h, k, l)|^2 + |F_r|^2]$ and, thus, will provide no information about phases. The mechanical stability required is much harder to achieve when the wavelength is 1 Å than it is when wavelength is 5,000 Å! For that reason, among others, both of the approaches used to phase macromolecular X-ray diffraction patterns experimentally use reference beams that are generated *inside* crystals rather than being supplied from the outside.

6.2 MACROMOLECULAR DIFFRACTION PATTERNS CAN BE PHASED BY ADDING HEAVY ATOMS TO CRYSTALS

Of the many contributions made to macromolecular crystallography by Max Perutz, by far the most important was his discovery of the first of the two experimental methods used to measure the phases of reflections in macromolecular diffraction patterns. His approach requires not only crystals of the macromolecule of concern, but also **isomorphous** crystals of derivatives of that same macromolecule that have small numbers of atoms of high atomic number bound at specific locations in each unit cell. Crystals are isomorphous if the structures of the contents of their unit cells are the same, save, in this case, for the presence or absence of a small number of heavy atoms, and their symmetries and unit cell dimensions are identical. The radiation scattered by the heavy atoms in derivatized crystals provides the reference beam that makes phase determination possible.

The simplest way to prepare a heavy atom derivative of some macromolecular crystal is to soak one of its crystals in a solution in which a heavy atom–containing compound of low molecular weight is dissolved. (NB: The solution used must be one in which the crystal is stable!) The outcome of such an experiment depends on: (1) the reactivity of the heavy atom compound used, (2) its concentration in the solution, (3) the length of time the crystal is exposed to the agent, and (4) the reactivity of the macromolecule itself. The reason derivatives can be prepared this way is that macromolecular crystals include solvent-filled channels that are so large that low molecular weight molecules can readily diffuse into their interiors.

However it is done, the success of any heavy atom derivatization procedure is judged by comparing the diffraction patterns of crystals that have been treated with heavy atom compounds with those of the corresponding native—underivatized—crystals. What the crystallographer hopes is: (1) that the lattice of the derivatized crystal will be identical to that of the native crystal to within some very small tolerance (e.g. $<1\%$ change in the lengths of unit cell vectors); (2) that the derivatized crystal will belong to the same space group as the native crystal; and (3) that there will be significant differences in the intensities of corresponding reflections between the derivatized crystal and the native crystal.

6.3 THE NUMBER OF HIGH-Z ATOMS PER UNIT CELL NEEDED FOR PHASING IS SMALL

Even Perutz may have been surprised to discover how small the number of heavy atoms it takes to produce useful changes in the intensities of the reflections produced by macromolecular crystals, but it is easy to understand why this is so using Wilson statistics. Hemoglobin, the protein Perutz spent his career studying, has a molecular weight of about 64,000, which means that for these purposes, it can be thought of as an assembly of $\sim 5,000$ carbon atoms ($Z = 6$). Thus, for crystals of hemoglobin that contain 1 molecule per unit cell, Wilsonian arguments indicate that the average value of $F(h, k, l)$ should be proportional to $\sim(5,000)^{1/2}6(= 424)$. If a single mercury atom ($Z = 80$) binds to, or reacts with, a unique site in each hemoglobin molecule in such a crystal, the amplitude of the contribution mercury makes to each reflection will be $\sim 1^{1/2}80 (= 80)$ on that same scale. Thus, if the phase of the mercury contribution to some reflection happened to be the same as the phase of the hemoglobin contribution to that same reflection, the amplitude of that reflection in the derivatized crystal's diffraction pattern would be $\sim 504 (= 424 + 80)$ instead of ~ 424, which implies an $\sim 41\% (504/424)^2$ increase in measured intensity! When we take into account the fact that the phases of the heavy atom contribution to the reflections of a derivatized crystal are seldom the same as the phases of the protein contributions, and we perform a more careful analysis, the following expression emerges:

$$\langle \Delta \mathrm{I} \rangle / \langle \mathrm{I_P} \rangle = (2N_{\mathrm{HA}}/N_{\mathrm{P}})^{1/2} \left(\mathrm{f_{HA}} / \langle \mathrm{f_P^2} \rangle^{1/2} \right), \tag{6.1}$$

where $\langle \Delta \mathrm{I} \rangle / \langle \mathrm{I_P} \rangle$ is the ratio of the average change in reflection intensity due to derivitization and the average reflection intensity of the native protein; N_{HA} and N_{P} are the number of heavy atoms and the number of macromolecular atoms per asymmetric unit, respectively; $\mathrm{f_{HA}}$ is the structure factor of the heavy atoms; and $\langle \mathrm{f_P}^2 \rangle$ is the average of the square of the structure factors of all the atoms in the macromolecule. For the hemoglobin example we just discussed, this formula indicates that the average change in reflection intensity produced by the derivatization in question should be $\sim 26\%$.

It is important to realize that even though the average change in the intensities of reflections caused by derivization is $\sim 26\%$, the average intensity of the reflections obtained from crystals of the hemoglobin-mercury complex under consideration will be only slightly different from the average intensity of the reflections obtained from otherwise similar crystals of underivatized hemoglobin. The average value of $|\mathrm{F_{HP}}|$ will be $\sim 432 (= [(5000)6^2 + 80^2]^{1/2})$, which means that the average value of $\mathrm{I_{HP}}$ will be only $\sim 3.6\%$ larger than the average value of $\mathrm{I_P} [= (432/424)^2]$. The reason that the addition of a single heavy metal atom to the protein has so little effect on average reflection intensities is that the contribution the metal atom makes to the intensity of any given reflection is almost as likely to decrease its value as to increase it.

6.4 THE HEAVY ATOM ISOMORPHOUS REPLACEMENT STRATEGY FOR PHASING REQUIRES THE COMPARISON OF INTENSITIES MEASURED FROM DIFFERENT CRYSTALS

Provided the locations of the heavy atoms in a derivatized crystal can be determined, a critical issue that we will address shortly, it is easy to extract the desired phase information once the diffraction patterns of both the native crystal and the derivatized crystal have been measured. For each reflection in the crystal's diffraction pattern, a picture can be drawn that is similar to figure 6.1, which shows something that should be obvious, namely that $\mathrm{F_{HP}}(h, k, l)$, the structure factor of the derivatized protein, is the (vector) sum of $\mathrm{F_P}(h, k, l)$, the structure factor of the underivatized protein, which can be written $|\mathrm{F_P}| \exp(i\alpha_{\mathrm{P}})$, and $\mathrm{F_H}(h, k, l)$, the structure factor of the heavy

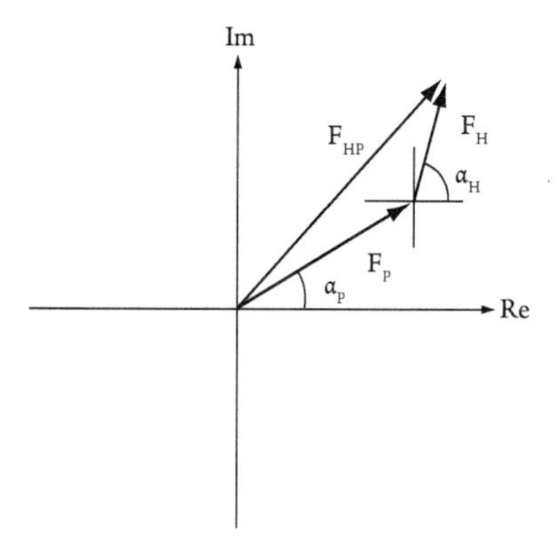

Figure 6.1 The Argand diagram for a typical reflection in the diffraction pattern of a protein crystal that contains some heavy atoms. The structure factor of the derivatized protein for that reflection is F_{HP}. The structure factor of the underivatized protein is F_P, and the heavy metal structure factor is F_H. α_P and α_H are the phase angles of the protein and heavy atom structure factors, respectively.

metal atom(s) in the unit cell, which is the same as $|F_H| \exp(i\alpha_H)$. The Pythagorean theorem assures us that:

$$|F_{HP}|^2 = [|F_h| \sin(\alpha_H - \alpha_P)]^2 + [|F_P| + |F_H| \cos(\alpha_H - \alpha_P)]^2,$$

and thus:

$$\cos(\alpha_H - \alpha_P) = \left(|F_{HP}|^2 - |F_H|^2 - |F_P|^2\right)/2\,|F_P|\,|F_H|. \qquad (6.2)$$

Because α_H and $|F_H|$ can both be calculated once the locations of the sites where heavy metal atoms bind have been determined (see section 6.8), and $|F_{HP}|$ and $|F_P|$ can be measured experimentally, the data will provide two estimates for α_P:

$$\alpha_P = \alpha_H \pm \cos^{-1}\left[\left(|F_{HP}|^2 - |F_H|^2 - |F_P|^2\right)/2\,|F_P|\,|F_H|\right].$$

Many years ago, D. Harker devised a graphical method for estimating phases that is still useful (figure 6.2). A circle of radius $|F_P|$ is drawn on an Argand diagram, the center of which is the origin. A vector is drawn on that same diagram of length $|F_H|$ in the direction $(\alpha_H + \pi)$, and then a circle of radius $|F_{HP}|$ is drawn centered on the end of that vector. Vectors drawn from the center of the Argand diagram to the intersections of the two circles represent the two solutions to the phase that are consistent with the data, F_P and F_P'. Their phases are α_P and α_P', respectively.

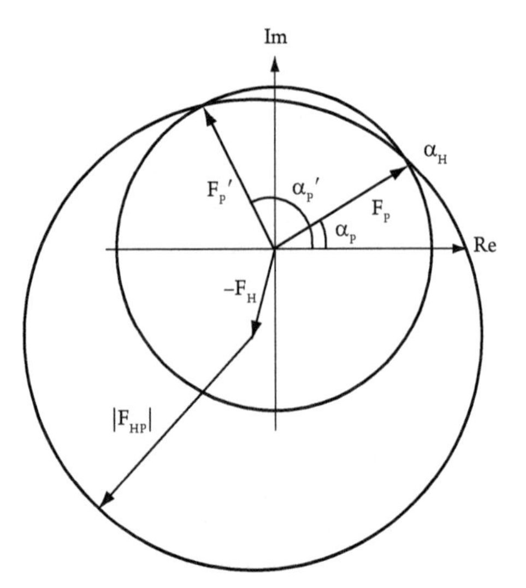

Figure 6.2 Harker's method for determining phases by isomorphous replacement.

The correct alternative can be identified by repeating this process using a derivative of the crystal in question that has heavy atoms bound at a different set of sites. In theory, for each reflection, one of the two values for α_P that was obtained from the first experiment will be the same as one of the two values for α_P that emerges from the second. Perutz's phasing strategy is called **multiple isomorphous replacement** (MIR), because it requires the crystallographer to obtain data from (at least) two different derivatives of the crystal under investigation.

6.5 ANOMALOUS DATA CAN ALSO PROVIDE PHASE INFORMATION

J. M. Bijvoet, the crystallographer who demonstrated that anomalous diffraction can be used to determine the absolute hand of chiral molecules crystallographically (section 5.5), also knew that anomalous data contain phase information. That information is to be found in the differences in measured intensities between reflections that bear the relationship (h, k, l) and $(-h, -k, -l)$. Pairs of reflections that are so related are called **Bijvoet pairs**, and their differences in amplitude are called **Bijvoet differences.**

In section 5.5, it was demonstrated that if the scattering factors of any of the atoms in a crystallized molecule have an appreciable $\Delta f''$ component then $I(h, k, l) \neq I(-h, -k, -l)$. The $\Delta f''$ component of the scattering factors of the vast majority of atoms found in biological macromolecules are so small at the X-ray wavelengths crystallographers generally use that they can be ignored, as we know, but this is often not the case for atoms that have higher atomic numbers. For example, for X-rays having wavelengths of 1.54 Å, the $\Delta f''$ of Hg ($Z = 80$) is ~7.6 electrons. It follows that the Bijvoet differences in the diffraction pattern of the crystals of some macromolecule will be dominated by the contributions made by the high-Z atoms it contains, the same way the differences in measured amplitudes between native crystals and isomorphously derivatized crystals are determined by heavy atom contributions.

To be more precise, for any given reflection obtained from a crystal that contains high atomic number atoms we can write:

$$F(h, k, l) = |F_{HP}(h, k, l)| \exp(i\alpha_{HP}) + i\left|F_H''(h, k, l)\right| \exp(i\alpha_H),$$

where $|F_{HP}(h, k, l)|$ and α_{HP} are the amplitude and phase of the contribution made by the real parts of the scattering factors of the protein and the heavy atom(s) bound to it to that particular $F(h, k, l)$, and $|F_H''(h, k, l)|$ and α_H are the amplitude and phase of the contribution made to $F(h, k, l)$ by the imaginary part of the scatter from those same heavy atoms. This being the case,

$$F(-h, -k, -l) = |F_{HP}(h, k, l)| \exp(-i\alpha_{HP}) + i\left|F_H''(h, k, l)\right| \exp(-i\alpha_H).$$

Thus:

$$I(h, k, l) = |F_{HP}(h, k, l)|^2 + \left|F_H''(h, k, l)\right|^2 + 2|F_{HP}(h, k, l)|\left|F_H''(h, k, l)\right| \sin(\alpha_{HP} - \alpha_H), \quad (6.3)$$

and:

$$I(-h, -k, -l) = |F_{HP}(h, k, l)|^2 + \left|F_H''(h, k, l)\right|^2 - 2|F_{HP}(h, k, l)| \left|F_H''(h, k, l)\right| \sin(\alpha_{HP} - \alpha_H). \quad (6.4)$$

It follows that the difference in intensity between reflections that are Friedel mates will be:

$$[I(h, k, l) - I(-h, -k, -l)] = 4|F_{HP}(h, k, l)|\left|F_H''(h, k, l)\right| \sin(\alpha_{HP} - \alpha_H). \quad (6.5)$$

If the heavy atoms responsible for the anomalous scattering observed can be located in the unit cell, $|F_H''(h, k, l)|$ and α_H can both be computed. In addition, if $|F_P| > |F_H''|$, which will often be true, then $[(1/2)[I(h, k, l) + I(-h, -k, -l)]]^{1/2}$ will be approximately $|F_{HP}(h, k, l)|$ (see appendix 6.1). Hence, it will be possible to obtain an estimate of $\sin(\alpha_P - \alpha_H)$ from every such Friedel pair.

Thus, in theory, there is as much phase information in the Bijvoet differences obtained from a single crystal that contains anomalously scattering atoms as there is in isomorphous heavy atom differences, which must be extracted from the data obtained from pairs of crystals. The fundamental distinction between the two kinds of phase information is that anomalous data provides estimates of sin $(\alpha_P - \alpha_H)$, whereas heavy atom data provides estimates of $\cos(\alpha_P - \alpha_H)$. [NB: If both $\cos(\alpha_P - \alpha_H)$ and $\sin(\alpha_P - \alpha_H)$ can be obtained experimentally for a reflection, $(\alpha_P - \alpha_H)$ will be uniquely determined because there will be only one value of α_P that will satisfy both.]

6.6 PATTERSON FUNCTIONS DISPLAY THE INTERATOMIC DISTANCES AND DIRECTIONS OF A CRYSTAL

There is nothing to be gained from measuring $\cos(\alpha_P - \alpha_H)$ and/or $\sin(\alpha_P - \alpha_H)$ unless α_H can be estimated, and that is possible only if we know where the heavy atoms are in the unit cell. The single most important tool for obtaining the positional information required to compute α_H s was provided by A. L. Patterson in 1934.

At the time, Patterson was interested in solving the crystal structures of small molecules, not macromolecules, and like everyone else, he knew that electron density distributions cannot be computed from diffraction data in the absence of phases. His contribution was to point out that useful information about molecular structure can be obtained from the functions that result when diffraction intensities are Fourier transformed directly in the complete absence of any phase information. The function that results is called a **Patterson function**, $P(x, y, z)$, and it is defined as follows:

$$P(x, y, z) = \sum_h \sum_k \sum_l I(h, k, l) \exp\left[2\pi i(hx + ky + lz)\right]. \qquad (6.6)$$

In the event that Friedel's law holds, the triple sum can be written:

$$2 \sum_h \sum_k \sum_l I(h, k, l) \cos\left[2\pi (hx + ky + lz)\right],$$

and then the sums then run from 0 to the maximum values of h, k, and l (with the $(0, 0, 0)$ term given a weight of 1.0 instead of 2.0).

From section 2.10, we know that Patterson functions must be the three-dimensional autocorrelation functions of the electron density distributions of crystals, but it may not be immediately obvious why Patterson functions are worth thinking about. Imagine a molecule that consists of point atoms. The Patterson function of that molecule will be nonzero at $\mathbf{r}$ *only* if there is at least one pair of atoms in the molecule so positioned that $\mathbf{r}_i - \mathbf{r}_j = \mathbf{r}$, where $\mathbf{r}_i$ and $\mathbf{r}_j$ are vectors specifying the positions of atoms i and j in the molecule relative to any convenient origin. At any such point, the value of $P(\mathbf{r})$ will be $\sum f_i f_j$ where the sum runs over all pairs of atoms that are separated by exactly $\mathbf{r}$, and f_i and f_j are the scattering factors of the two atoms in each such pair. If two atoms are separated by $\mathbf{r}$, they are also separated by $(-\mathbf{r})$ because the sense of $\mathbf{r}$ depends only on which atom in the pair is taken (arbitrarily) as being the reference atom. Thus, $P(\mathbf{r})$ is certain to be centrosymmetric; $P(\mathbf{r}) = P(-\mathbf{r})$. In addition, the Patterson function will have an origin peak the value of which is $\sum f_i^2$, because every atom is connected to itself by a vector of length 0.

Figure 6.3 (top right) shows what the two-dimensional Patterson function of a planar molecule consisting of three point atoms looks like (top left), and it also shows what the Patterson function of a two-dimensional crystal composed of those

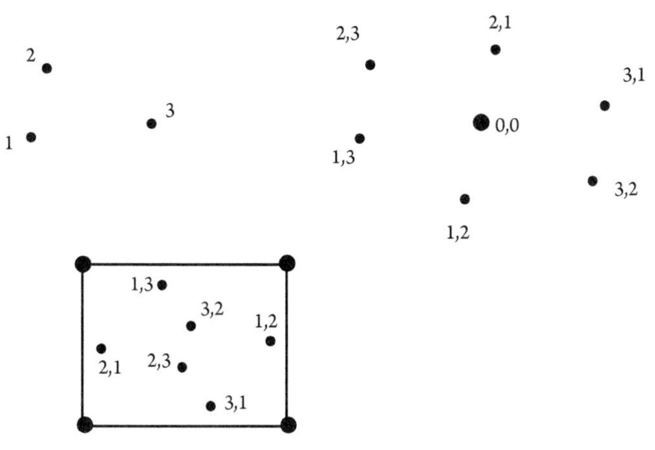

Figure 6.3 The structure of a three-atom, planar molecule (upper left) compared to its Patterson function (upper right), and to one unit cell's worth of the Patterson function that would be obtained if the intensity data from a planar crystal of those molecules that has the unit cell shown were Fourier transformed.

molecules might look like (bottom). Note that the Patterson function of a crystal is the convolution of a real-space lattice, which is identical to the lattice of the parent crystal, with the Patterson function of the contents of one of that crystal's unit cells. Because this is so, each unit cell in a crystal's Patterson function contains peaks corresponding to all the atom–atom vectors in the Patterson function of the contents of its unit cells, even though the volume occupied by the Patterson function of the contents of a crystal's unit cell is larger than that of its unit cells by roughly a factor of 2 in every direction. The same effect can also be seen in two dimensions, as a comparison of figure 6.3, upper right, with figure 6.3, lower left, shows.

> *Comment: It is easy to understand why the Patterson functions of crystals can be described mathematically as motifs convoluted with lattices, and why the lattice of the Patterson function of a crystal is the same as that of the crystal itself. For every atom at* $\mathbf{r}$ *in the unit cell chosen as the origin unit cell, there will be a corresponding atom at* $(\mathbf{r} + n\mathbf{a} + m\mathbf{b} + o\mathbf{c})$ *in the unit cell that is displaced from the origin by* $(n\mathbf{a} + m\mathbf{b} + o\mathbf{c})$*, where* $\mathbf{a}$*,* $\mathbf{b}$*, and* $\mathbf{c}$ *and the unit cell axes, and n, m, and o are integers. Thus,* $\mathrm{P}(0) = \mathrm{P}(n\mathbf{a} + m\mathbf{b} + o\mathbf{c})$ *for all n, m, and o.*

At first glance, it is not obvious which of the several lattice points associated with some unit cell is the origin peak for any given Patterson peak inside that unit cell, but it is not difficult to figure out. For a two-dimensional lattice like the one shown in figure 6.3, if the origin for the peak found at fractional coordinates (x, y), measured from the lower left lattice point, really is the lower left lattice point, then there must be a peak at $(a - x, b - y)$, which is the $(-x, -y)$ peak associated with the lattice point in the upper right corner of the unit cell. If a peak is not found at that position, then the origin for the (x, y) peak must either be the lattice point at the upper left corner of the unit cell or its mate at lower right corner of the unit cell. The same reasoning works as well in three dimensions as it does in two. Once Patterson peaks have been associated with their "parent" lattice points this way, it is a simple matter to reconstruct the molecular Patterson function (figure 6.3, upper right) from the Patterson function of the crystals it forms (figure 6.3, lower left).

6.7 MACROMOLECULAR CRYSTAL STRUCTURES CANNOT BE SOLVED USING PATTERSON FUNCTIONS ALONE

The peaks in the Patterson functions of real molecules are not points. In fact, they can be no smaller than the peaks that would be obtained by convoluting the electron density distribution functions of the two atoms they represent, and temperature factor effects will make them even bigger than that. The reason the size of individual peaks in Patterson functions is so important is that, apart from the large peak at the origin, the number of peaks in the Patterson function of a molecule that contains N atoms would be $N(N-1)$ if there were no accidental overlaps. (Hydrogen atoms scatter so weakly that their contributions to Patterson functions can usually be ignored.) Thus, the Patterson function of a small protein that might contain 1,000 nonhydrogen atoms will have 10^6 peaks in it. In any Patterson function that densely populated, accidental overlap of peaks will be the rule rather than the exception, and it will be impossible to identify any of its features with specific pairs of atoms in the protein.

6.8 HEAVY ATOM SITES IN DERIVATIZED CRYSTALS CAN BE LOCATED USING DIFFERENCE PATTERSONS

The reason Patterson functions are useful for determining the locations of heavy atoms in derivatized crystals is that it is experimentally feasible to do a reasonably good job of estimating the amplitude of the contribution heavy atoms make to each of the reflections in the diffraction pattern of a derivatized crystal. The Patterson function obtained when that component of the data is transformed is likely to be reporting on the positions of perhaps 10 atoms or less, rather than 1,000, and thus is likely to be simple enough to understand. If MIR phasing is being done, the intensity assigned to each reflection in the data set used for heavy atom Patterson function computations, $\mathrm{I_D}(h, k, l)$, is:

$$\mathrm{I_D}(h, k, l) = \left[\mathrm{I_{HP}}(h, k, l)^{1/2} - \mathrm{I_P}(h, k, l)^{1/2}\right]^2. \qquad (6.7)$$

As appendix 6.1 shows, if the heavy atom contribution to the overall diffraction of a derivatized crystal is modest, as it usually is, for many reflections, $\mathrm{I_D}(h, k, l) \approx \mathrm{I_H}(h, k, l)$, as required. (It is easy to show that the estimate for $\mathrm{I_D}(h, k, l)$ that emerges if $\mathrm{I_D}(h, k, l)$ is set equal to $[\mathrm{I_{HP}}(h, k, l) - \mathrm{I_P}(h, k, l)]$, which might seem to be the more natural definition, will be less accurate than the one obtained using equation (6.6) (see question 1).)

As the discussion in appendix 6.1 also makes clear, the quality of the estimates for $\mathrm{I_H}(h, k, l)$ obtained using equation (6.7) will vary a lot from one reflection to the next. Hence, the data used to compute these Patterson functions, which are called "heavy atom difference Pattersons", are not ideal. The reason heavy atom difference Pattersons are as useful as they have proven to be is that the number of heavy atom peaks that need to be visualized is usually orders of magnitude smaller than the number of measured intensities available to determine their locations. Because Fourier transformation is a least squares process, they are quite insensitive to quasi-random error. Thus, heavy atom peaks are usually the dominant features in difference Patterson maps, even though there are may be a lot of noise features present also.

Atoms that are strong anomalous scatterers (i.e., heavy atoms) can also be located using difference Patterson techniques. In this case, the data that should be transformed, $I_{anom}(h, k, l)$, are:

$$I_{anom}(h, k, l) = \left[I(h, k, l)^{1/2} - I(-h, -k, -l)^{1/2}\right]^2. \tag{6.8}$$

The relationship between $I_{anom}(h, k, l)$ and anomalous contributions to measured intensities can be worked out using the same reasoning that leads to the conclusion that $I_D(h, k, l) \approx I_H(h, k, l)$ for MIR data (see appendix 6.1), but the outcome is less satisfying:

$$I_{anom}(h, k, l) \approx 2|F_H''(h,k,l)|^2 \left[1 - \cos 2(\alpha_{HP} - \alpha_H)\right].$$

Thus, Patterson functions computed using $I_{anom}(h, k, l)$ include not only a component that is proportional to $I_H''(h, k, l)$, as desired, but also a contribution that is weighted by $\cos[2(\alpha_H - \alpha_P)]$. This second contribution tends to obscure the first and, hence, is, effectively, noise.

6.9 ATOMIC COORDINATES CAN BE DEDUCED FROM THE HARKER SECTIONS OF THE PATTERSON FUNCTIONS

The structures of small molecules can be deduced from the Patterson functions computed using the the diffraction data their crystals provide by procedures that usually depend on chemically derived information about the way their atoms are bonded together. Unhappily, no such information is available to help crystallographers interpret heavy atom difference Patterson maps. The heavy atoms in a derivatized crystal can be anywhere in its unit cell, and there may be many of them. What then?

This problem was partially solved by D. Harker shortly after Patterson introduced his function to the crystallographic community. Harker pointed out that the Patterson functions of crystals that have nontranslational symmetry often include special planes in which the positions of peaks are directly related to atomic coordinates. These special planes in three-dimensional Patterson functions are called **Harker sections**. The only symmetry elements macromolecular crystallographers need to consider are rotational axes and screw axes because all the other symmetry operations that give rise to Harker sections invert chiral centers.

To understand what a Harker section is, consider a crystal that has a single, two-fold axis coincident with the **c** axis of its unit cells. If there is an atom in the asymmetric unit of such a crystal at $(x\mathbf{a} + y\mathbf{b} + z\mathbf{c})$, the unit cell will also contain a second atom of exactly the same kind at $(-x\mathbf{a} - y\mathbf{b} + z\mathbf{c})$. The vectors connecting that pair of atoms will thus be $(2x\mathbf{a} + 2y\mathbf{b} + 0\mathbf{c})$ or $(-2x\mathbf{a} - 2y\mathbf{b} + 0\mathbf{c})$, depending on which atom is chosen as the origin. Thus, in the $(x, y, 0)$ plane of the Patterson function of this crystal, which will be its Harker section, every atom in the asymmetric unit will be represented by two peaks, one at $(2x, 2y, 0)$ and the other at $(-2x, -2y, 0)$ (see section 6.6). Moreover, only symmetry-related atoms pairs are likely to contribute peaks to that plane because only by bad luck will a pair of atoms that are not two-fold related have identical z coordinates. Thus, in theory, except for a $(+x, +y)/(-x, -y)$ uncertainty, it should be possible to determine the x and y coordinates of every atom in the asymmetric unit of such a crystal from the positions of the peaks in this single plane of its three-dimensional Patterson function. [For a two-fold screw axis aligned with **c**, the Harker section will be the $(x, y, 1/2)$ plane.] If some crystal has two perpendicular two-fold axes, which many do, it will be possible to obtain estimates of all three of the coordinates of every one of the heavy atoms in its asymmetric unit from the locations of peaks in the two perpendicular Harker sections obtained from its difference Pattersons to within a $(+x, +y, +z)/(-x, -y, -z)$ ambiguity. In that case, the information about heavy atom positions gleaned from a Patterson function should be enough to determine heavy atom intensities and phases.

The absolute hand of the MIR electron density maps obtained when any set of experimental amplitudes and phases is first Fourier inverted is invariably determined by an arbitrary choice that got ratified at the time the positions of the heavy atoms used for phasing were determined. The reason an arbitrary choice has to be made is that there is no way of knowing whether the coordinates assigned to any heavy atom binding site on the basis of Harker section data should be (x, y, z) or $(-x, -y, -z)$. In this regard, it should be noted that if there are heavy atoms at *more* than one site per asymmetric unit, once this $(+x, +y, +z)/(-x, -y, -z)$ choice has been made for one of them, all the other sites must conform because otherwise the peaks observed *outside* the Harker sections of the heavy atom/anomalous difference Patterson function, which report on the distances and directions relating atoms that are not symmetry mates, will not be consistent with observation. Another way of putting it is that if there are n heavy atoms in some derivatized crystal, instead of there being 2^n ways of placing them in the unit cell, which would be the case if the only information available about their locations was the information contained in Harker sections, there will be only two, and those two will be mirror images of each other.

We end this discussion by noting that over the years, a number of computational methods have been developed for solving crystal structures using ordinary, nonanomalous diffraction data as the sole source of information. These approaches, which are referred to collectively as **direct methods**, are far more powerful than the Patterson approach outlined in previous sections, and the complexity of the structures that can be solved by direct methods continues to increase. At this point, the crystal structures of small biological macromolecules are within reach. Direct methods can also be used to determine the positions of heavy atoms in the unit cells of macromolecular crystals, and they are particularly useful when the number of heavy atom positions that must be determined is large. The input data used for this purpose is the same as that used to compute heavy atom/anomalous difference Pattersons.

The model for the placement of heavy atoms in the unit cell that emerges from either Patterson and/or direct methods analyses of difference data must indicate not only where heavy atoms are found in the unit cell, but if there are several such sites, it must also specify the differences in the degree to which those sites are occupied by heavy atom compounds, if they exist. It is easy to understand why there might be such differences in crystals that have been derivatized by the soak-in method. The surfaces of biological macromolecules are chemically heterogeneous, and they are liable to include a variety of sites capable of reacting/interacting to some degree with any given, heavy atom–containing compound. Because these sites will differ in binding constant, at any given concentration of a heavy atom compound, some may be saturated, others may be partially saturated, and still others may be empty. For the purposes of computing either heavy atom or anomalous contributions to scattered intensities, partially occupied sites must be taken properly into account. Their contribution to $F_H(h, k, l)$ will be $(f_H \exp[-2\pi i(hx + ky + lz)])$ times a coefficient called the **occupancy**, which is a number between 0 and 1.0. (Remember that crystallographic electron density distributions are distributions averaged the whole crystal.)

6.10 MULTIPLE-WAVELENGTH ANOMALOUS DIFFRACTION COMBINES ANOMALOUS AND HEAVY ATOM PHASE DETERMINATION IN A SINGLE EXPERIMENT

For decades the macromolecular community paid comparatively little attention to phasing by anomalous diffraction, a technique given equal billing here. The reason was that the X-ray sources then available generated radiation efficiently at only single wavelengths. Often the K_α radiation of copper ($\lambda = 1.54$ Å) was used, and at that wavelength $\Delta f''$ is small for many elements, even those having quite high atomic numbers.

Starting in the early 1980s, macromolecular crystallographers began using the X-radiation produced by synchrotrons. Operationally, synchrotron light sources differ from laboratory X-ray generators in two important respects. First, the beams they generate are orders of magnitude brighter than those produced by laboratory X-ray generators, which implies a corresponding, and highly welcome, increase in the rates of data collection. Second, instead of producing X-rays at only a single wavelength, synchrotrons generate X-rays having wavelengths that vary continuously over a wide range. X-ray beams having the narrow range of wavelengths required for diffraction experiments—"monochromatic X-rays"—are extracted from this white radiation using monochromators of one kind or another, and they are adjustable. Thus, the users of synchrotrons can set the wavelength of the X-rays they use to any value they like over a considerable range. It should not be surprising that crystallographers interested in exploiting anomalous diffraction effects might prefer to use X-rays having the wavelength at which $\Delta f''$ is at its maximum for one or another of the heavy atoms that their crystals contain.

At about the time synchrotron radiation was coming to the fore, it was discovered that many/most macromolecular crystals can be flash frozen without disordering them too seriously and that freezing markedly reduces the rate at which crystals decay due to radiation damage. In addition, efficient area detectors became available, which further accelerated the rate at which data can be collected, and so for the first time it became practical to collect complete diffraction data sets from single crystals of macromolecules. Prior to that, most macromolecular data sets were generated by merging partial data sets collected from (often) scores of different crystals.

These developments are all taken advantage of by a phasing technique called **multiple-wavelength anomalous diffraction** (MAD), which simultaneously uses both the heavy atom and the anomalous diffraction methods to estimate phases. MAD experiments can only be done on macromolecular crystals that include heavy atoms that have X-ray absorption edges in the wavelength range that is convenient for crystallographic data collection, in other words, between ~2Å and ~0.5Å.

As is explained in chapter 3, both $\Delta f'$ and $\Delta f''$ change dramatically at atomic absorption edges. As wavelength decreases, $\Delta f'$ passes through a sharp minimum while $\Delta f''$ rises from a low basal value, passes through a sharp peak, and then assumes a new, higher "plateau" value. In a MAD experiment, a full data set is collected from a single crystal at the wavelength at which $\Delta f''$ is maximum, and that data set is used for anomalous phase determination in the usual way. Two additional data sets are collected, one at the wavelength at which $\Delta f'$ is minimum, and the second at a "remote" wavelength, which is usually chosen to be a wavelength shorter than the other two. (It is easy to determine experimentally which wavelengths should be used for MAD experiments on a particular crystal, but the details are of no concern here.)

The second pair of data sets are combined to generate a **dispersion** data set for determining the positions of anomalous scatterers, $|F_d(h, k, l)|$:

$$
\begin{aligned}
|F_d(h, k, l)| = (1/2)\left\{\left[I_{min}(h, k, l)^{1/2} + I_{min}(-h, -k, -l)^{1/2}\right]\right. \\
\left. - \left[I_{rem}(h, k, l)^{1/2} + I_{rem}(-h, -k, -1)^{1/2}\right]\right\}.
\end{aligned}
$$

By averaging Bijvoet pairs this way, the contribution made by the anomalous component of each reflection is minimized, and $|F_d(h, k, l)|$ is determined primarily by the difference in $\Delta f'$ between λ_{min} and λ_{rem}. Thus, $|F_d(h, k, l)|$ is approximately the same as the signal that would be observed in a conventional heavy atom isomorphous replacement experiment if there were no atoms at the position of the anomalously scattering atoms in the native crystal, and atoms that have scattering factors equal to $\left(\Delta f'_{min} - \Delta f'_{rem}\right)$ at those same positions in the derivatized crystal.

Because the heavy atom positions implied by both the anomalous data and the dispersion data should be the same, their difference Pattersons should be identical, save for noise. For that reason, MAD data sets are usually analyzed en masse, rather than analyzed one data set at a time. The objective is to find the single arrangement of heavy atoms in the unit cell that best satisfies *all* the data in a statistically appropriate sense. Because estimates emerge for both $\cos(\alpha_P - \alpha_H)$ and $\sin(\alpha_P - \alpha_H)$ for each reflection, the phase of every reflection in a diffraction pattern can be uniquely determined using the data gleaned from a single MAD experiment. In theory, there should be no need to prepare a second derivative.

MAD experiments are often done on crystals of proteins in which the sulfur-containing amino acid methionine has been replaced by selenomethionine. Selenium has an X-ray absorption edge at $\lambda \approx 0.98$ Å, and once appropriate selenium-labeled crystals have been obtained, there is no need to indulge in heavy atom soak-in trials, which tend to be hit or miss.

6.11 EXPERIMENTAL ERROR COMPLICATES THE EXPERIMENTAL DETERMINATION OF PHASES

Experimental error makes both isomorphous replacement and anomalous phasing methods challenging to execute. Data errors will necessarily lead to errors in phase estimates, and not surprisingly, electron density maps computed with inaccurate phases are hard to interpret. Furthermore, the accuracy of experimental phase estimates is bound to vary from one reflection to the next because the measurement errors that affect those estimates do, and if decent electron density maps are to result, the Fourier transformations done to compute them should be error-weighted to ensure that the well-measured data dominate.

The errors that affect the accuracy of the phases extracted from anomalous data are not the same as those that affect phase estimates obtained by conventional isomorphous replacement methods, and both approaches have their advantages and disadvantages in this regard. The chief advantage of MIR phasing is that MIR difference signals are almost always larger than anomalous difference signals, which makes them easier to measure. Its disadvantage is that MIR phasing depends on the comparison of data sets obtained from different crystals, and this gives rise to data scaling issues that those determining phases by anomalous diffraction generally do not confront because (often) all the data necessary for anomalous phasing can be obtained from a single crystal.

When data from different crystals must be compared or merged, two kinds of problems are encountered that can contribute to error: imperfect isomorphism, and scaling. It is an unhappy fact that different crystals of the same macromolecule, even those obtained at the same time from the same preparation, always differ at least a little in unit cell dimensions; perfect isomorphism is never achieved. When the unit cell dimensions of two crystals differ, the locations in the reciprocal space sampled by their reciprocal lattices will differ also, and thus, even if every other aspect of the measurement process is ideally executed, $F(h, k, l)$ for crystal A will still not equal $F(h, k, l)$ for crystal B. Unit cell variation of this sort is particularly likely to be observed when data from crystals that have not been exposed to heavy atom compounds are compared to data from crystals that have been. The general rule of thumb is that one should not use data sets for MIR phasing that differ in unit cell dimensions by more than 1%, but beyond this, little can be done to control errors due to lack of isomorphism.

Data set scaling is also a problem because just as it is impossible to obtain two crystals that have exactly the same lattice, it is also impossible to obtain two crystals that are so similar in size, shape, and B-factor that for all (h, k, l), the measured intensities obtained from one can be meaningfully compared to the measured intensities obtained from the other. For this reason, before data sets are compared or merged they must be **scaled** to minimize the impact of such systematic differences, and over the years this kind of scaling has been done in many different ways. For example, the data obtain from each of the two crystals of interest can be sorted into thin shells of constant $|\mathbf{S}|$, and then the average value of I computed for both data sets, shell by shell. The data from the data set that extends to the larger value of $|\mathbf{S}|$ can be scaled to match the data in the data set that extends to the lower value of $|\mathbf{S}|$ by multiplying each of the reflections from the higher resolution data set that fall in a given shell in reciprocal space by $\langle I_{low} \rangle / \langle I_{high} \rangle$, where $\langle I_{low} \rangle$ is the average intensity of the data from the lower resolution data set within that shell, and $\langle I_{high} \rangle$ is the corresponding average for the higher resolution data set. This kind of scaling, or variants of it, is useful even when the data sets compared come from crystals that differ in heavy atom content because the effect of heavy atoms on average intensities is much smaller than their effect on the differences between intensities, as we saw earlier.

A second kind of scaling is important for both MIR and anomalous phasing. Phases can be obtained from both MIR and anomalous data (see equations (6.2) and (6.5)) only if you know what the value of $|F_H|$ (or $|F_H''|$) would be if it were measured in the same crystals that produced the data that are being processed. As we already know, that information is obtained, ultimately, by combining information about heavy atom positions and occupancies with the appropriate atomic scattering factors. However, this information alone is not enough. Among other things, an a priori calculation of amplitudes and phases would have to take into account the sizes, and shapes of the crystals of concern, which are not fully known, as well as the properties of the apparatus used for data collection. It is much easier to computed $|F_H|$ and/or $|F_H''|$ using occupancies, heavy atom positions, and atomic scattering factors alone, and then process the experimental data assuming that the $|F_H|$ and/or $|F_H''|$ contribution to each reflection can be written $\beta|F_H(h, k, l)|_{calc}$ or $\beta|F_H''(h, k, l)|_{calc}$, where β is a scale factor that should have the same value for *all* reflections. Because the typical macromolecular data set includes thousands of reflections, it is not hard to find the value of β that best explains the data.

Outright measurement error also contributes to phase errors. The detectors used to measure the intensities of reflections, however they do it, register either the number of X-ray photons that reached their surfaces at some location in some interval of time, or something proportional to that. Thus, measured $I(h, k, l)$s include a contribution from random, statistical error that ensures that their standard deviations will never be better than the square root of the number of photons that reached the detector in the time interval of concern. In fact the standard errors of individual observations are always worse than that because of the background that must be subtracted from crude, measured intensities, which are themselves known only to within the same sort of statistical uncertainty. The standard deviation associated with a specific $I(h, k, l)$ in a data set, σ_I, can be reduced either by measuring that reflection for a longer period of time, or by averaging multiple observations of the same reflection.

The importance that measurement error might have for subsequent computations becomes evident when it is realized that by custom, macromolecular crystallographers use all the data they measure from crystals out to the value of $|\mathbf{S}|$ where the average value of the intensities at that $|\mathbf{S}|$ divided by the average standard deviation of those intensities is 2.0, which implies that the average value of F is only 4 times the average value of its standard deviation. Data that poorly measured are certain to be noisy, and sums and differences computed using data of that quality will be noisier still.

Finally, we have to realize that the accuracy of the techniques used to determine heavy atom locations and occupancies is also limited by experimental error. It follows that the information available about both $|F_H(h, k, l)|$ and α_H is imperfect, quite apart from any scaling inaccuracies, simply because the positions of heavy atom binding sites and their occupancies are never perfectly determined.

6.12 EXPERIMENTAL PHASE DATA SPECIFY PHASE PROBABILITY DISTRIBUTIONS

The measurement issues discussed above are all sources of error, and the impact of error on the information about phases contained in MIR experiments is easily understood using Harker diagrams (figure 6.4). If all measurement error is assumed to be errors in the estimates for $|F_{HP}|$ and $|F_P|$, which is an oversimplification, but not too seriously so, the circles of radius $|F_{HP}|$ and $|F_P$, whose intersections determine phases in a Harker phase diagram can be replaced by annuli having widths equal, perhaps, to twice the standard errors of the two measurements. The area where the two annuli overlap will then define one or two ranges of angles within which the true phase for some reflection is likely to fall, confirming the obvious. Real data do not determine phases exactly.

A more sophisticated way to analyze the data would be to replace the circles for $|F_{HP}|$ and $|F_P|$ with annuli that have radial profiles that correspond to Gaussians of the appropriate standard deviation, each profile having its maximum on the measured radii for $|F_{HP}|$ and $|F_P|$. Using this approach, a number can be assigned to every point in the Argand diagram that is the product of the values of the $|F_{HP}|$ and $|F_P|$ Gaussians appropriate for that point. That number is proportional to the probability that $|F_{PH}|$, $|F_P|$, and α_P might have the corresponding values. If such a probability plot is integrated along radii of constant α_P, and normalized appropriately, a new plot will emerge that shows how the probability of α_P varies with α_P. Figure 6.5 shows what such a plot might look like.

Several useful ideas emerge when phase data are approached this way. First, each experiment done to measure the phase of some reflection can be represented as phase probability distribution, $p(\alpha)$, that will often be bimodal, but need not be. (If $I_P = I_{HP}$ for some reflection, $p(\alpha)$ will be a constant for that reflection, and nothing will have been learned about its phase from that particular isomorphous replacement experiment.) Second, each heavy atom, isomorphous replacement experiment

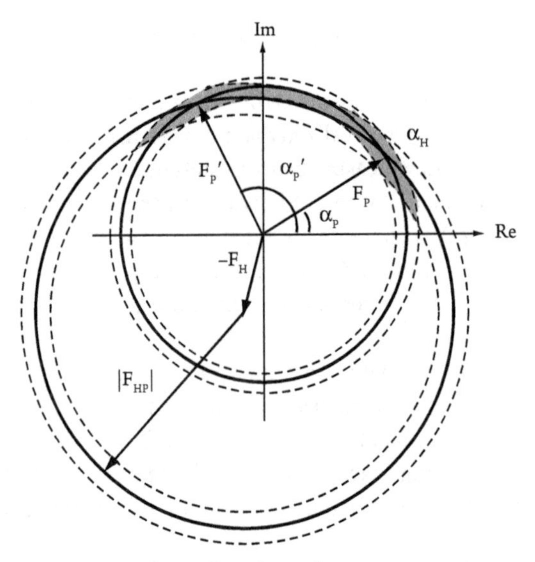

Figure 6.4 The Harker phase diagram appropriate for the data presented in figure 6.2 if the errors estimated for the measurements of both $|F_{HP}|$ and $|F_P|$ were both $\sim$10%. The dotted circles have radii that are $|F_{HP}| \pm 10\%$ and $|F_P| \pm 10\%$. The shaded area is the region where the two annuli intersect, and it is the region within which the true value of α_P is likely to fall.

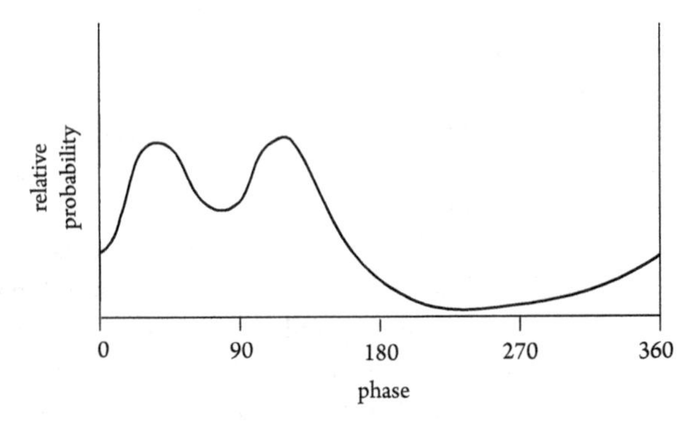

Figure 6.5 A sketch of the relative probability of that the value of α_P has some particular value versus α_P that crudely corresponds to the data shown in figure 6.4.

done with some crystal will produce a plot of this sort for each reflection in its diffraction pattern. Clearly, the totality of the phase information available for each reflection will be the product of all of the experimental phase distributions available for it, normalized so that the area under the curve is 1.0, in other words,

$$p_{total}(\alpha) = \prod p_i(\alpha) \Big/ \int \left[\prod p_i(\alpha)\right].$$

Crystallographers hope that when it is all over, the $p_{total}(\alpha)$ distributions for all the reflections in their data sets will be single, narrow peaks, but the real world is seldom so kind.

The software used by macromolecular crystallographers today incorporates all of the concepts described herein, and spares them from having to look at Harker diagrams or phase probability diagrams, let alone make decisions about what phases should be assigned individual reflections.

6.13 THE IMPACT OF PHASE ERRORS ON ELECTRON DENSITY MAPS CAN BE CONTROLLED

If the final phase probability distribution available for some reflection is not a single, narrow peak, what phase should be used for that reflection when it comes time to compute electron density functions? We could use the phase for which $p_{total}(\alpha)$ reaches its maximum value, which is the most probable phase, but if the phase distributions of different reflections vary in width, which they will, this approach will lead to electron density distributions in which reflections that have poorly determined phases (i.e., have broad $p_{total}(\alpha)$ peaks), contribute as much as those that have well-determined phases (i.e., have narrow $p_{total}(\alpha)$ peaks), and that would be perverse.

In the 1950s, D. Blow and F. Crick pointed out that if the objective is to produce experimental electron density maps in which, averaged over the whole unit cell, the deviation of the computed map from the actual electron density distribution in the crystal is the smallest it can be, which seems sensible, a different choice of phase should be made. In this case, the best phase to use for any reflection is the centroid value of its probability distribution:

$$\alpha_{ave} = \int_0^{2\pi} \alpha p_{total}(\alpha)\, d\alpha.$$

Furthermore, rather than use $|F(h, k, l)|$ as the amplitude for each reflection, we should use $m|F(h, k, l)|$, where m is a number between 0 and 1.0 called the **figure of merit**.

The figure of merit of a reflection is the component of the $p_{total}(\alpha)$ distribution in the direction of the Argand diagram specified by α_{ave}. Thus:

$$m = \int_0^{2\pi} [\cos(\alpha_{ave})\cos(\alpha) + \sin(\alpha_{ave})\sin(\alpha)] p_{total}(\alpha)\, d\alpha$$

$$= \int_0^{2\pi} \cos(\alpha_{ave} - \alpha)\, p_{total}(\alpha)\, d\alpha$$

If $p_{total}(\alpha)$ for some reflection is a Dirac delta–like single peak, its figure of merit will be 1.0, and the phase assigned to it will be α_{max}. However, if $p_{total}(\alpha)$ has two peaks of similar shape that differ in average phase by 180°, or is a constant independent of α, its figure of merit will be 0, which makes sense because if $p_{total}(\alpha)$ looks either way, it will be impossible to decide which phase to use for it, and if that it so, it is better to omit that reflection from computations of electron density maps than to include it. Electron density maps computed this way are called **best Fouriers**.

It is perfectly possible for the phase probability distribution for some reflection obtained from a single heavy atom derivative or a single anomalous diffraction observation to have either one peak that is highly dominant or two peaks that are close together. This will happen whenever $F_H(h, k, l)$ and $F_P(h, k, l)$ are roughly parallel or antiparallel, which they will be for all reflections in the centrosymmetric planes in the reciprocal space of a crystal. (In centrosymmetric planes, Fs have to be either parallel or antiparallel because the phases of all reflections must either be 0° or 180°.)

Comment: To understand how the diffraction pattern of a chiral molecule like a protein could have centrosymmetic planes, consider crystals that have two-fold axes aligned with **c**. *For reflections in the $(h, k, 0)$ plane of the reciprocal space of such a crystal,*

$$\begin{aligned} F(h,k,0) &= \sum f_i \left(\exp\left[-2\pi i\left(hx+ky\right)\right] + \exp\left[-2\pi i\left(hx+ky\right)\right]\right) \\ &= \sum f_i \cos\left[2\pi\left(hx+ky\right)\right], \end{aligned}$$

where the sum is over all atoms in a single asymmetric unit. Thus, $F(h, k, 0)$ will be real. [This is the reciprocal space corollary of the observation that the $(x, y, 0)$ plane of the Patterson functions of crystals with two-fold axes aligned with **c** *are Harker sections.]*

It follows that while the data obtained from a single phasing experiment will never uniquely determine the phases of all reflections in a macromolecular diffraction pattern, the amount of phase information it yields is much greater than 0. For that reason, it can be useful to compute best Fourier electron density maps using the data from single derivatives. A map of this kind is called a **single isomorphous replacement** map, or a **single-wavelength anomalous diffraction** map. For example, single wavelength anomalous diffraction phase information from one heavy atom derivative can be used to compute electron density maps from difference data obtained with a second heavy atom derivative and, thereby, locate the heavy atom sites in the second derivative. It should be noted in passing that single isomorphous replacement and single wavelength anomalous diffraction maps can be very useful in the early stages of any experimental phase determination process because they can be used to prove that the heavy atom data being obtained are self-consistent and to find low occupancy heavy atom sites.

6.14 THE LIKELIHOOD THAT THE EXPERIMENTS DONE TO PHASE A DIFFRACTION PATTERN HAVE PRODUCED RELIABLE DATA CAN BE ASSESSED STATISTICALLY

Over the years, several statistical measures have been developed to help crystallographers assess the quality of their experimental phases. We will discuss two of them here: **average figure of merit**, and **phasing power**.

As we already know, the figure of merit of a single reflection, m, is the value of the cosine of the difference between $(\alpha_{ave} - \alpha)$ averaged over the entire phase probability distribution of that reflection. The average figure of merit of an entire diffraction data set is that cosine value averaged over all reflections in the data set. Clearly, if the average value of m is 1.0, the experimental data have perfectly specified the phase of every reflection, and if its average value is 0, an electron density map computed using those phases is likely to be pure noise.

Experimental electron density maps computed using phase sets that have average phase uncertainties large enough to astonish the uninitiated are often useful. Average figures of merit around 0.5 are common, which implies an average phase uncertainty of $\pm 60°$!. Electron density maps with phase errors that large are useful for two reasons. First, to at least a first approximation, phase errors are randomly distributed among reflections, and Fourier synthesis is a least-squares mathematical operation (see chapter 2). Second, phase errors tend to be less for intense reflections, which are the reflections that make big contributions to electron density maps, than they are for weak reflections, which make lesser contributions to electron density maps, because it is easier to make the measurements required for phasing when reflections are intense.

For isomorphous replacement experiments, phasing power is defined as follows:

$$\left[\sum |F_H(h,k,l)_{calc}|^2 / \sum (|F_{HP}(h,k,l)| - |F_{HP}(h,k,l)_{calc}|)^2\right]^{1/2}.$$

In this expression, the sums run over all reflections, and $F_H(h, k, l)_{calc}$ is the appropriately scaled contribution of the heavy atom(s) to the (h, k, l) reflection calculated using the experimentally determined heavy atom distribution. $|F_{HP}(h, k, l)|$ is the experimentally observed value for the amplitude of the (h, k, l) reflection in the data set obtained from the heavy atom–derivatized crystal of concern, and $|F_{HP}(h, k, l)_{calc}|$ is the amplitude computed for that reflection using the measured amplitude of its protein component, $|F_P(h, k, l)|$; the phase experimentally determined for $|F_P(h, k, l)|$; and the phase and amplitude computed for the heavy atom contribution to that same reflection. $(|F_{HP}(h, k, l)| - |F_{HP}(h, k, l)_{calc}|)$, thus, measures the magnitude of inconsistency that may exist between what has been learned about the lengths and orientations of two sides of the phase triangle and the (measured) length of its third side, $|F_{HP}(h, k, l)|$. It is called the **lack of closure error**. If all of the data were perfect, $(|F_{HP}(h, k, l)| - |F_{HP}(h, k, l)_{calc}|)$ would be 0 for all reflections, and the phasing power of the data would be infinite. Thus, big numbers for phasing powers are good. In practice, we hope that, on average, $|F_H(h, k, l)_{calc}|$ will be larger than the lack of closure error, but in fact, many macromolecular crystal structures have been solved using data that had a phasing power less than 1.0.

6.15 DIFFRACTION PATTERNS CAN BE PHASED BY MOLECULAR REPLACEMENT

Two important facts about biomolecules that are relevant to their crystallography have been understood for many years: (1) that the unit cell dimensions and space group of the crystals a macromolecule forms can vary with crystallization conditions, and (2) that macromolecules having similar sequences generally have similar three-dimensional structures. Both phenomena have led crystallographers to develop methods for using crystal structures that have already been determined to solve the structures of related crystals. The approach is called **molecular replacement**, and its importance as a method for solving crystal structures grows with every passing year.

Suppose some macromolecule forms two kinds of crystals that differ in unit cell size and symmetry, and that its structure has been solved in one of those crystals. Although it could be that the conformation of the macromolecule in the second crystal is very different from its conformation in the crystals used to solve its structure, it is more likely to be similar. If that is so, it should be possible to generate a usefully accurate model for the structure of the second crystal by placing the structure obtained from the first crystal in the unit cell of the second crystal in the appropriate location(s) and orientation(s) computationally.

Once such a model has been obtained, it ought to be possible to compute phases for the diffraction pattern of the second crystal that are accurate enough so that a reasonable approximation to the electron density map of the unknown structure emerges when the amplitudes of its diffraction pattern are Fourier-inverted using those phases. That structure can then be refined in the usual way (see chapter 7) to obtain the final result. The same approach will work even when the molecule used to produce trial phases has a sequence that is not the same as that of the molecule in the crystal of unknown structure, provided their three-dimensional structures are similar.

The fundamental problem in molecular replacement is determining how to position trial structures in the unit cells of unsolved crystals. In theory, this problem could be solved by brute force, that is, by trial and error. Diffraction patterns could be computed for some test structure positioned in the unit cell of the second crystal form in every conceivable location and orientation, and those computed diffraction patterns compared with the measured data, one by one. The packing that best explained the measured data would become the point of departure for solving the structure of the crystal of unknown structure.

Even though the trial-and-error process just described would work, it has a serious practical drawback. When molecules are "moved around" in unit cells computationally as rigid bodies, their atoms all move in a highly correlated way, and for that reason, movements of an entire molecule that are small compared to $1/|\mathbf{S}|_{max}$ for the relevant diffraction data set will produce big changes in the values computed for the intensities of reflections. Thus, the six-dimensional space (three angular coordinates plus three positional coordinates) that must be explored in a search of this sort must be searched on a very fine grid to ensure that the correct answer is found. A six-dimensional search on a grid that fine is liable to be prohibitively expensive computationally.

6.16 MOLECULAR REPLACEMENT SEARCHES CAN BE DIVIDED INTO A ROTATIONAL PART AND A TRANSLATIONAL PART

Molecular replacement would still be seldom used to solve crystal structures if it were not possible to separate the rotational component of the search process just described from its translational part. The rationale for that separation is provided by the shift theorem (chapter 2), which shows that changes in the locations of objects in real space that are unaccompanied by changes in their orientation alter the phases of their transforms but not the amplitudes. The advantage gained by making that separation is obvious. For example, it might reduce a search of 1,000,000 translation-s-orientations to a rotational search involving 1,000 different orientations followed by a translational search that involves 1,000 positions.

Rotational searches are usually done by comparing Patterson functions, rather than by comparing Fourier coefficients, although the two kinds of comparisons are closely related. The degree of similarity between the Patterson function of some crystal, $P_t(\mathbf{r})$, and the Patterson function of the molecular structure being used to search for an initial molecular model, appropriately rotated and sampled, $P_s(\mathbf{R}^{-1}\mathbf{r})$, where $\mathbf{R}$ is the matrix specifying a particular rotation of the search molecule, is assessed by evaluating the following overlap integral:

$$R(\mathbf{R}) = (1/V) \int P_R(\mathbf{r})\, P_S\left(\mathbf{R}^{-1}\mathbf{r}\right) dV_r.$$

$R(\mathbf{R})$ is the **rotation function**. It can be shown that this expression is equivalent to:

$$R(\mathbf{R}) = \left(1/V^2\right) \sum_{h,k,l} \sum_{h',k',l'} I(h,k,l)\, I\left(h',k',l'\right) G_{h,k,l,h',k',l'},$$

where $I(h, k, l)$ is the intensity of the (h, k, l) reflection in the crystal of unknown structure. $I(h', k', l')$ is the intensity of the (h', k', l') reflection of the rotated test structure, computed on some other (rotated) lattice that is appropriate for it. It will almost always be true that the rotated lattice used to compute the structure factors of the test molecule does not superimpose on the lattice of the unknown structure. The $G_{h,k,l,h',k',l'}$ coefficients deal with the interpolation that has to be done in reciprocal space to evaluate the magnitude of the Patterson function of the test structure on the lattice of the unknown structure. Those coefficients are determined entirely by the geometry of the two lattices and their rotational orientations, and they are easily computed.

Rotational searches are more challenging when there are several copies of the test molecule in the unit cell, even if that molecule constitutes the entire asymmetric unit. The reason is that the Patterson function obtained from the data such a crystal provides includes intermolecular vectors as well as the intramolecular vectors that convey information about rotational orientation. However, intramolecular vectors are more likely to be represented at small values of $|\mathbf{r}|$ than intermolecular vectors are. Hence, if the range of $|\mathbf{r}|$ over which $R(\mathbf{R})$ is evaluated runs from $|\mathbf{r}| = 0$ to some fixed value of $|\mathbf{r}|$ that is much less than a, b, or c, the lengths of the unit cell vectors, $R(\mathbf{R})$, will be determined primarily by the orientations of interest. The value for the upper bound to $|\mathbf{r}|$ is a variable the crystallographer may have to adjust to obtain a useful result, and big values for $R(\mathbf{R})$ are good.

Once the rotation problem has been solved, the location of the center of gravity of the test macromolecule in the unit cell must then be determined. (NB: There is no center of gravity issue for crystals that have primitive triclinic unit cells.) Translational searches are done in many different ways, and rotation functions remain a useful way to assess the quality of models that emerge. The difference is that in this instance, the overlap integral *must* be evaluated over the entire unit cell so that long, intermolecular vectors are taken into account.

It is simpler to solve the structures of crystals that have many copies of their asymmetric units in the unit cell by molecular replacement than the preceding discussion might lead you to believe. It will invariably be found that only a comparatively small subset of the set of all conceivable placements of the test molecule in the unit cell corresponds to a packing of molecules in the unit cell that is physically plausible. Thus, many of the models tested will fail before much computation has been done to explore them simply because they imply that neighboring molecules overlap, which cannot be true.

Molecular replacement procedures do not always work, even when it is found subsequently that the test structure is not dissimilar to the structure of the molecule in the crystal of concern. Because this is so, those trying to solve structures by molecular replacement must sometimes deal with a difficult psychological problem. Am I failing because my model is bad, or am I failing because I am not using it correctly? Nevertheless, a lot of structures have been solved by molecular replacement, and some remarkable triumphs have been obtained. There have been crystal structures solved by molecular replacement using starting structures that accounted for less than 10% of the macromolecular material in the unit cell. In addition, the number of macromolecular structures that have been solved, which is already in the tens of thousands, continues to grow rapidly. Our ability to identify pairs of amino acid or nucleotide sequences that are likely to adopt similar three-dimensional structures continues to improve also. Thus, the probability that a crystallographer will find a structure in the Protein Data Bank that can be used to solve some other structure by molecular replacement is rising all the time. There are some who think that within a decade or so, most new protein crystal structures will be solved that way.

PROBLEMS

1. Show that the quantity $[I_{HP}(h, k, l) - I_P(h, k, l)]$ does not correspond as well to $I_{HA}(h, k, l)$ as $[I_{HP}(h, k, l)^{1/2} - I_P(h, k, l)^{1/2}]^2$. (See appendix 6.1.)
2. Mercury (Hg; $Z = 80$) is often the heavy atom component of compounds used to prepare isomorphous heavy atom derivatives of protein crystals because mercurials react preferentially with the sulfhydryl groups of cysteines, of which most proteins contain only a modest number. Suppose your objective is to ensure that the average intensity change that results from derivatization of some proteins you are working with is not less than

20% of the average intensity of the reflections the crystals of those protein produce in the absence of heavy atoms.

a. What would the molecular weight be of the largest protein containing a single reactive cysteine that you could work with that would meet that criterion? (To make calculations easy, assume that all the atoms in protein molecules are carbon atoms.)

b. Would the outcome of this computation be any different if there was more than one copy of one of these proteins in the unit cells of the crystals available?

3. Consider a single, isolated benzene molecule. Draw a sketch showing what the Patterson function of that molecule looks like. Indicate the relative amplitudes of each of its peaks. Ignore its hydrogen atoms.

4. a. Show that in general, $FT(I(\mathbf{S})) = [FT^{-1}(I(\mathbf{S}))]^*$.

b. If $\rho(\mathbf{r})[= FT^{-1}(F(\mathbf{S}))]$ is real, what will the relationship be between $FT(I(\mathbf{S}))$ and $[FT^{-1}(I(\mathbf{S}))]^*$?

c. If $\rho(\mathbf{r})$ is real for some crystal, will it make any difference if its Patterson function is computed by forward Fourier transformation of the intensity data the crystal yields, or by backward Fourier transformation?

APPENDIX 6.1 HEAVY ATOM DIFFERENCE PATTERSONS AND ANOMALOUS DIFFERENCE PATTERSONS

Equation (6.7) describes the way isomorphous replacement data are treated before they are used to compute heavy atom difference Patterson functions. Here, we show why this method of differencing leads to useful estimates for $I_H(h, k, l)$. Equation (6.7) can be written:

$$I_D(h, k, l) = I_{HP}(h, k, l) + I_P(h, k, l) - 2\,|F_{HP}(h, k, l)|\,|F_P(h, k, l)|\,.$$

The terms in this equation can be evaluated using equation (6.2), and when this is done, we obtain:

$$\begin{aligned} I_D(h, k, l) &= 2F_P^2 + F_H^2 + 2F_HF_P\cos(\alpha_H - \alpha_P) \\ &\quad - 2F_P\left[F_P^2 + F_H^2 + 2F_HF_P\cos(\alpha_H - \alpha_P)\right]^{1/2}. \end{aligned}$$

The second term on the right-hand side of this equation can be rearranged as follows:

$$\begin{aligned} &2F_P\left(F_P^2 + F_H^2 + 2F_HF_P\cos(\alpha_H - \alpha_P)\right)^{1/2} \\ &= 2F_P^2\left(1 + (F_H/F_P)^2 + 2(F_H/F_P)\cos(\alpha_H - \alpha_P)\right)^{1/2}. \end{aligned}$$

If $2|F_P\cos(\alpha_H - \alpha_P)| > F_H$, which will often, but by no means always be the case, then:

$$2F_P\left(F_P^2 + F_H^2 + 2F_HF_P\ \cos\left(\alpha_H - \alpha_P\right)\right)^{1/2}$$
$$\approx 2F_P^2\left(1 + 2\left(F_H/F_P\right)\ \cos\left(\alpha_H - \alpha_P\right)\right)^{1/2}.$$

and using the first terms of the power series expansion for expressions of the form $(1 + x)^{1/2}$ to obtain an approximate expression for the square root term, we get:

$$2F_P^2\left(1 + 2\left(F_H/F_P\right)\ \cos\left(\alpha_H - \alpha_P\right)\right)^{1/2} \approx 2F_P^2\left(1 + \left(F_H/F_P\right)\ \cos\left(\alpha_H - \alpha_P\right)\right),$$

and hence, under these circumstances:

$$I_D\left(h,\ k,\ l\right) \approx F_H^2 = I_H\left(h,\ k,\ l\right).$$

It is important to point out that if $I_{HP}(h,\ k,\ l) \approx I_P(h,\ k,\ l)$, little can be learned about $I_H(h,\ k,\ l)$ from isomorphous differences because this observation could mean that $|F_H(h,\ k,\ l)|$ is small, but it could also mean that $|F_H(h,\ k,\ l)|$ is large, but that the phase of the heavy metal contribution is such that $F_P(h,\ k,\ l)$, $F_{HP}(h,\ k,\ l)$ and $F_H(h,\ k,\ l)$ form an isosceles triangle.

A similar approach can be used to work out the implications of equation (6.8), the expression that describes the way anomalous diffraction data are differenced to obtain the information about $|F_H''|$, that is, $[I(h,\ k,\ l)^{1/2} - I(-h,\ -k,\ -l)^{1/2}]$. From equation (6.3), we know that:

$$|F\left(h,\ k,\ l\right)| = \Big[|F_{HP}\left(h,\ k,\ l\right)|^2 + \left|F_H''\left(h,\ k,\ l\right)\right|^2$$
$$+ 2\ |F_{HP}\left(h,\ k,\ l\right)|\ \left|F_H''\left(h,\ k,\ l\right)\right| \sin\left(\alpha_{HP} - \alpha_H\right)\Big]^{1/2}.$$

This expression can be written:

$$|F\left(h,\ k,\ l\right)| = |F_{HP}\left(h,\ k,\ l\right)| \Big[1 + \left|F_H''\left(h,\ k,\ l\right)\right|^2 / \left|F_{HP}\left(h,\ k,\ l\right)\right|^2$$
$$+ 2\left(\left|F_H''\left(h,\ k,\ l\right)\right| / |F_{HP}\left(h,\ k,\ l\right)|\right) \sin\left(\alpha_{HP} - \alpha_H\right)\Big]^{1/2}$$

and if $|F_H''(h,\ k,\ l)|^2/|F_{HP}(h,\ k,\ l)|^2$ is small compared to 1.0, which it often will be, then:

$$|F\left(h,\ k,\ l\right)| \approx |F_{HP}\left(h,\ k,\ l\right)| \left[1 + 2\left(\left|F_H''\left(h,\ k,\ l\right)\right| / |F_{HP}\left(h,\ k,\ l\right)|\right)\right.$$
$$\left.\sin\left(\alpha_{HP} - \alpha_H\right)\right]^{1/2},$$

and:

$$|F\left(h,\ k,\ l\right)| \approx |F_{HP}\left(h,\ k,\ l\right)| + \left|F_H''\left(h,\ k,\ l\right)\right| \sin\left(\alpha_{HP} - \alpha_H\right).$$

When the same reasoning is applied to equation (6.4), finds:

$$|F(-h, -k, -l)| \approx |F_{HP}(h, k, l)| - \left|F_H''(h, k, l)\right| \sin(\alpha_{HP} - \alpha_H).$$

It follows that:

$$\left[I(h, k, l)^{1/2} - I(-h, -k, -l)^{1/2}\right] = 2\left|F_H''(h, k, l)\right| \sin(\alpha_{HP} - \alpha_H),$$

and, thus, the data that should be used to compute anomalous difference Patterson functions, $I_{anom}(h, k, l)$, is:

$$\begin{aligned} I_{anom}(h, k, l) &= \left[I(h, k, l)^{1/2} - I(-h, -k, -l)^{1/2}\right]^2 \\ &= 4\left|F_H''(h, k, l)\right|^2 \sin^2(\alpha_{HP} - \alpha_H). \end{aligned}$$

From elementary trigonometry, we know that $\sin^2(\alpha_H - \alpha_P) = (1/2)(1 - \cos[2(\alpha_{HP} - \alpha_H)])$, thus:

$$I_{anom}(h, k, l)\, 2\left|F_H''(h, k, l)\right|^2 [1 - \cos 2(\alpha_{HP} - \alpha_H)].$$

Note that just as isomorphous differences provide little or no information about $|F_H|$ when $I_{HP}(h, k, l) \approx I_P(h, k, l)$, anomalous differences are uninformative about $|F_H''|$ when $I(h, k, l) \approx I(-h, -k, -l)$. The (approximate) equality of the intensities of two reflections that are Bijvoet mates could mean that $|F_H''|$ is small, but it could also mean that $(\alpha_{HP} - \alpha_H) \approx 0$.

7

Electron Density Maps and Molecular Structures

Once estimates of the phases of the reflections in a crystal's diffraction pattern have been obtained, it is easy to compute the electron density distribution of its unit cells. What do these electron density maps look like, and how are they interpreted?

7.1 EXPERIMENTAL ELECTRON DENSITY MAPS DISPLAY THE VARIATION IN ELECTRON DENSITY WITHIN THE UNIT CELL WITH RESPECT TO THE AVERAGE

Most experimental electron density maps are computed using all the diffraction amplitudes and phases available for a crystal out to the spherical shell in reciprocal space within which the average signal-to-noise ratio of the measured intensities drops to 2.0, or some other, arbitrarily chosen, small value. The $(0, 0, 0)$ reflection is *always* missing from the data because it is physically impossible to measure, and other low angle reflections may be missing as well because the beam stop that protected the detector from the direct beam was so large that it kept them from reaching the detector.

The absence of a value for $F(0, 0, 0)$ from experimentally determined diffraction data sets has a profound effect on the electron density maps derived from them. Rayleigh's theorem (section 2.11) tells us that the average electron density in the unit cell of an electron density map computed using a data set that lacks a value for $F(0, 0, 0)$ must be 0. Thus, rather than depicting $\rho(\mathbf{r})$, all such electron density maps will be displays of $\delta(\mathbf{r})$, the difference between $\rho(\mathbf{r})$ and the average electron density. In regions of the unit cell that are occupied by well-ordered molecules, $\delta(\mathbf{r})$ maps will have strong positive features at locations where atoms are always found, and strong negative features between atoms, where atoms are never found. However, in disordered regions, such as regions occupied by solvent, $\delta(\mathbf{r})$ will be less variable, and have an average value less positive than the maxima in regions where ordered structure is present, but more positive than the minima in those same regions. (The consequences of the omitting other low angle reflections from the data are discussed in section 7.3.)

7.2 ELECTRON DENSITY MAPS ARE CONTOURED IN UNITS OF SIGMA

It follows that crystallographic electron density maps are three-dimensional arrays of both positive and negative numbers that are proportional to $\delta(\mathbf{r})$. They are invariably displayed as contour maps, and the contour levels are usually multiples of the root mean square average value of $\delta(\mathbf{r})$ within a single unit cell, which is identical to the standard deviation of the $\rho(\mathbf{r})$ over the unit cell. Because statisticians use the Greek letter sigma, σ, to designate standard deviations, a crystallographer might describe some map as being contoured at $+1$ sigma, for example. A map of this sort, which would display no negative features, would show the outlines of the regions in the unit cell where $\delta(\mathbf{r}_i) \geq 1.0\,\sigma$, where:

$$\sigma = \left[(1/N)\sum \delta\,(\mathbf{r}_i)^2\right]^{1/2}.$$

N is the number of points in the unit cell where the electron density was computed, and the sum is over all points. If the values of $\delta(\mathbf{r})$ in such a map were normally distributed, and there is no reason why they should be, the volume(s) enclosed by the 1σ contour level would fill about 15% of the unit cell.

It is important not to forget that the absolute electron density that corresponds to the $+1\sigma$ contour in an electron density map depends on the Bragg spacing of the highest angle reflections in the data set used to compute that map. Rayleigh's theorem tells us that, if everything else were the same about two crystals except the scattering angle at which their diffraction patterns become too weak to measure effectively, $+1\sigma$ would correspond to a larger absolute fluctuation in electron density in the map of the crystal that diffracts to higher angle than it does in the map of the crystal that diffracts to lower angle.

7.3 THE POINT-TO-POINT RESOLUTION OF AN ELECTRON DENSITY MAP IS ROUGHLY $1/|S|_{MAX}$

The interpretability of an electron density map is critically dependent on its **resolution**, a concept that will be introduced here, but is so important that it will be revisited several more times as we proceed. As we already know, the average intensity of reflections in macromolecular diffraction data sets falls as $|\mathbf{S}|$ increases both because of thermal disorder and because the amplitudes of atomic scattering factors decrease with scattering angle. The reciprocal of the value of $|\mathbf{S}|$ of the shell within which the data in some diffraction data set is deemed to have become too weak to be worth measuring, which we will call $|\mathbf{S}|_{max}$, is the resolution of that data set. It is also the length scale of the smallest structural features that can be revealed by electron density maps computed using that data set, as we have noted before.

Comment: The adjectives "high" and "low" are commonly used to describe resolutions, and their usage in this context can be confusing. A high-resolution diffraction data set is one that extends to a (comparatively) large value of $|\mathbf{S}|$ (e.g., has a large $|\mathbf{S}|_{max}$). The electron density maps derived from a high-resolution data set are also described as high-resolution because the smaller $1/|\mathbf{S}|_{max}$ is, the smaller the structural details those maps may reveal (i.e., the more highly resolved the map is). A "low" resolution data set is just the opposite in this regard. Because $|\mathbf{S}|_{max}$ is small for such a data set, $1/|\mathbf{S}|_{max}$ is large, and only gross features will be evident in the (low-resolution) electron density maps that are computed from it. Thus, an electron density map in which features having length scales smaller than, say, 10 Å, cannot be discerned is a "low-resolution map", but an electron density map that reveals features having length scales of the order of 2 Å is "high resolution", even though 10 is bigger than 2.

In optics, the word *resolution* has a more intuitive meaning. The resolution of an image produced by a microscope is the smallest separation a pair of points can have in the object being imaged and not be mistaken for a single point in its image. Although it might not be obvious at first glance, the crystallographic definition of resolution and its optical definition are closely related.

The connection between these two definitions of resolution can be made using concepts introduced in chapter 2. There it is shown that if the inverse Fourier transformation of some (one-dimensional) function in reciprocal space is computed over a limited range of s, the real-space function obtained will be the function that would have been obtained if the range of s had not been truncated, convoluted with the inverse Fourier transform of the square wave of amplitude 1.0 that specifies the limits of that truncation. The same concept is valid in three dimensions. If $\rho(\mathbf{r})$ is the three-dimensional electron density distribution that would result from inverse Fourier transformation of the relevant diffraction data starting at $|\mathbf{S}| = 0$ and ending at $|\mathbf{S}| = \infty$, and $\rho'(\mathbf{r})$ is the electron density distribution obtained when that same data are inverse Fourier transformed over the range $|\mathbf{S}| = 0$ to $|\mathbf{S}| = |\mathbf{S}|_{\mathrm{max}}$, then:

$$\rho'(\mathbf{r}) = \rho(\mathbf{r}) * \mathrm{FT}^{-1}\left(\mathrm{A}\left(|\mathbf{S}|\right)\right),$$

where $\mathrm{A}(|\mathbf{S}|) = 1.0$ for all $\mathbf{S}$ such that $|\mathbf{S}| \leq |\mathbf{S}|_{\mathrm{max}}$, and for all $\mathbf{S}$ outside that range, $\mathrm{A}(|\mathbf{S}|) = 0$.

The only question that remains is what $\mathrm{FT}^{-1}(\mathrm{A}(|\mathbf{S}|))$ looks like. In most cases, $\mathrm{A}(|\mathbf{S}|)$ will be a sphere in reciprocal space that has a value of 1.0 from $|\mathbf{S}| = 0$ to $|\mathbf{S}| = |\mathbf{S}|_{\mathrm{max}}$, where $|\mathbf{S}|_{\mathrm{max}}$ is the (crystallographic) resolution limit of the data used to compute $\rho'(\mathbf{r})$. Although the algebra required to compute $\mathrm{FT}^{-1}(\mathrm{A}(|\mathbf{S}|))$ is messy (appendix 7.1), the result is gratifyingly simple. Appropriately normalized:

$$\mathrm{FT}^{-1}\left(\mathrm{A}\left(|\mathbf{S}|\right)\right) = \left(3/\left(2\pi\,|\mathbf{r}|\,S_{\max}\right)^{3}\right)\left[\sin\left(2\pi\,|\mathbf{r}|\,S_{\max}\right) - \left(2\pi\,|\mathbf{r}|\,S_{\max}\right)\cos\left(2\pi\,|\mathbf{r}|\,S_{\max}\right)\right].$$

Not surprisingly, this three-dimensional function resembles sin x/x. It has a strong symmetrical peak centered on $r = 0$ that is surrounded by weaker, spherically symmetric ripples that rapidly die in amplitude as $|\mathbf{r}|$ increases (figure 7.1). Thus, in $\rho'(\mathbf{r})$, every point in $\rho(\mathbf{r})$ is replaced by one of these functions, and the larger the value of $|\mathbf{S}|_{\max}$, the more delta like $\mathrm{FT}^{-1}(\mathrm{A}(|\mathbf{S}|))$ becomes. [We note for future reference that $\mathrm{FT}^{-1}(\mathrm{A}(|\mathbf{S}|))$ is the transfer function, or point-image function (see section 10.2) of the crystal-diffractometer system being used to determine $\rho(\mathbf{r})$. It is what a point atom embedded in the crystal would look like in the corresponding electron density map.]

How far apart must two points be in $\rho(\mathbf{r})$ in order that the image of the two of them, which is the sum of two adjacent point spread functions in $\rho'(\mathbf{r})$, have a shape that can be clearly differentiated from the shape of an image of a single point? How different must *different* be? If we take the view that the word *resolved* implies that an image of two adjacent points must have a clear minimum between their respective maxima, then the optical resolution of an X-ray structure is:

$$|\mathbf{r}|_{\mathrm{res}} \approx 0.9/\,|\mathbf{S}|_{\max}\,, \tag{7.1}$$

as you can readily verify by making plots of the sums of pairs of sphere transforms separated by different distances. [Optical resolution is discussed further in chapter 9.] We should not forget that in the derivation of equation (7.1), it is implicitly assumed that the phases and amplitudes used to compute electron density maps are all perfectly accurate, which is never the case. Low-resolution data are almost always measured more accurately than high-resolution data simply because low-resolution intensities are stronger. Thus, the point-to-point resolution

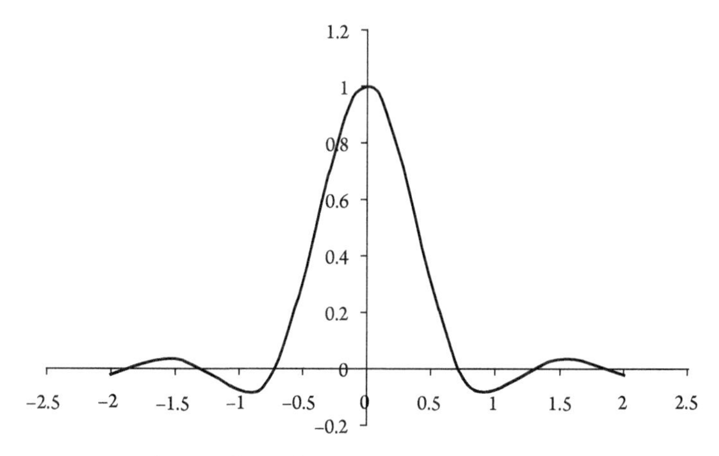

Figure 7.1 The transform of a sphere. Normalized amplitude is plotted as a function of the value of $(|\mathbf{r}||\mathbf{S}|_{\mathrm{res}})$.

of a real crystallographic electron density map is never as good as equation (7.1) implies.

The arguments just made not withstanding, by including all the diffraction data out to $|\mathbf{S}|_{max}$ in their electron density map computations, macromolecular crystallographers are taking advantage of almost all the information their crystals afford. For that reason, the truncation of the data at $|\mathbf{S}|_{max}$ has only a modest impact on the appearance of most crystallographic electron density maps. In effect, the maps are going to be that poorly resolved even if $|\mathbf{S}|_{max}$ is set at a larger value because the data past $|\mathbf{S}|_{max}$ are nearly 0, and noisy in the bargain. The image of an atom in crystallographic electron density map computed using all the data out to $|\mathbf{S}|_{max}$ is the convolution of the inverse Fourier transform of its temperature factor Gaussian with the inverse Fourier transforms of its atomic form factor (see chapter 5). The value of $|\mathbf{S}|_{max}$ makes a difference only if $|\mathbf{S}|_{max}$ is set at a value so low that the data still are quite strong at that resolution, in other words, when the data are truncated prematurely. In this case, the images of atoms that would have been obtained if all the data had been included are replaced by the convolutions of those images with $FT^{-1}(A(|\mathbf{S}|))$, which has negative as well as positive fringes (see figure 7.1). Interference between adjacent fringe patterns in a truncated image generates image artifacts called **series termination errors** that are liable to be particularly blatant around the boundaries of its features.

Now that we know about the transforms of spheres, it is easy to deal with a problem that was raised in section 7.1, but not addressed there, namely the effect of omitting low-resolution reflections other than $F(0, 0, 0)$ from electron density map calculations. The effect is easily described; it changes $A(|\mathbf{S}|)$ from a solid sphere of radius $|\mathbf{S}|_{max}$ into spherical shell that has the same outer radius and an inner radius of $|\mathbf{S}|_{min}$. Using the approach outlined in appendix 7.1, it is straightforward to compute $FT^{-1}A(|\mathbf{S}|)$ for a hollow sphere, but there is little insight to be gained from pursuing it. If $|\mathbf{S}|_{max} >> |\mathbf{S}|_{min}$, which it usually is, the shape of the central peak of $FT^{-1}A(|\mathbf{S}|)$ will hardly differ at all from that of the transform of the solid sphere that has the same outer radius. It is only in the fringes of $FT^{-1}A(|\mathbf{S}|)$, where amplitudes are small in any case, that the transforms of the spherical shell and the solid sphere differ much. Thus, the absence of very low-resolution reflections from some data set will have comparatively little effect on the appearance of the images obtained from it.

However, if $|\mathbf{S}|_{max}$ and $|\mathbf{S}|_{min}$ differ by only a little, for example, by a factor of 2, or less, the omission of the inner part of the data may have a visible effect. In that case, not only will the contribution of $F(0, 0, 0)$ be missing from the electron density map, but so also will be all the $F(h, k, l)$ terms that define the coarser features in the unit cell. The high-angle reflections in diffraction data sets modulate the shapes of those long wavelength features in images; they make them crisper, better defined, and more nuanced. Not surprisingly, therefore, the effects of high-resolution reflections tend to concentrate at the edges of features, where the first derivative of the electron density with respect to position is large. Thus, when only the higher resolution data are included in a Fourier synthesis, edges will be emphasized, and rather than looking like a solid of some size and shape, each feature in such an electron density map will look like a hollow shell of that same size and shape.

7.4 HOW HIGH IS HIGH ENOUGH?

The goal of the macromolecular crystallographer is to produce electron density maps that are well enough resolved so that their features can be interpreted chemically; thus, it is reasonable to wonder how high the resolution of an electron density map must be to make that possible. Clearly, an accurate atomic model for the structures of the molecules in a crystal ought to emerge from electron density maps that have resolutions high enough so that individual atoms are resolved. Because the distance between pairs of carbon atoms joined by single bonds is 1.54 Å, electron density maps having resolutions higher than ~1.7 Å should suffice (see equation (7.1)), and in practice, as you would expect, electron density maps having resolutions that high are easily interpreted, even though real resolutions are never as high as nominal resolutions.

What can be done with electron density maps that have (nominal) resolutions lower than 1.7 Å? One way to address this question is to compare the number of reflections in the diffraction patterns of crystals with the number of parameters that must be determined in order to place atoms in their unit cells. That comparison is instructive because the number of independent parameters that can be extracted from a data set cannot exceed the number of independent observations it reports. Suppose that the unit cell of some nucleic acid is a cube 100 Å on edge, and that 50% of its volume is occupied by solvent. Because the density of nucleic acid is about 1.50×10^3 kg/m^3, the molecular weight of the average nucleotides is about 330, and there are about 24 nonhydrogen atoms per nucleotide, there will be about 1,200 nucleotides, or about 29,000 atoms in each of that crystal's unit cells. (NB: We ignore hydrogen atoms here because they have so few electrons that they cannot be visualized in electron density maps having resolution much lower than 1 Å.) Using this information, we can estimate the number of reflections (see table 7.1) the diffraction pattern of these crystals will contain out to some limiting resolution and compare it to the number of atomic coordinates needed to specify the structure of its unit cell, which is $\sim 87,000$ ($= 3 \times 29,000$).

It follows that if the location of every nonhydrogen atom in this hypothetical crystal had to be determined using *only* the information provided by its diffraction pattern, an atomic resolution interpretation of its electron density maps would not be possible unless the resolution of the data set used to compute its electron density map was higher than ~2.8 Å, and even that is an underestimate. No molecular model will account for the diffraction data obtained from some crystal unless it includes

Table 7.1 DEPENDENCE OF THE NUMBER OF REFLECTIONS ON RESOLUTION

Resolution	No. of reflections	Reflections/coordinate
10 Å	2,094	0.024
5 Å	16,755	0.193
4 Å	32,724	0.376
3 Å	77,570	0.891
2 Å	261,794	3.01
1 Å	2,094,395	24.07

estimates for the temperature factors of its atoms. Thus, the number of parameters that need to be extracted from this hypothetical data set might be as large as 115,000, which implies that the breakeven resolution might be ~2.6 Å, and it could be higher if some of the solvent is visible.

This argument suggests that the only way the structure of a macromolecular crystal might be solved (i.e., an accurate, all-atom model proposed for it), using an electron density map having a resolution worse than ~2.6 Å is by feeding independently acquired information about macromolecular structures into the interpretation process. This is, in fact, what macromolecular crystallographers do all the time. Most of that outside information comes from high-resolution crystal structures of the monomers of which biopolymers are composed—amino acids and nucleotides. Most macromolecular electron density maps are commonly interpreted by fitting entire monomer structures into electron density maps, not by fitting individual atoms into them one at a time.

Monomer fitting dramatically reduces the amount of information that needs to be gleaned from the electron density maps of macromolecules. When models are being created this way, monomer bond lengths and bond angles are held fixed, and only the torsions angles of rotatable bonds are allowed to vary. Thus, the parameters that must be specified to place a single monomer in the map are the values for the relevant torsion angles, the location of its center of mass, and its orientation. To understand the impact this modus operandi might have, consider the hypothetical nucleic acid crystal we have been talking about. The conformation of each of its 24-atom nucleotide monomers is determined once 7 torsion angles have been set, and, thus, only 13 parameters have to be determined for each one: 7 torsion angles, 3 coordinates for the location of its center of mass location, and 3 angles to specify its orientation. Thus, the total number of parameters to be determined, including individual atomic B-factors, drops from about 115,000 to about 44,000, and that reduces the resolution required from about 2.6 Å to about 3.6 Å. If we were to "cheat" on B-factors by applying single, collective B-factors to the base, sugar, and phosphate group atoms of each nucleotide rather than attempting to estimate B-factors to each atom individually, the number of parameters to be determined would drop to about 19,000, and the minimum resolution would fall to 4.8 Å.

7.5 MACROMOLECULAR ELECTRON DENSITY MAPS HAVING RESOLUTIONS WORSE THAN ~3.5 Å ARE DIFFICULT TO INTERPRET CHEMICALLY

The theoretical arguments just made indicate that the lowest workable resolution for macromolecular electron density maps is much lower than the resolution at which individual atoms can be resolved. What happens in practice? Figure 7.2 shows what the electron density distribution map of a small region in the large ribosomal subunit from the archaebacterium *Haloarcula marismortui* looks like at two different resolutions: 5 Å (left), and 3.1 Å (right). The model for that part of the structure, which is based on a higher resolution electron density distribution,

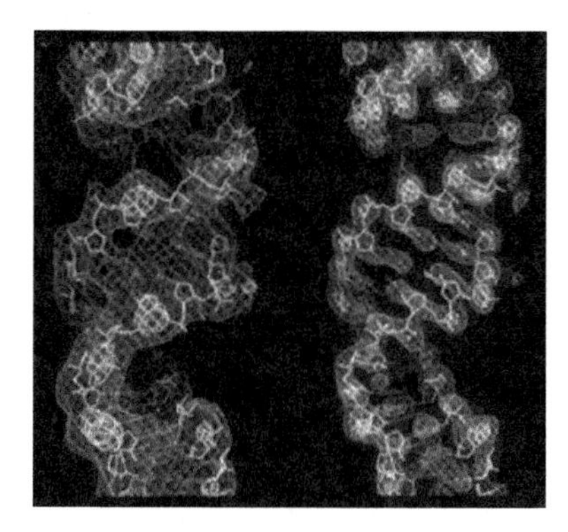

Figure 7.2 Two electron density maps of the same segment of the 23S rRNA of the *Haloarcula marismortui* ribosome. The left-hand electron density is taken from a 5 Å resolution map. The electron density on the right comes from a 3.1 Å resolution map (see Ban, N., Nissen, P., Hansen, J., Moore, P. B., and Steitz, T. A. [2000], The complete atomic structure of the large ribosomal subunit at 2.4 A resolution, *Science* 289, 905–920). The model for that part of the structure is superimposed.

is superimposed on both maps, and it clearly rationalizes them both. However, if the lower resolution map were the only one available, it would be a challenge to fit an RNA helix into it in proper register, let alone to distinguish one base pair from another. Thus, the real reason why reliable nucleic acid structures cannot be obtained from electron density maps having resolutions around 4.8 Å is not that there is insufficient information embedded in those maps, but rather the difficulty of identifying monomers and following polymer chains correctly in electron density maps that have resolutions that low. In practice, electron density maps of proteins become interpretable at resolutions a little lower than 3 Å, and because nucleotides are larger than amino acids, the interpretability threshold is crossed for nucleic acids at lower resolution, ~3.5 Å.

It should be noted that an increasing number of papers are being published that offer atomic resolution interpretations of electron density maps that have resolutions considerably worse than the thresholds just given. Almost all such interpretations are based not on the fitting monomer structures into electron density, but rather on the fitting of independently determined, atomic resolution structures of entire segments of macromolecules into electron density. It is a stretch to call the product of such exercises "structures", because in regions where those pieces of predetermined structure do not fit the electron density well, there is no way of determining how the map should be interpreted. Models of this sort are best thought of as working hypotheses.

Comment: A structure is an atomic resolution model for a macromolecule that is supported by a body of data that includes more independent observations than the number of independent parameters required to specify it.

7.6 SOLVENT MAY BE VISIBLE IN MACROMOLECULAR ELECTRON DENSITY MAPS

One reason why electron density maps having resolutions substantially better than 4 or 5 Å are required to solve macromolecular structures is that there is no way of being sure in advance what is going to be visible in them. The crystals formed by small organic and inorganic molecules are usually samples of a pure substance: They contain only a single kind of molecule. Thus, the information available about the chemical structure of a small molecule tells crystallographers exactly how many atoms they ought to see in its electron density maps. Macromolecular crystals are different. As we already know, the gaps between adjacent macromolecules in macromolecular crystals are large enough to accommodate large amounts of water, as well as solutes of modest molecular weight. In fact, the fraction of the total volume of a macromolecular crystal accounted for by solvent is typically $\sim$50%, and although it can be larger than that, it is seldom much less.

As we already know, crystallographic electron density distributions are *average* electron density distributions, where the average is over both time and all the unit cells in a crystal. Thus, at every location in the unit cell of a macromolecular crystal where the time-/space-averaged probability of finding an atom is reasonably high, say $>$30%, the image of an atom will appear in its electron density map. Because macromolecular crystals depend on macromolecular interactions for their existence, most of the macromolecular material in their unit cells is likely to be imaged, but there is no guarantee that it will all be visible. Regions in a macromolecule that are not involved in crystal contacts may be flexible, and if the B-factors of the atoms in those regions are large enough, they may not show up at all in maps contoured at $+1\sigma$, and maps contoured at a lower level, say $+0.5\sigma$, may be so noisy that nothing can be made of the weaker features they contain. In addition, there will be electron density for water molecules or low-molecular weight solute molecules at sites on the surfaces of macromolecules where those molecules bind with high probability. Nothing will be seen in the regions of the crystal where the solvent structure varies substantially from one unit cell to the next. Thus, crystallographers cannot be sure what they are looking at in a macromolecular electron density map until a lot of effort has been made to interpret it, and clearly, the higher the resolution of the map, the higher the probability that it will be interpreted correctly.

7.7 INITIAL MODELS MUST BE REFINED

Computer programs exist that greatly simplify the task of fitting monomers into macromolecular electron density maps. Indeed, this kind of model building may be completely automated before long. However, it is important to recognize that the models for macromolecules obtained by the initial fitting of monomer structures into electron density maps are, at best, useful first approximations to the truth, not the final word. They must all be thoroughly **refined** before they are made public.

The input for most crystallographic refinement programs includes: (1) an initial model for the contents of the unit cell that may or may not include estimates of atomic B-factors, (2) the experimentally measured $F(h, k, l)$ reflections for the crystal of concern, (3) information about the covalent geometry of the relevant monomers, and (often) (4) information about the energetic costs of distorting monomer geometries, as well information about interaction energies. During refinement, the positions of the atoms in a model are adjusted in ways that improve the correspondence between measured diffraction amplitudes and those implied by the model, while keeping the overall energy of the model as low as possible. The energy constraints ensure that individual bond lengths and bond angles will remain close to accepted values, and that the interactions between nonbonded atoms implied by the model will be chemically sensible. The output should be a chemically acceptable model for the molecule(s) of concern that is more consistent with the measured diffraction data than its antecedent.

Comment: It is at the refinement stage that the ratio of parameters to observations becomes critical. Refinement programs do not work well, if at all, if the number of parameters being refined exceeds the number of experimental data.

Several general points should be made about electron density maps, initial models, and refinement. First, because of local disorder in crystals, the ease with which experimental electron density maps can be interpreted can vary a lot from one region of the unit cell to the next. Second, the **radius of convergence** of *all* refinement programs—the size of the errors in atomic positions they will correct—is limited. Refinement programs *do not* make silk purses out of sows' ears, and all of them have a distressing tendency to output models similar to the models fed into them, rather than significantly different models that are objectively better. Third, because experimental phases are obtained by procedures that require the differencing of measured data, the quality of the phase information on which experimental electron density maps are based always drops faster with increasing $|\mathbf{S}|$ than the ratio of average intensity to the average standard error of intensities. However, there are a number of ways to use molecular models that are barely satisfactory to improve initial phase estimates, especially at high resolution. Thus, once a reasonable first model has been obtained for some molecule, we can often correct the errors it contains that are too gross to be eliminated by refinement by appropriate back-checking and, thus, extend its resolution to the limit of the data (see section 7.12).

7.8 R-FACTORS ARE USED TO MEASURE THE CONSISTENCY OF MOLECULAR MODELS WITH MEASURED DIFFRACTION DATA

By tradition, the crystallographic community assesses the correspondence between molecular models and the data from which they derive using a global statistic called the **R-factor**, which is defined as follows:

$$R = \sum_{h,k,l} ||F(h, k, l)_{obs}| - |F(h, k, l)_{calc}|| / \sum_{h,k,l} |F(h, k, l)_{obs}|, \qquad (7.2)$$

where both sums run over all measured reflections, $|F(h, k, l)_{obs}|$ is the measured amplitude of the (h, k, l) reflection, and $|F(h, k, l)_{calc}|$ is the amplitude computed for that reflection using the atomic model *du jour* for the contents of the unit cell. Small values of R imply that a model agrees well with the data, and large values mean that the agreement is poor. The objective of all refinement procedures is to adjust models in ways that drive the value of R down.

R-factors are a statistical disaster. Their principal shortcoming is that they treat data that have been well measured the same way they treat data that have been measured poorly. In a better world, crystallographers would compute a figure of merit for their molecular models that summed the discrepancies between observed data and calculated data in such a way that discrepancies that were large compared to estimated error counted for more than those that are small. However, the crystallographic community has used R-factors for so long that there is little chance they will ever be replaced by something better. It follows that we need to know what value of R to expect if the model obtained for some crystal is perfectly accurate, and we also need to know what the value of R would be if the model were perfect nonsense.

We start by considering the case where the model is perfect. When that is the case, by definition, measurement error in the $|F(h, k, l)|_{obs}$, which we will call $\delta(h, k, l)$, must account for R, and so:

$$R_{perfect} = \sum_{h,k,l} ||F(h, k, l)_{calc} + \delta(h, k, l)| - |F(h, k, l)_{calc}|| / \sum_{h,k,l} |F(h, k, l)_{calc} + \delta(h, k, l)|$$

$$\approx \sum_{h,k,l} |\delta(h, k, l)| / \sum_{h,k,l} |F(h, k, l)_{obs}|.$$

The second equation is obtained from the first, assuming that the errors in the measured Fs are statistically distributed, and for that reason, the sum of all measured Fs is likely to be very close to the sum of all calculated Fs. $\delta(h, k, l)$ will be of the order of the standard error of $|F(h, k, l)|_{obs}$, $\sigma_F(h, k, l)$. We can get a sense of the likely magnitude of $R_{perfect}$ from the tables summarizing data statistics that are often found in crystallographic papers, which usually include estimates of the average values of the standard errors of the measured amplitudes or intensities. For well-measured macromolecular data sets, $R_{perfect}$ might be 0.05 to 0.10. R_{awful} can be estimated using Wilson statistics (see chapter 5), and it is not too hard to show that the R-factor for a model of some macromolecule that is totally incorrect will be ~0.586 (appendix 7.2). The range of values possible for R is thus quite small.

It is important to realize that initial models that are good enough to lead investigators to correct structures can have R-values in the 40% range. It is also important to realize that the values of R that we can expect from carefully refined models depend on resolution. Refined, ~1 Å resolution crystal structures may have R-factors below 10%, and 5% is not unusual for small-molecule crystal structures solved at resolutions in this range. Refined, ~3 Å resolution macromolecule crystal structures, on the other hand, are unlikely to have R-factors smaller than 25%.

Sixty years ago, V. Luzzatti derived an expression that relates the standard error of the atomic coordinates in a macromolecular structure to both the resolution of the data used to determine it and its R-factor. That derivation, which makes heavy use of Wilson statistics and assumes that the sole source of error in diffraction data sets is counting error, is too complicated to replicate here. Suffice it to say, the higher the resolution of the data and the lower the R-factor of the model, the more accurate its atomic coordinates should be. Using Luzzati's approach, one estimates that for well-refined models, the (average) standard error for individual atomic coordinates will be roughly a quarter of the resolution of the data used to determine them. On more than one occasion, two different laboratories have independently solved the structure of the same macromolecular crystal. Comparison of atomic coordinates between pairs of independently determined structures indicates that experimental coordinate errors are larger than those Luzzati's theory predicts, but not dramatically so.

7.9 FREE-R IS USEFUL TOOL FOR VALIDATING REFINEMENTS

Both the amplitudes and the phases in measured diffraction data sets include noise, and in addition, measured data sets are seldom complete. Thus, inevitably, the electron density maps produced by inverting real data are inaccurate, and their inaccuracies have two consequences. First, they distort both the shapes and the apparent locations of the electron density features that correspond to atoms or groups of atoms. Noise of this sort limits the accuracy with which atomic positions can be determined. Second, noise can produce features in images that look real, but are not. In solvent regions, for example, noise features can easily be mistaken for water molecules and/or solvent ions.

Refinement programs are liable to deal with accuracy problems of the first kind in a rational way, but they provide no protection whatever from accuracy problems of the second kind. Biochemists seldom know in advance how many water molecules and/or ions bind tightly to some macromolecule under crystallization conditions, let alone where they bind. In addition, at resolutions worse than about 1 Å, only the oxygen atom in a water molecule can be seen in electron density maps, and so waters look like monatomic ions. They are single peaks in the electron density, just like noise peaks. Worse than that, the occupancies of bound water molecules and/or ions can always be less than 1.0 and their B-factors need not be the same as the B-factors of neighboring atoms in the macromolecule. Thus, almost any peak adjoining electron density assigned to macromolecular material could be a bound water molecule or ion.

It is hard to resist the temptation to identify noise peaks close to the surfaces of macromolecules in electron density maps as water molecules or ions, especially if the macromolecule provides charged groups and/or hydrogen bond donors or acceptors nearby that might explain why a water molecule or an ion would be found there. When water or ions are added to structural models at the positions where they really do bind to macromolecules, R-factors go down. Unhappily, R-factors also go down when noise peaks are interpreted as waters or ions because each ion or water molecule added to a model increases the number of degrees of freedom that can be refined to improve the fit of models to data. Furthermore, each such addition to a model, in effect, tells the refinement program that some component of the data that in fact is noise should be treated as though it is signal.

Overinterpretation of electron density maps can be minimized using a self-validation technique devised by A. Brunger. The first step in the procedure is the division of the measured diffraction data into two unequal sets, one that includes about 95% of the data, and the other, the remaining 5%, chosen at random. Refinement is done using the 95% part of the data, but progress is assessed not by improvements in the R-factor computed using the 95% portion of the data, which is called the **working R-factor**, but rather by improvements in the R-factor computed using the 5% part of the data, which is called **free-R**. Because the noise in the 5% part of the data should be uncorrelated with the noise in the 95% part of the data, and because the positions, B-factors, and occupancies of all atoms contribute to the value calculated for every $F(h, k, l)$, adjustments to a model that address its real shortcomings should reduce both its working R-factor and its free-R. Adjustments made in response to noise are likely to reduce the working R-factor while leaving the free-R unchanged or even increased in value. Refinement is stopped when free-R bottoms out, and at that point, it is normally observed that the working value of R is a few percentage points smaller than free-R.

7.10 PHASES RULE

As we already know, experimental electron density maps (images) are computed using measured amplitudes and phases. What is more important for the quality of the image that emerges, the amplitudes or the phases? The answer to this question is well-known; phases are more important than amplitudes. Figure 7.3 shows a conventional photograph, and two images derived from it by Fourier computation. The middle panel was obtained by inversion of the Fourier amplitudes obtained by transformation of the image on the left, setting all phases to 0. It is incomprehensible; it looks like a Patterson function. By contrast, the right panel, which was computed using the correct phases, but with all amplitudes set to 1.0, is recognizably similar to the parent image. The message could not be simpler. If the phases are right, a useful image will emerge. If the phases are wrong, the image will be wrong.

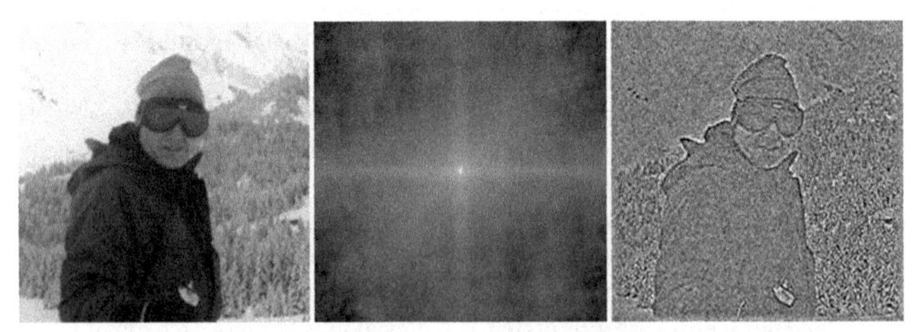

Figure 7.3 Phases versus amplitudes. The left panel is a photograph of a skier on a mountainside. The middle panel is an image generated by back Fourier transformation of the coefficients obtained by Fourier transformation of the image in the left panel with all phases set to 0. The right panel is an image computed by back Fourier transformation of the coefficients obtained by Fourier transformation of the image in the left panel with all amplitudes set to 1.0. (Figure reproduced with the permission of its creator: Professor Chris Jacobsen, Argonne National Laboratory.)

Earlier, we noted that the errors in experimentally determined phase estimates are often shockingly large, which is why it is important to compute experimental electron density maps using figure-of-merit weighting. If the molecular model obtained by the interpretation of an experimental electron density map is any good at all, however, the average difference between the phases computed using that model and their true values is likely to be much smaller than the average error of the experimental phases used to obtain the parent map. Furthermore, a map computed using observed Fs and model phases will always be "prettier" than its parent, experimental map. Features will be better defined, and the noise level will appear to be lower. The problem is that because phases rule, the structure evident in maps of the latter kind is guaranteed to be identical to the model structure, no matter what. Thus, crystallographers interested in improving their models must overcome a serious problem. How do you identify errors in an initial model? How do you eliminate them so that a better, more correct model can be obtained? How do you keep the phases of your first model from making that model the only model you ever get?

7.11 REGIONS WHERE MODELS DO NOT CORRESPOND TO ELECTRON DENSITY MAPS CAN BE IDENTIFIED USING DIFFERENCE ELECTRON DENSITY MAPS

The difference Fourier transformation is the single most powerful tool available for identifying regions in electron density maps where models and data are inconsistent and, thus, for eliminating such discrepancies. Difference Fouriers are easily computed:

$$\rho_d(x, y, z) = \sum_{h,k,l} \left(|F_{obs}(h, k, l)| - |F_{calc}(h, k, l)|\right) \times \exp\left[2\pi i\left(hx + ky + lz\right) + i\alpha_{calc}(h, k, l)\right],$$

where $F_{obs}(h, k, l)$ is the observed amplitude of a reflection, and $F_{calc}(h, k, l)$ and $\alpha_{calc}(h, k, l)$ are the amplitude and phase computed for that reflection using the model to be tested. It is easy to show (see appendix 7.3) that:

$$\rho_d(x, y, z) \approx (1/2)\left[\rho(x, y, z) - \rho_{mod}(x, y, z)\right], \tag{7.3}$$

where $\rho(x, y, z)$ is the true electron density distribution, and $\rho_{mod}(x, y, z)$ is the electron density distribution that corresponds to the current model. Thus, the difference electron density distribution will have positive features where there is something present in the unit cell that is not accounted for by the model and negative features where the model includes something that should not be there. In short, difference electron density maps can identify the places in the unit cell where models need to be corrected and indicate what needs to be done. It should be obvious that, averaged over the entire unit cell, the root mean square value of $\delta(\mathbf{r})$, σ, will be much smaller in a difference map than it is in a regular electron density map. For that reason, difference maps are commonly contoured at intervals of 3σ to 4σ instead of 1σ, and for obvious reasons, both positive and negative contours are examined. (In difference maps contoured at $\pm 1\sigma$, it can be very difficult to distinguish significant differences from noise. For the record, difference electron density maps are often computed using amplitude differences that are not computed the in the manner described here, but these variants all work the same way.)

Two points need to be made at this juncture.

1. It is essential that difference Fouriers be computed from time to time as crystallographically derived macromolecular models are refined. The radius of convergence of refinement algorithms is small, as has already been pointed out. Thus, erroneous models will emerge from refinements if care is not taken to identify the regions in less-refined models where the correspondence between the model and data is poor and then to correct them by hand.

2. Many of the macromolecules studied crystallographically interact with low--molecular weight substrates, substrate analogs, or ligands of other kinds. It is often possible to add these small molecules to macromolecular crystallization mixtures, either before or after crystallization, and then to collect data from the crystals that emerge. These crystals are often isomorphous with the crystals that form under the same conditions in the absence of ligands. When this is so, difference Fouriers can be used to find where ligands bind to these macromolecules and to determine their conformations in the bound state. The amplitudes used are the differences between the Fs measured with the ligand present and the Fs measured when the ligand and the phases that should be used (i.e., α_{calc}) are those appropriate for the refined model of the macromolecule in question with no small molecules bound.

7.12 EXPERIMENTAL ELECTRON DENSITY MAPS CAN BE IMPROVED BY PHASE MODIFICATION AND EXTENSION

As we already know, experimental phases tend to be inaccurate, and in general, in any given data set, phase errors increase with $|\mathbf{S}|$ faster than the value of average intensity divided by average standard error of intensity diminishes. Thus, not only are experimental electron density maps less accurate than we would like, but their effective resolutions are also worse than those implied by diffraction limit of the $F(h, k, l)$'s on which they are based, because the contributions made to electron density maps by high-angle reflections, which as a group have poorly measured phases, are diminished by their low figures of merit.

The crystallographic community has devoted a lot of energy to devising techniques both for improving the accuracy of phase estimates and for extending the phasing information available for data sets out to their resolution limits. Impressive progress has been made, and we will consider only one of these procedures in detail. It is called **solvent flattening**.

The idea behind solvent flattening is easy to understand. A large fraction of the volume of the unit cell of a macromolecular crystal is filled with solvent, and within those regions, $\rho(\mathbf{r})$ should be approximately constant. It follows that the fluctuations in density observed in the solvent regions of experimental electron density maps are (mostly) noise, and because phases are harder to measure than amplitudes are, that noise is likely to represent phase error. What can be done about this?

The first step in any solvent-flattening computation is delineation of the solvent-filled regions in the unit cell. Even in rather crudely phased experimental electron density maps, it is usually possible to distinguish solvent regions, where the fluctuations in electron density tend to be modest, from regions occupied by macromolecular material, where the fluctuations in electron density are larger. (A number of computer programs are available that will find solvent boundaries automatically; the details are of no concern here.) In the second step, the density on the solvent side of the solvent–macromolecule boundary is set equal to 0, that is, the solvent regions are "flattened", and on the macromolecular side of the boundary, the electron density is left alone. The modified electron density distribution that emerges is then used to compute a new set of phase estimates. In the third step, these solvent-flattened phase estimates are combined with the experimentally measured phase estimates available in a statistically sensible way so that solvent-flattened phases make large contributions to the estimates that emerge for phases that were poorly determined experimentally, but have a small impact on the phase estimates that were well-determined experimentally. This procedure can be carried out in several different ways, each of which has its own advantages and disadvantages, but again the details are unimportant. The electron density map obtained by inverse Fourier transformation of the measured amplitudes using these improved phase estimates should have a flatter solvent region and should also be more interpretable in its macromolecular regions than its purely experimental predecessor was. It is often useful to iterate this process several times.

When solvent flattening is used for phase extension, the procedure is somewhat different. In this case, the sine qua non is an atomic model for the structure of the contents of the unit cell; an electron density map is not enough. Using that model, phase estimates can be computed for *all* the reflections in a data set no matter what their resolution. The problem is how to combine those computed phases with the fact that the solvent region should be flat to arrive at usefully accurate estimates for the phases of the high-resolution reflections, for which there is little or no experimental phase information available, so that new information emerges about the structure. To get good results from this process, phases must be extended in very small increments of resolution. In each cycle, a thin shell of high-resolution data for which no experimental phases are available is added to the lower resolution, experimentally phased data, and maps are calculated with the phases of those high-resolution data that the most recent model of the structure requires. Many cycles of structure refinement and solvent flattening must then be done to ensure that the solvent region is as featureless as possible and the R-factor and free-R are as low as they can be before the next shell of unphased, high-resolution data is added. In this way, the model can be (slowly) forced to respond appropriately to each new increment of amplitude information, and the resolution of the electron density map extended to the edge of the available data.

7.13 SOLVENT FLATTENING AND THE NYQUIST THEOREM

In section 4.5, it was pointed out that because neighboring molecules in a crystal touch each other, but do not overlap, the number of Fourier components that can be accessed experimentally in the diffraction pattern of a crystal is the minimum necessary to compute the electron density distribution of its unit cells out to the resolution of the data. However, Fourier coefficients are complex, and because only their amplitudes can be measured experimentally, formally, at least, the intensity data we measure for a crystal provide only half the information required to determine its structure.

However, as we also argued in section 4.5, and illustrated in an entirely different way in section 7.4, this reasoning substantially exaggerates the magnitude of the phase problem that needs to be solved. The number of reflections in the diffraction pattern of a crystal can greatly exceed the number of parameters required to describe the locations of all the atoms in its unit cell. It follows that at resolutions higher than the minimum required to solve structures, which is estimated in table 7.1, the amplitudes and phases of reflections cannot be independent of each other; they must correlate. All direct methods used for determining crystal structures exploit these correlations to determine phases, as do all of the techniques used for extending and improving phases.

It is not hard to understand why the presence of solvent regions in macromolecular crystals leads to correlations in the diffraction data obtained from them. Consider a one-dimensional crystal, the unit cell of which has a length a. The relationship between the electron density distribution in its unit cell and its one-dimensional Fourier transform is:

$$\rho\,(r_i) = (1/a) \sum |\mathrm{F}\,(h)| \exp\left[i\alpha\,(h)\right] \exp\left(2\pi i r_i h\right),$$

where $\rho(r_i)$ is the electron density at the point in the unit cell where the fractional coordinate is r_i, and the sum runs from $-h_{\max}$ to $+h_{\max}$, h being the index of each reflection. If the data extend to some resolution $|\mathbf{S}|_{\max}(= h_{\max}/a)$, it will suffice to determine the value of ρ at a regular series of points within the unit cell that are separated by a distance of $(1/2|\mathbf{S}|_{\max})$, as per Nyquist. (The data can be used to compute ρ at more points than that, but nothing new will be learned by so doing because the values of ρ at those additional points will be linear combinations of the values of ρ we already know.) Interestingly, under these circumstances, the number of points where ρ is determined exactly equals the number of reflections in the diffraction pattern. Thus, if we knew the value of ρ at all these points, we would be able to compute all the F_j out to the limiting resolution in all directions, but no further.

What benefits would accrue if we were able to solvent flatten under these circumstances? The previous equation for electron density can be written in the following simplified form:

$$\rho_i = (1/a)\sum F_h E_{ih},$$

where $E_{ih} = \exp(2\pi i r_i h)$. For each ρ_i in the solvent region of the unit cell, we know that:

$$\rho_i = (1/a)\sum F_h E_{ih} = \langle \rho_{\text{solv}} \rangle.$$

Assuming all the $|F_h|$ have been measured experimentally, every such equation available could be used to determine the relationship between the phase of one reflection and the phases of all other reflections. If half the volume of the unit cell were solvent, and ρ was real, the number of phases that needed to be determined would equal the number of solvent equations available, and in theory, at least, there should be no need to measure any phases at all (see section 8.3).

Other techniques for phase improvement and extension exploit correlations the same way. For example, if the molecule in some crystal has internal symmetry elements that do not coincide with the symmetry elements of the crystals in which it is embedded, that is, if it has **noncrystallographic symmetry**, this fact can be exploited to improve phase estimates. Once the boundary of the molecule has been delineated and the orientation of its noncrystallographic symmetry elements determined, perhaps using an initial, experimentally phased electron density map, we can insist that the electron density in the unit cell have the noncrystallographic symmetry required by averaging the starting experimental electron density appropriately. The phases obtained by computation from that averaged map should be better than experimental phases, and thus form a useful basis for further refinement. In this instance, we are implicitly taking advantage of the fact that the size of the structure that needs to be solved to establish the structure of the unit cell is smaller than the crystallographic asymmetric unit. Thus, the ratio of data to unknowns is more favorable than it would be otherwise. Another, subtler, strategy for phase enhancement is called **histogram matching**. It takes advantage of the observation that in protein structures, the probability of encountering electron density features of different relative magnitudes has distribution (a histogram) that is invariant from one protein to the next. This fact is exploited by adjusting the electron density distributions in experimental electron density maps of proteins so that the distribution of electron densities in protein-containing regions conforms to the expected histogram. Once again, phases computed using histogram-corrected electron density distributions are more accurate than experimental phases are.

7.14 LET THE BUYER BEWARE

Not all crystal structures are right. Crystal structures are *interpretations* of electron density maps. They are *not* primary data. Furthermore, the path that leads from measured diffraction intensities to a crystal structure is a long one, and at many steps along the way, the crystallographer may have to exercise judgment. Not surprisingly, mistakes have been made in the past, and there is every reason to believe they will continue to be made. Some structures that have been deposited in the Protein Data Bank were/are seriously wrong, and included in this unhappy group are some that were published in high visibility journals such as *Science* and *Nature*.

The structural biology community has been struggling for decades with the fact that on the computer screen, any all-atom model for a macromolecule looks as authentic as any other, even models that have been fabricated out of whole cloth. How is anyone to know whether the structural model shown in some computer-generated illustration is reliable or not? Sadly, there is no foolproof way to answer this question, but there are ways "consumers" can test the validity of crystal structures, and those who use them owe it to themselves to do so. It pays to look the gift horse in the mouth.

The first question and most important test a structure must pass is biochemical. Does the structure make sense biochemically? By the time the structures of most macromolecules get solved, biochemists and geneticists will usually have been working on them for years. Valid macromolecular structures invariably rationalize, or are consistent with >95% of the biochemical/genetic information available on the relationship between their structures and their functions. Red flags should go up when this is not so. There are a few well-documented examples of macromolecules that crystallize in conformations distinctly different from those they assume when performing their biological functions. Crystal structures of that sort may flunk the "sense test" for cause. However, most of the crystal structures that flunk it do so because serious errors were made in either the generation or the interpretation of their electron density maps.

Papers describing new crystal structures invariably include a table of statistics that summarizes the quality of the data used to solve them and the quality of the structure itself. That information needs to be scrutinized because it also bears on reliability. The most important statistic is resolution. The higher the resolution of the data used to compute an electron density map, the easier that map should have been to interpret. Electron density maps that have poor resolution present all kinds of problems to their interpreters. For example, it is hard to identify amino acid side chains in electron density maps of proteins that have resolutions above 3.0 Å, and correspondingly, it is easy to fit peptide chains into them out of register. Registration errors are hard to detect in low-resolution maps, because even in high-resolution electron density maps, there may be parts of a protein that are not well visualized due to local disorder. Thus, there is no way to know in advance how many residues will be visible in the electron density map of a protein, let alone which ones they will be. In addition, when the resolution is poor, it is easy to interpret electron density belonging to disulfide bonds as polypeptide backbone, and the protein models that emerge when this mistake is made may not even have the correct fold topology. And there are other ways to get into trouble! Structures based on electron density maps that have resolutions worse than 4 Å should be treated with suspicion unless they were interpreted using accurately determined structures of closely related macromolecules.

If the resolution of a structure is reassuring, the consumer should then look at its R-factors. In this case, it is hard to be specific about what to look for because the R-factors of correct structures vary a lot, with resolution being an important determinant, but not the only one. What can be said with some certainty is that the free-R value for a structure should always be a few percentage points worse than its working R-factor value. If it is not, something odd is likely to have happened during refinement. For example, it may be that the set of *F*s used to assess R-free were not independent of the working set the way they were supposed to be. As for the actual values of R-free, any structure that has a free-R value much above 30% should be treated with caution. The only way a structure that is basically correct can have a free-R that high is if it has been inadequately refined, and if that is so, at the very least there may be a lot of small errors in its coordinates, which is not helpful. Finally, structures that contain serious errors often fail to refine well because the adjustments of coordinates required to reverse those errors are outside the radius of convergence of the programs used to refine them. In effect, refinements of wrong structures get "stuck", and one symptom of this may be a high R-free value.

The covalent geometry of the structure also deserves attention. Structures solved at moderate resolutions should have bond lengths that, on average, are within ± 0.01 Å of standard values, and bond angles that, on average, are less than 2° away from standard values. In addition, they should have very few nonbonded contacts less than ~ 4 Å. It is not that exceptions to these rules do not occur in nature, but they are rare, and there is no way they can be identified reliably in an electron density map that has a resolution worse than ~ 2.0 Å. At the very least, structures that have bad covalent geometries have not been properly refined. Bad geometry can also result when a fundamentally incorrect model is refined with loose constraints on covalent geometry.

It can also be useful to look at the header text that precedes the coordinate listings in Protein Data Bank files. Outlying bond lengths, bond angles, and close contacts are usually cataloged there. In addition, Ramachandran plots are useful tools for assessing protein structures. Almost all the torsion angles of the peptide backbones of well-refined, high-resolution protein structures fall in the allowed regions of such plots, with the overwhelming majority in the preferred region. A protein structure that does not have this property cannot be accurate.

Finally, bad structures commonly often have strange B-factors. The only way to find out about this is by inspecting the coordinate files that the Protein Data Bank supplies for structures. All refinement algorithms modify models for structures so as to reduce the difference between the diffraction data implied by those models and the measured data to the maximum degree possible, consistent with whatever external constraints are applied. The only variables in play are the coordinates of atoms and their B-factors. Because coordinates are constrained in most refinements by requiring that bond lengths and angles conform to prior knowledge, the only parameters that can be freely adjusted during refinement are the temperature factors. For that reason, the errors in models tend to accumulate in their B-values. If a structure has been poorly determined, the B-factors assigned to its atoms are liable to vary erratically from one residue to the next, and even from atom to atom within the same residue. In effect, the refinement program is trying to compensate for coordinate inadequacies by adjusting B-factors.

If a published structure you are interested in does not emerge from your examination of its credentials with a clean bill of health, you should consult with a knowledgeable crystallographer about it. Far better to confront its shortcomings before you have done a lot of structure-based experiments on it than after you have done so.

PROBLEMS

1. Either by plotting the sum of two crystallographic point transfer functions separated by a distance in real space of $1/|\mathbf{S}|_{max}$, or by computing that sum, confirm that at that separation there is a definite minimum between the two maxima, as the text claims there should be (see equation (7.1)).
2. Why should the R-free values associated with refined crystal structures be less than their working R-values? [Note: There are sound statistical reasons why this should be so; the explanation has nothing to do with crystallography per se.]
3. Suppose also that you have solved the structure of a crystal of some macromolecule, and that you have obtained an isomorphous crystal of that same macromolecule with a substrate analog bound. The cross R-factor between the two data sets is 0.10, where R_{cross} is given by the following expression:

$$R_{cross} = \sum_{h,k,l} \left| \left| \mathrm{F}\,(h,\ k,\ l)_{+\mathrm{ligand}} \right| - \left| \mathrm{F}\,(h,\ k,\ l)_{-\mathrm{ligand}} \right| \right| \Big/ \sum_{h,k,l} \left| \mathrm{F}\,(h,\ k,\ l)_{-\mathrm{ligand}} \right|.$$

   Using Rayleigh's theorem, estimate what the magnitude of σ will be in the difference electron density map computed using these two data sets relative to what it is in the electron density map computed using the data from the crystal that has no ligand bound.
4. Many of the area detectors used to collect X-ray diffraction data are square, and using such detectors, we could imagine collecting a data set the boundaries of which form a cube instead of a sphere. Why would it be a bad idea to compute electron density maps using data sets that had cubic (or rectangular) boundaries in reciprocal space?

APPENDIX 7.1 THE INVERSE FOURIER TRANSFORM OF THE SPHERICAL APERTURE FUNCTION

Consider $\mathrm{A}(|\mathbf{S}|)$, the function that equals 1.0 for all $\mathbf{S}$ such that $|\mathbf{S}| \leq |\mathbf{S}|_{res}$, and equals 0 for all $\mathbf{S}$ such that $|\mathbf{S}| > |\mathbf{S}|_{res}$. What is the inverse Fourier transform of that function? Because $\mathrm{A}(|\mathbf{S}|)$ is spherical, this question is best addressed using spherical polar coordinates. In spherical polar coordinate systems, there are three coordinates: a radius r, which is the distance from the center of the coordinate system to any point, and two angles: θ, the polar angle, and φ, the azimuthal angle. The polar angle is basically the latitude of some point, and its azimuthal angle is its longitude. Using these coordinates in both real space and reciprocal space, we can write the following expression for the inverse transform of $A(|\mathbf{S}|)$:

$$\mathrm{FT}^{-1}(\mathrm{A}(|\mathbf{S}|)) = \int_0^{S_{max}} \int_{\Theta=0}^{\pi} \int_{\Phi=0}^{2\pi} \exp\left[2\pi i|\mathbf{S}|\left(\begin{matrix}\sin\theta\,\sin\Theta(\cos\varphi\,\cos\Phi + \sin\varphi\,\sin\Phi)\\ +\cos\theta\,\cos\Theta\end{matrix}\right)\right] dV_s,$$

where $dV_S = |\mathbf{S}|^2 \sin\Theta\, d\Theta\, d\Phi\, d|\mathbf{S}|$, and Θ and Φ are the polar angles and azimuthal angles of a particular point in reciprocal space, respectively.

Because the function we are trying to evaluate is spherically symmetric, we do not need to evaluate the integral required at every single point in real space in order to understand what it looks like. It is enough to evaluate it along a single radial line in real space, secure in the knowledge that it will be just the same along every other radial line by reason of symmetry. The line to choose is the line along which θ and ϕ are both 0, because along that line the integral that must be evaluated is as simple as it gets:

$$\mathrm{FT}^{-1}(\mathrm{A}(|\mathbf{S}|)) = \int_0^{S_{max}} \int_{\Theta=0}^{\pi} \int_{\Phi=0}^{2\pi} \exp[2\pi i r|\mathbf{S}|\cos\Theta]|\mathbf{S}|^2 \sin\Theta\, d\Theta\, d\Phi\, d|\mathbf{S}|$$

Integrating over first Φ and then Θ we find:

$$\mathrm{FT}^{-1}(\mathrm{A}(|\mathbf{S}|)) = (2/r) \int_0^{S_{max}} |\mathbf{S}| \sin(2\pi r|\mathbf{S}|) d|\mathbf{S}|.$$

[Remember that $d(\cos\Theta) = -\sin\Theta\, d\Theta$.] The integral in the expression for $\rho_a(\mathbf{r})$ is a standard form the solution to which can be found in any table of integrals.

$$\mathrm{FT}^{-1}(\mathrm{A}(|\mathbf{S}|)) = (1/2\pi^2 r^3)[\sin(2\pi r S_{max}) - (2\pi r S_{max})\cos(2\pi r S_{max})],$$

and the expression is usually normalized by dividing by the volume of the sphere in reciprocal space, $(4/3)\pi S_{max}^3$. Thus:

$$\mathrm{FT}^{-1}(\mathrm{A}(|\mathbf{S}|))_{norm} = (3/(2\pi r S_{max})^3)[\sin(2\pi r S_{max}) - (2\pi r S_{max})\cos(2\pi r S_{max})].$$

APPENDIX 7.2 ESTIMATING THE R-FACTOR OF CRYSTAL STRUCTURES THAT ARE PERFECT NONSENSE

Wilson statistics provide a method for estimating the R-factor expected if the model proposed for some crystal structure bears no relation whatever to the true structure.

Wilson statistics tell us that the probability of observing a reflection whose amplitude is F, P(F), is:

$$\mathrm{P(F)} = \left(2\mathrm{F}/\sum f_i^2\right)\exp\left(-\mathrm{F}^2/\sum f_i^2\right).$$

This implies that the average probability of encountering a situation where the observed amplitude of some reflection is F_{obs} and the model amplitude is F_m is:

$$\langle|F_{obs}-F_m|\rangle = \int_0^{F_{obs}} (F_{obs}-F_M)\,P(F_M)dF + \int_{F_{obs}}^{\infty} (F_M-F_{obs})\,P(F_M)dF.$$

The reason the average has to be evaluated this way is that $|F_{obs} - F_m|$ is always positive. In order to evaluate these integrals, you have to remember that:

$$\int_0^{\infty} P(F)\,dF = \int_0^{F_{obs}} P(F)\,dF + \int_{F_{obs}}^{\infty} P(F)\,dF = 1.$$

It is also useful to recall that:

$$\left(2/\sum f_i^2\right)\int_0^{F_{obs}} F\exp\left(-F^2/\sum f_i^2\right)dF = 1-\exp\left(-F_{obs}^2/\sum f_i^2\right).$$

When it is all over, we find:

$$\langle|F_{obs}-F_{mod}|\rangle = F_{obs} - 2\int_0^{F_{obs}} \exp\left(-F_{obs}^2/\sum f_i^2\right)dF_{obs} + (1/2)\left(\pi\sum f_i^2\right)^{1/2}.$$

Now, the averaging has to be repeated so that we can obtain a value for $|F_{obs} - F_{mod}|$ averaged over all possible values of F_{obs}:

$$\langle\langle|F_{obs}-F_{mod}|\rangle\rangle = \int_0^{\infty} F_{obs}P(F_{obs})dF_{obs}$$

$$-2\int_0^{\infty}\left[\int_0^{F_{obs}} \exp\left(-F_{obs}^2/\sum f_i^2\right) dF_{obs}\right]$$

$$P\left(F_{obs}\right) dF_{obs} + (1/2)\left(\pi\sum f_i^2\right)^{1/2}\int_0^{\infty} P\left(F_{obs}\right) dF_{obs}.$$

This expression is not hard to evaluate, provided we recognize that the second integral on the right-hand side of the equation must be integrated by parts. (NB: If u is defined as the integral inside the square brackets of the second term, then du/dF_{obs} at any value of F_{obs} is $\exp\left(-{F_{obs}}^2/\sum f_i^2\right)$.) Using all this information, we find:

$$\langle\langle|F_{obs} - F_{mod}|\rangle\rangle = \left[\pi^{1/2} - (\pi/2)^{1/2}\right]\left(\sum f_i^2\right)^{1/2}.$$

By definition:

$$\begin{aligned} R_{awful} &= \langle\langle|F_{obs} - F_{mod}|\rangle\rangle/\langle F\rangle \\ &= \left[\pi^{1/2} - (\pi/2)^{1/2}\right]\left(\sum f_i^2\right)^{1/2} / (1/2)\left(\pi\sum f_i^2\right)^{1/2} \\ &= 2 - 2^{1/2} \approx 0.586. \end{aligned}$$

APPENDIX 7.3 THE DIFFERENCE FOURIER

Difference electron density maps are computed as follows:

$$\rho_d\left(x, y, z\right) = \sum_{h,k,l}\left[|F_{obs}(h, k, l)| - |F_{calc}(h, k, l)|\right]$$

$$\exp\left[2\pi i\left(hx + ky + lz\right) + i\alpha_{calc}(h, k, l)\right],$$

where $F_{obs}(h, k, l)$ is the observed amplitude of a reflection, and $F_{calc}(h, k, l)$ and $\alpha_{calc}(h, k, l)$ are the amplitude and phase computed for that same reflection using the model to be tested.

Using the generalized form of the Pythagorean theorem, and omitting (h, k, l) indices for convenience, we obtain (see figure 7.3.1):

$$F_{obs}^2 = \left[F_{calc} + \Delta F\cos\left(\Delta\alpha - \alpha_{calc}\right)\right]^2 + \left[\Delta F\sin\left(\Delta\alpha - \alpha_{calc}\right)\right]^2$$

$$F_{obs}^2 = F_{calc}^2 + \Delta F^2 + 2F_{calc}\Delta F\cos\left(\Delta\alpha - \alpha_{calc}\right).$$

In these equations, ΔF is the difference between F_{obs} and F_{calc}, and $\Delta\alpha$ is its phase. Hence:

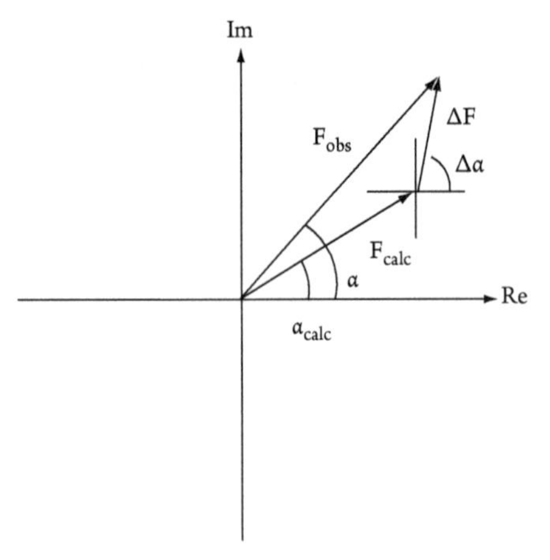

Figure 7.3.1 Argand diagram identifying the quantities that must be taken into account when computing difference electron density maps.

$$F_{\mathrm{obs}}^2 - F_{\mathrm{calc}}^2 = \Delta F^2 + 2F_{\mathrm{calc}}\Delta F\ \cos\left(\Delta\alpha - \alpha_{\mathrm{calc}}\right),$$

and thus:

$$|F_{\mathrm{obs}}| - |F_{\mathrm{calc}}| = \Delta F^2/\left(|F_{\mathrm{obs}}| + |F_{\mathrm{calc}}|\right) \\ +2F_{\mathrm{calc}}\Delta F\ \cos\left(\Delta\alpha - \alpha_{\mathrm{calc}}\right)/\left(|F_{\mathrm{obs}}| + |F_{\mathrm{calc}}|\right).$$

A useful expression for $[|F_{\mathrm{obs}}(h, k, l)| - |F_{\mathrm{calc}}(h, k, l)|]\exp[i\alpha_{\mathrm{calc}}(h, k, l)]$ can be extracted from this equation if we remember that $\cos(x) = [\exp(ix) + \exp(-ix)]/2$. Thus:

$$\begin{aligned}\left(|F_{\mathrm{obs}}| - |F_{\mathrm{calc}}|\right)\exp\left(i\alpha_{\mathrm{calc}}\right) = {} & \left[\Delta F^2/\left(|F_{\mathrm{obs}}| + |F_{\mathrm{calc}}|\right)\right]\exp\left(i\alpha_{\mathrm{calc}}\right) \\ & + \left[F_{\mathrm{calc}}\Delta F/\left(|F_{\mathrm{obs}}| + |F_{\mathrm{calc}}|\right)\right]\exp\left(i\Delta\alpha\right) \\ & + \left[F_{\mathrm{calc}}\Delta F/\left(|F_{\mathrm{obs}}| + |F_{\mathrm{calc}}|\right)\right] \\ & \quad \exp\left[i\left(\Delta\alpha + 2\alpha_{\mathrm{calc}}\right)\right].\end{aligned}$$

The physically relevant component of $(|F_{\mathrm{obs}}| - |F_{\mathrm{calc}}|)\exp(i\alpha_{\mathrm{calc}})$ is the second term to the right of the equal sign. Its phase is $\Delta\alpha$, the phase of ΔF, and if ΔF is small, which it will be if the model is reasonably accurate, $|F_{\mathrm{obs}}| \approx |F_{\mathrm{calc}}|$. Hence, the amplitude associated with the second term will be $\sim\Delta F/2$. Thus, the electron density this component contributes to $\rho_d(x, y, z)$ when the difference maps are computed will be $(1/2)[\rho(x, y, z) - \rho_{\mathrm{mod}}(x, y, z)]$.

The other two components of $\rho_d(x, y, z)$ are comparatively unimportant. If ΔF is small, the contribution of the first term to the right of the equal sign will be much weaker than the second because it is second order in ΔF. Furthermore, it will be a noisy version of the model for the macromolecule because its Fourier components have the same phases as that model, but inappropriate amplitudes (see figure 7.2). The third term, which has amplitudes the same size as the second, will contribute noise to $\rho_d(x, y, z)$ because its phases are uncorrelated with the underlying structure. Thus:

$$\rho_d\left(x, y, z\right) \approx (1/2)\left[\rho\left(x, y, z\right) - \rho_{\text{mod}}\left(x, y, z\right)\right].$$

PART THREE

Noncrystallographic Diffraction

8

Diffraction from Noncrystalline Samples

Biological materials that are not crystalline scatter radiation just like anything else, and as always, their scattering patterns reflect their structures. This chapter addresses two different kinds of scattering experiments that can be done with noncrystalline samples, one of which has still to prove itself, and the other of which is well-established. The more speculative of the two approaches may someday provide structural biologists with high-resolution, two-, or even three-dimensional X-ray images of noncrystalline biological specimens. The other technique, which has been in use for about 70 years, yields useful, low-resolution information about the conformations of macromolecules in solution.

8.1 X-RAY MICROSCOPY CAN BE DONE WITHOUT LENSES

A new method for imaging noncrystalline biological specimens with X-rays, which is called **coherent diffractive imaging** (CDI), has attracted great interest lately. In theory, CDI can yield two-dimensional images of biological specimens that resemble electron micrographs. Even though it is far too soon to predict what the impact of CDI is going to be on biology, it is worth discussing here because its relationship to better established imaging techniques is so interesting. CDI depends on the surprising fact that under certain circumstances, images can be recovered from experimentally measured scattering patterns in the absence of any prior information about the structure of the object from which the data were obtained, or of any experimental phase information.

In theory, CDI experiments are comparatively simple to carry out. Samples can be prepared for CDI the same way they are prepared for electron microscopy, namely by deposition on thin carbon films that are supported mechanically by a fine metal mesh. A sample of this sort is then exposed to a beam of monochromatic X-rays with the plane of the supporting film normal to the beam, and the scattered radiation is recorded on an area detector.

If useful images are to be obtained from the diffraction patterns of objects the size of whole cells, the transverse coherence lengths of the X-ray beams used to produce them must be unusually large, for example, tens of microns. This requirement is one reason why most of the CDI experiments reported so far have been done using long wavelength X-rays (e.g., $\sim$20 Å). The longer the wavelength of the X-rays used, the larger the angular divergence of the X-ray beam that can be tolerated for a given transverse coherence length (see section 1.9). The other reason is that the longer the wavelength of the radiation used, the larger the angular size of the small $|\mathbf{S}|$ part of a sample's diffraction pattern, which is where most of the information about the internal structures of cell-sized objects will be found. The larger that range of angles, the easier it is to measure the small $|\mathbf{S}|$ data.

CDI is a child of the synchrotron light source. Synchrotrons readily produce beams of long wavelength X-rays that are so brilliant that even after they have been subjected to the stringent collimation required to obtain the large transverse coherence lengths CDI experiments require, the flux of photons they deliver to samples is large enough to make data collection practical.

Before considering how images are recovered from CDI data, it is worth thinking about what that data will look like. In the first place, because the sample is held in a fixed orientation in the beam during data collection, the data represent a portion of a single, spherical shell of the three-dimensional Fourier transform of the sample. Furthermore, the part of the sphere of reflection that is close to $|\mathbf{S}| = 0$ will not be included in the data because of the beam stop, which for CDI experiments must be made as small as possible. The data are likely to weaken dramatically as $|\mathbf{S}|$ increases, and at some value of $|\mathbf{S}|$, which is unlikely to be very large, it will become indistinguishable from noise. Finally—and this is vitally important—because the specimens used for CDI experiments are not crystalline, the scattering data will be a continuously varying function of $\mathbf{S}$, rather than a set of discrete reflections.

8.2 THE FOURIER TRANSFORM OF A PROJECTION IS A CENTRAL SECTION

What would you learn about the three-dimensional structures of an object if you could Fourier transform the amplitudes gleaned from a scattering data set of this sort? If the data extend only to small values of $|\mathbf{S}|$ that question has a simple answer. In the region where scattering angles are small, the surface of the sphere of reflection is indistinguishable from the plane in reciprocal space that is tangent to that sphere at $|\mathbf{S}| = 0$. Any plane in reciprocal space that includes the point $|\mathbf{S}| = 0$ is called a **central section**. Thus, the image generated by inverse Fourier transformation of low-resolution CDI data will resemble the image that would result if the data from the corresponding central section were inverse Fourier transformed.

If the electron density distribution of the specimen is described using an orthogonal axis system (x, y, z), the z-axis of which is parallel to the incident X-ray beam, then the natural coordinate system for its transform will be X, Y, and Z, which are the axes in reciprocal space parallel to x-, y-, and z-axes, respectively. In this coordinate system, points in the central section of the transform of the object that is perpendicular to the beam and tangent to the Ewald sphere at $|\mathbf{S}| = 0$ will have reciprocal space coordinates $(X, Y, 0)$. Because:

$$\mathrm{F}(X, Y, Z) = \int_{-\infty}^{+\infty}\int_{-\infty}^{+\infty}\int_{-\infty}^{+\infty} \rho(x, y, z) \exp\left(-2\pi i(xX + yY + zZ)\right) dx\, dy\, dz,$$

it follows that:

$$\mathrm{F}(X, Y, 0) = \int_{-\infty}^{+\infty}\int_{-\infty}^{+\infty}\left[\int_{-\infty}^{+\infty} \rho(x, y, z) dz\right] \exp\left(-2\pi i(xX + yY)\right) dx\, dy. \quad \mathbf{(8.1)}$$

The term in square brackets in equation (8.1) is the **projection** of the electron density of the object onto the (x, y) plane in real space, and hence, the image generated when $\mathrm{F}(X, Y, 0)$ is inverse Fourier transformed will be that projection, which should resemble the images produced by microscopes of specimens so thin that they are in focus through their entire thicknesses.

Clearly as $|\mathbf{S}|$ increases, the distance between the surface of the sphere of reflection and the central section tangent to the sphere at $|\mathbf{S}| = 0$ will increase also, and at some scattering angle, the difference between the diffraction data measured in a CDI experiment and the data of the corresponding central section will become too large to ignore. The value of $|\mathbf{S}|$ beyond which the central section approximation ceases to be valid is determined by two parameters: (1) the wavelength of the X-rays used, and (2) the thickness of the specimen. The distance in reciprocal space between the surface of the Ewald sphere and the surface of the central section tangent to it at $|\mathbf{S}| = 0$, $\delta(2\theta)$, has the following dependence on 2θ:

$$\delta(2\theta) \approx (1/\lambda)(1 - \cos 2\theta).$$

The thickness of an object—its extent in z—determines the maximum rate at which its three-dimensional Fourier transform can vary as a function of Z. It follows that CDI data will deviate from central section data significantly when scattering angles exceed the scattering angle where $\delta(2\theta) \sim t^{-1}$, where t is the thickness of the sample. Because $(1 - \cos 2\theta) = 2 \sin^2 \theta$, it follows that $|\mathbf{S}|_{\text{limit}} \sim (2/t\lambda)^{1/2}$.

What does this equation mean in practice? Suppose a sample 1μ thick is being imaged by CDI using X-rays having a wavelength of 20 Å. The preceding formula indicates that the diffraction data obtained from that sample at scattering angles less than 3.6° will be central section data for all intents and purposes. At that scattering angle, the Bragg spacing is about 300 Å. Thus, if the measured data were truncated at that scattering angle—assuming it is intense enough to measure beyond that scattering angle—the image obtained by inversion of the truncated data would correspond to a projection image that had a (crystallographic) resolution of $\sim$300 Å. Because conventional optical microscopes have resolutions of the order of 2,000 Å (see section 9.15), we can understand why it might be interesting to image cells by CDI.

8.3 CONTINUOUS TRANSFORMS CAN BE INVERTED BY SOLVENT FLATTENING

The reason CDI data sets can be Fourier inverted is that they are oversampled. The transform of an object of a nonrepeating structure is continuous, and thus, in theory, the scattering profile produced by such an object can be measured at all locations in reciprocal space, not just at the small number of locations that would be permitted if the object were embedded in the unit cell of a crystal. Detectors necessarily generate sampled versions of continuous transforms because their surfaces are divided into pixels of finite size. However, it is (usually) a simple matter to position an area detector far enough from a sample so that the interval in reciprocal space represented by the separation between adjacent pixels on its face is small compared to L^{-1}, where L is the maximum linear dimension of the area illuminated by the X-ray beam. That length might be of the order of 20 μ, or 2×10^{-5} m, and it is the length that determines what the transverse coherence length of the X-ray beam would have to be to image an object that large properly.

If the detector used for a CDI experiment is placed so that its pixels sample the continuous transform of some object at intervals of $(2L)^{-1}$ in both dimensions, the data it provides will be proportional to what would have been obtained from a two-dimensional crystal of the object being imaged, the unit cells of which are $(2L \times 2L)$ squares. Inside each such hypothetical unit cell there will be an $(L \times L)$ square within which the entire image of the object of interest is contained, and the rest of the unit cell must have an electron density of 0. The density of measured data in reciprocal space thus exceeds the density required to specify the structure of the $(L \times L)$ square that contains the object by a factor of 4, and because Fourier coefficients are complex, that means that the density of the measured data in reciprocal space exceeds the Nyquist minimum by a factor of 2. The problem is overdetermined.

A number of procedures have been devised for solving overdetermined Fourier inversion problems such as this one (e.g., see Fienup, 1982), and they are all closely related to the solvent-flattening procedures crystallographers use to improve and extend phases (see sections 7.12 and 7.13). One way to begin is by computing a two-dimensional electron density map for the object of concern using measured $|F|$s and randomly assigned phases. The image that emerges will be total nonsense, but the next step in the procedure is to set the electron density in that image to 0 everywhere except in its middle quarter. Improved estimates for the phases can be obtained by Fourier transformation of this corrected image, and a second trial image can be computed using that new set of phases and, again, the measured |F|s. That new image is modified as previously described, and the cycle iterated as many times as it takes to make the electron density in the region being set to 0, stabilize at 0. When that happens, the structure of the structured part of the image will be found to have stabilized also, and the problem will have been solved. In other words, phases will have been found for the set of measured amplitudes that yield an electron density distribution for the unit cell that is consistent with prior knowledge—that is structured in its central region—and 0 everywhere else. Computational tests have shown that these procedures yield accurate images in both two and three dimensions, and that the set of |F|s used must be oversampled by not less than a factor of 2 in each dimension to ensure convergence.

8.4 CAN THE STRUCTURES OF MACROMOLECULES BE SOLVED AT ATOMIC RESOLUTION BY X-RAY IMAGING?

Advances in synchrotron technology may make it possible to determine the three-dimensional structures of macromolecules at atomic resolution using CDI-like methods. Conceptually, the strategy is simple. A fine spray is prepared of some macromolecular solution that has, on average, less than one macromolecule per droplet, and those droplets are passed through a pulsed beam of X-rays with the timing arranged so that a single-pulse, "snap-shot" diffraction pattern is recorded from each drop. Each diffraction pattern obtained from a drop that contains a macromolecule will provide scattering data that falls on the surface of a single Ewald sphere in the macromolecule's reciprocal space. It will correspond to what crystallographers call a "still". Figure 8.1 shows what such a still would look like if the macromolecule being investigated were a small protein. Provided the relative orientations of the shells of data obtained from successive drops can be determined, which is a problem electron microscopists have already solved in principle (see Chapter 13), an over-sampled version of $F(\mathbf{S})$ should emerge when the data obtained from many drops are merged. Once a complete data set has been amassed, it can be inverted using CDI methods.

The obvious advantages of this approach to determining molecular structure are that it bypasses the need to prepare crystals and that it should require only tiny quantities of material. Can it be executed in practice? For a long time, the biggest barrier to progress was the lack of suitable X-ray sources. Single macromolecules scatter X-rays so weakly that the total number of X-ray photons to which they must be exposed in order to obtain a decently accurate scattering pattern from them is orders of magnitude more than the number required to vaporize them. This being the case, the only way the effects of radiation damage can be avoided is if all the photons necessary to record such a diffraction pattern are delivered in a single X-ray pulse, the duration of which is much shorter than the picoseconds ($\sim 10^{-12}$ s) it takes for the chemical events to occur that will destroy the molecule. Synchrotron X-ray beams are naturally pulsed, but the pulses of radiation produced by the most powerful of today's synchrotrons are not intense enough for this application. However, a new

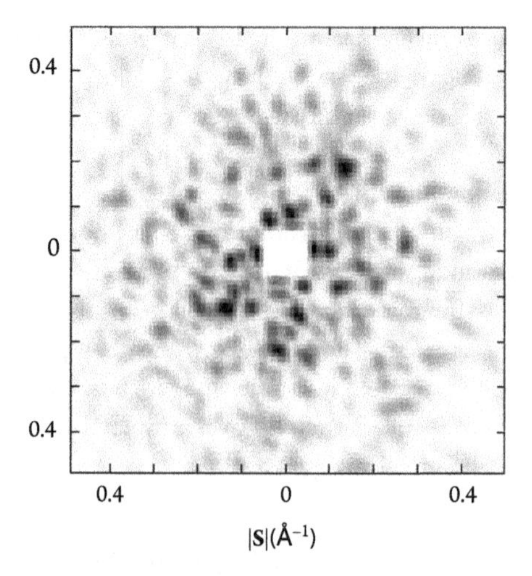

Figure 8.1 Computed appearance of a single still from the diffraction pattern of a small protein. I(**S**) is shown for a single shell of data that corresponds to the diffraction pattern of lysozyme, a protein having a molecular weight of about 14,700. Black features correspond to large I(**S**). In the computation done to produce this figure, it was assumed that the molecule was in the gas phase.

class of X-ray generators, called **free-electron lasers**, is under development that can produce femtosecond ($\sim 10^{-15}$ s) pulses containing the requisite number of photons.

Proof of principle experiments published as this book was going to press show that structures having resolutions at least as high as 1 nm can be produced using single-pulse data obtained using free-electron lasers. If the technology advances as expected, a revolution in macromolecular structure determination could result. However, it would be premature to give up on crystallography just yet. There is overwhelming evidence that the conformations of all macromolecules vary with time to at least some degree, and conformational variability will complicate the task of determining macromolecular structures by this single molecule method. The data merging this approach requires will certainly be based on the assumption that all the partial data sets obtained for some molecule were obtained from objects that all had the same conformation (before they were vaporized). It may turn out that in solution, the temperature factors of many macromolecules are so high that atomic resolution electron density maps cannot be produced for them this way. Thus, it may also turn out that crystals make macromolecular structure determination easier not only by amplifying scattering intensities, but also by limiting the number of conformational states those molecules can access.

8.5 SOLUTION-SCATTERING PATTERNS PROVIDE ROTATIONALLY AVERAGED SCATTERING DATA

For decades, biophysicists have obtained low-resolution information about macromolecules from the scattering observed when solutions of macromolecules interact with radiation. Thus, unlike CDI, macromolecular solution scattering is a well-established discipline that is backed by a rich body of theory, the basic elements of which will be explored here. The data sought in a solution-scattering experiment is the difference between the scattering profile of a solution of some macromolecule, and the scattering profile of the solvent in which that molecule is dissolved, with due allowance being made for the fact that the volume fraction of solvent in a macromolecular solution is less than 1.0.

In dilute solution, macromolecules, like all other solutes, behave as though they were components of an ideal gas. The key properties of an ideal gas are not only that the positions and rotational orientations of its constituent molecules are uncorrelated, but that they change rapidly on time scales much shorter than measurement times. For that reason, in the ideal limit, solute scatter is unaffected by intermolecular interference effects; the intensities of the radiation scattered by individual macromolecules add, not their scattering amplitudes.

Just as is the case with real gases, macromolecular solutions display ideal behavior only in the limit of zero concentration. Even in comparatively low concentration solutions, macromolecular positions may correlate appreciably, or, more accurately, anticorrelate, due to excluded volume effects. The higher the concentration, the more pronounced correlations of this sort become, and the more likely that other correlation-inducing phenomena such as aggregation will occur. These effects all have their greatest impact on the lowest scattering angle regions of solution-scattering curves, which is unfortunate because, as we shall see, this is the part of the solution-scattering profile of a macromolecule that is most likely to be useful. Thus, in theory, macromolecular solution-scattering profiles should always be measured at several concentrations, and the data extrapolated to zero concentration, point by point, to eliminate concentration effects.

If the structures of all the macromolecules in some solution are identical, and the radiation beam incident on the solution is a thin, cylindrical pencil, the data will be radially symmetric; it will depend only on $|\mathbf{S}|$, which is to say only on the scattering angle. Furthermore, it will be the sum of the scattering produced by molecules that are in every possible orientation. Thus, $\mathrm{I}(\mathbf{S})$ will be $\mathrm{I}(|\mathbf{S}|)$, and:

$$\mathrm{I}\,(|\mathbf{S}|) = N\langle \mathrm{I}_{\mathrm{mol}}(\mathbf{S})\rangle.$$

In this equation, N is the number of solute molecules illuminated by the beam, and $\langle \mathrm{I}_{mol}(\mathbf{S})\rangle$ is the rotationally averaged diffraction pattern of a single solute molecule.

The proportionality of $\mathrm{I}(|\mathbf{S}|)$ to N has an important practical consequence. By comparison with crystalline diffraction, solution-scattering intensities are very weak. Everything else being equal, at any given $\mathbf{S}$, the scattering produced by N macromolecules in solution will be N-fold weaker than the intensity of the reflections obtained from a crystal containing N molecules. (There are same caveats associated with this assertion, but qualitatively, it is correct.)

The relationship between the structure of a macromolecule and its rotationally averaged diffraction pattern is straightforward. From chapter 2, we know that for a single molecule in a specific orientation:

$$\mathrm{F}(\mathbf{S}) = \sum_{\text{all atoms}} \mathrm{f}_j \exp\left(-2\pi i \mathbf{r}_j \cdot \mathbf{S}\right),$$

where $\mathbf{r}_i$ is a vector describing the position of the jth atom, the scattering factor of which is f_i (equation (2.7)). It follows that:

$$\mathrm{I}(\mathbf{S}) = \sum_{\text{all}} \sum_{\text{atoms}} \mathrm{f}_i \mathrm{f}_j \exp\left[-2\pi i \left(\mathbf{r}_i - \mathbf{r}_j\right) \cdot \mathbf{S}\right].$$

The rotationally averaged value of $\mathrm{I}(\mathbf{S})$ will be designated $< \mathrm{I}(\mathbf{S}) >$, or $\mathrm{I}(|\mathbf{S}|)$, or $\mathrm{I}(S)$, depending on circumstances, and it is easy to show that:

$$\mathrm{I}(S) = \sum_{\text{all}} \sum_{\text{atoms}} \mathrm{f}_i \mathrm{f}_j \sin\left(-2\pi \left|\mathbf{r}_i - \mathbf{r}_j\right| S\right) / \left(-2\pi \left|\mathbf{r}_i - \mathbf{r}_j\right| S\right). \quad (8.2)$$

(see appendix 8.1). This equation, which is attributed to Debye, describes both the scatter produced by gases, and with small adjustments that will be described later, the contribution macromolecular solutes make to the scattering of solutions.

8.6 SOLUTION-SCATTERING EXPERIMENTS DETERMINE LENGTH DISTRIBUTIONS AND VICE VERSA

In equation (8.2), the only properties of pairs of atoms that count are their scattering factors and the distance between them. For that reason, the double sum over atoms in the Debye equation can be simplified using a function, $\mathrm{p}(r)$, which is defined as follows:

$$\mathrm{p}(r) \equiv \sum \mathrm{f}_i \mathrm{f}_j,$$

where the sum is over all pairs of atoms in a macromolecule that are separated by distances that lie between r and $r + dr$. Using $\mathrm{p}(r)$, $\mathrm{I}(S)$ can be written in integral form:

$$\mathrm{I}(S) = \int_0^{r_{\max}} \mathrm{p}(r) \left[\sin(2\pi r S)/2\pi r S\right] dr, \quad (8.3)$$

where $r_{\max}$ is the maximum separation between atoms in the molecule. [NB: Because $\mathrm{p}(r) = 0$ for all $r > r_{\max}$, equation (8.3) would be valid if $r_{\max}$ were replaced by ∞.]

The function $p(r)$ is called the **length distribution**, or **distance distribution** of a molecule because if the scattering factors of all the atoms in a molecule were the same, a plot of $p(r)$ versus r, suitably scaled, would display the number of interatomic vectors in that molecule having a length between r and $r + dr$ as a function of r; it would be that molecule's interatomic distance spectrum. $p(r)$ is the scattering factor–weighted version of the distance distribution of a molecule.

Equation (8.3) can be rewritten as follows:

$$SI(S) = (1/2\pi) \int_0^\infty [p(r)/r] \sin(2\pi rS)\, dr,$$

and once this is done, it becomes obvious that $[SI(S)]$ is the Fourier sine transform of $[p(r)/r]$, and that implies that $[p(r)/r]$ is the inverse Fourier sine transform of $[SI(S)]$:

$$p(r)/r = 8\pi \int_0^\infty [SI(S)] \sin(2\pi rS)\, dS.$$

It follows that once $I(S)$ has been measured for some molecule, its length distribution can be computed directly. It also follows that the length distribution of a

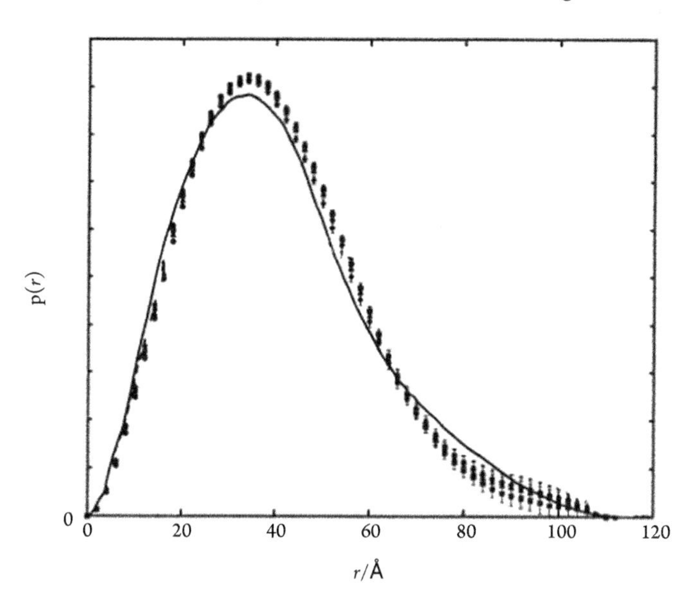

Figure 8.2 Typical length distributions. The length distribution computed for the protein elongation factor G using its crystal structure (solid line) is compared to the length distributions measured for it under several different solvent conditions (data points with error bars). $p(r)$ is given in arbitrary units. (Adapted from Czworkoski, J., Moore, P. B. [1997], The conformation properties of elongation factor G and the mechanism of translocation, *Biochemistry* 36, 10327–10334, with the permission of the American Chemical Society.)

molecule is a real-space representation of all the information about its structure that can be recovered from one of its solution-scattering profiles. The $p(r)$ profile in figure 8.2 is typical of those recovered from the solution-scattering data produced by solutions of globular proteins.

As we already know, the Patterson function of a molecule displays both the lengths and directions of all of its interatomic vectors. Furthermore, length distributions, like Patterson functions, are obtained by Fourier transformation of intensity data. Thus, length distributions and Patterson function ought to be related, and they are. $p(r)dr$ is the (weighted) sum of all the peaks in a molecule's three-dimensional Patterson function that are located between r and $r + dr$ from the origin.

It follows that the information in three-dimensional diffraction patterns that is missing from solution-scattering profiles has to do with the directions of interatomic vectors. That said, if the length distribution of a macromolecule could be assigned, which is to say if all pairs of atoms in that molecule for which $|\mathbf{r}_i - \mathbf{r}_j|$ falls between r and $r + dr$ could be identified for all r, it would be possible to recover the three-dimensional structure of the macromolecule from its length distribution by triangulation. Furthermore, if this were done, the only aspect of the molecule's structure left undetermined would be its absolute hand, assuming it is chiral. Because nothing better would emerge if the peaks in the three-dimensional Patterson function of the same molecule were assigned, we might think that the solution-scattering profile of a macromolecule is every bit as valuable as its three-dimensional diffraction pattern, which is certainly not so. Nevertheless, in a small number of cases, solution-scattering experiments have been done that made it possible first to assign features in length distributions, and then to generate three-dimensional models for structures from such data by triangulation.

8.7 MOLECULAR WEIGHTS AND RADII OF GYRATION ARE EASILY EXTRACTED FROM SOLUTION-SCATTERING PROFILES

The low angle part of a macromolecular solution-scattering curve contains useful information about molecular size and shape. This fact becomes evident when equation (8.2) is expanded as a power series:

$$\mathrm{I}(S) = \sum_{\text{all}} \sum_{\text{atoms}} \mathrm{f}_i \mathrm{f}_j \left[1 - (1/3!) \left(-2\pi \left|\mathbf{r}_i - \mathbf{r}_j\right| S\right)^2 + \ldots\right].$$

The scatter of some macromolecule in the $S = 0$ direction, that is, its **forward scatter**, $\mathrm{I}(0)$, which can never be directly observed because of the beam stop, equals $\sum\sum \mathrm{f}_i \mathrm{f}_j$, which is simply the square of the total number of electrons in the molecule. Thus:

$$\mathrm{I}(S) = \mathrm{I}(0) \left[1 - (1/3!) \left(2\pi |\mathbf{r}_i - \mathbf{r}_j| S\right)^2 + \ldots\right].$$

This equation applies to a single molecule. For a sample containing N such molecules, the forward scatter will be $N\mathrm{I}(0)$, and the forward scatter produced by some macromolecular solution in some particular experimental apparatus can be written as follows:

$$I(0) = k\left[(electrons/mass)\,(MW/No)\right]^2 (C/MW).$$

In this expression, k is an instrumental constant that depends on factors such as the size of the volume illuminated by X-rays, the X-ray flux, and the measurement time, and $(electrons/mass)$ is the number of electrons per unit mass of the macromolecule in question, which can be obtained from its elemental composition. C is the weight concentration of the macromolecule in the solution; MW is its molecular weight; and N_o is Avogadro's number. Rearranging this expression slightly, we find:

$$I(0)/C = k[(electrons/mass/No)]^2(MW).$$

Thus, the molecular weight of the macromolecule responsible for the scattering measured is proportional to $I(0)/C$, and, hence, molecular weights can be obtained from solution-scattering data.

The easiest way to obtain molecular weight estimates from solution-scattering methods is to compare the $[I(0)/C]$s of macromolecules of unknown molecular weight with the $[I(0)/C]$s of macromolecules of known molecular weight that belong to the same chemical class, using the same instrument. This approach is convenient because the number of electrons per unit mass varies only slightly within any given chemical class—all proteins are alike in this regard—and because it is often a nuisance to measure/estimate k directly.

Forward scatter cannot be measured directly, but its value can be estimated by extrapolation, provided solution-scattering data are measured at values of 2θ that are small enough. As the power series expansion discussed previously indicates, at small angles:

$$I(S) \approx I(0)\left\{1 - \left[(4\pi^2/6I(0)) \sum_{\text{all}} \sum_{\text{atoms}} f_i f_j \left|\mathbf{r}_i - \mathbf{r}_j\right|^2\right] S^2\right\}.$$

For convenience, call quantity inside the square brackets A, and recall that when $\exp(-x^2)$ is expressed as a power series, its first two terms are $(1 - x^2)$. Thus, at small S, $I(S)$ should resemble a Gaussian that has the following functional form:

$$I(S) = I(0)\exp(-AS^2).$$

It follows that at very low angles, solution-scattering data should fall on a straight line when plotted as $\ln[I(S)]$ versus S^2, which makes extrapolation to $S = 0$ easy. The intercept at $S = 0$ will be $\ln[I(0)]$, and slope of the line will be $(-A)$.

The quantity A is interesting in its own right because it reports on a useful, averaged property of the structure of a macromolecule. As demonstrated in appendix 8.2,

$$A = (4\pi^2/3)\left(\sum f_i r_i^2\right)/\sum f_i,$$

where r_i is the distance between the ith atom and the center of scattering mass of the molecule. The quantity $\left[\sum f_i r_i^2/\sum f_i\right]^{1/2}$ is called the molecule's **radius of gyration**. It gets its name because of its similarity to a quantity of the same name encountered in the mechanics of solid objects. In mechanics, the radius of gyration of an object, R_g, is:

$$R_g = \left[\sum m_i r_i^2/\sum m_i\right]^{1/2},$$

where m_i is the mass of the ith mass element in the object, and the origin of the coordinate system is located at the center of mass of the object. It follows that the low-angle regions of the solution-scattering profile of a macromolecule should obey the following equation:

$$\mathrm{I}(S) \approx \mathrm{I}(0)\exp\left[-(4\pi^2/3)Rg^2S^2\right].$$

In this expression, R_g is the scattering radius of gyration, not the mechanical radius of gyration. This equation is called **Guinier's law** after the scientist who discovered it.

Because atomic masses of low Z atoms are roughly proportional to atomic number, the radius of gyration of a macromolecule determined by scattering methods will be almost the same as its mechanical radius of gyration. Thus, if the mass and density of a molecule are known, comparison of the volume of the molecule to its radius of gyration affords a sense of its shape. The smallest radius of gyration an object can have is that of the sphere of the same volume, and it is easy to show that the radius of gyration of a sphere of uniform density of radius a is $(3/5)^{1/2}a$. If the measured radius of gyration is close to that minimum value, the object must be quasi-spherical, if it is much larger, the molecule must be elongated.

Both $\mathrm{I}(0)$ and R_g can be obtained from $\mathrm{p}(r)$. The forward scattering is the area under the length distribution curve of a macromolecule:

$$\mathrm{I}(0) = \int_0^\infty \mathrm{p}(r)dr$$

and:

$$R_g^2 = (1/2)\left[\int_0^\infty r^2\mathrm{p}(r)dr/\int_0^\infty \mathrm{p}(r)dr\right].$$

Thus, R_g^2 is half of the second moment of a macromolecule's length distribution. For the record, two other statistical parameters that characterize the structure of a macromolecule can be extracted directly from $p(r)$. The first is the average interatomic vector length, which is simply the first moment of its length distribution curve, and the second is r_{max}, which is the largest distance between a pair of atoms in that molecule. Beyond r_{max}, $p(r) = 0$.

8.8 SOLUTION-SCATTERING PROFILES ARE STRONGLY AFFECTED BY THE SCATTERING LENGTH DENSITIES/ELECTRON DENSITIES OF SOLVENTS

In the theory presented in the preceding section, the solvent in which macromolecules are dissolved is completely ignored; it is treated as though it were a vacuum. However, the electron densities of solvents are not 0, and solvent electron densities alter the contribution that macromolecules make to the total scattering from macromolecular solutions.

A useful, first approximation description of the effects solvents have on macromolecular scatter can be obtained by treating the volumes in solutions that are filled by solvent as volumes in which the electron density is constant. The electron density distribution of the macromolecules in some solution can be divided into two parts, a constant part equal to the average electron density of the macromolecule, $\langle\rho\rangle$:

$$\langle\rho\rangle = (1/V)\int \rho(\mathbf{r})dV = (1/V)\mathrm{I}(0),$$

where V is the volume of the macromolecule, and the integral runs over the entire volume of the molecule, and a variable part, $\delta(\mathbf{r})$, which is defined as $(\rho(\mathbf{r}) - \langle\rho\rangle)$. [NB: $\int \delta(\mathbf{r})dV = 0$.] The macromolecular component of the scattering profile of a macromolecular solution is generated by the perturbations in the electron density distribution of those solutions caused by their presence, $\mathrm{D}(\rho(\mathbf{r}))$, and because solvent cannot occupy volumes that are filled by macromolecules, to first approximation:

$$\mathrm{D}\left(\rho\left(\mathbf{r}\right)\right) = \left(\langle\rho\rangle - \rho_{sol}\right) + \delta\left(\mathbf{r}\right).$$

Using this expression for $\mathrm{D}(\rho(\mathbf{r}))$, and writing equation (8.2) in integral form, we obtain:

$$\mathrm{I}(S) = \int\int \left[\langle\rho\rangle - \rho_{sol} + \delta(\mathbf{r}_i)\right]\left[\langle\rho\rangle - \rho_{sol} + \delta(\mathbf{r}_j)\right] \left[\sin(2\pi|\mathbf{r}_i - \mathbf{r}_j|S)/(2\pi|\mathbf{r}_i - \mathbf{r}_j|S)\right] dV_i dV_j.$$

If the shape of the volume occupied by a macromolecule is described using a function, $\mathrm{V}(\mathbf{r})$, that has a value of 1.0 for all $\mathbf{r}$ inside the boundary of the macromolecule, and a value of 0 everywhere else, this equation can be recast as follows:

$$\begin{aligned} I(S) = & \Delta\rho^2 \int\int V(\mathbf{r}_i)V(\mathbf{r}_j)\left[\sin(2\pi|\mathbf{r}_i - \mathbf{r}_j|S)/(2\pi|\mathbf{r}_i - \mathbf{r}_j|S)\right] dV_i dV_j \\ & + 2\Delta\rho \int\int V(\mathbf{r}_i)\delta(\mathbf{r}_j)\left[\sin(2\pi|\mathbf{r}_i - \mathbf{r}_j|S)/(2\pi|\mathbf{r}_i - \mathbf{r}_j|S)\right] dV_i dV_j \\ & + \int\int \delta(\mathbf{r}_i)\delta(\mathbf{r}_j)\left[\sin(2\pi|\mathbf{r}_i - \mathbf{r}_j|S)/(2\pi|\mathbf{r}_i - \mathbf{r}_j|S)\right] dV_i dV_j, \end{aligned} \quad (8.4)$$

where $\Delta\rho$, which is called the **contrast**, equals $(\langle\rho\rangle - \rho_{sol})$. The first term on the right side of equation (8.4) is the scattering that would be observed if the electron density of the macromolecule was constant inside its boundary. The third term is scatter generated by fluctuations in electron density within the volume of the macromolecule. The second term represents interference between shape scatter and fluctuation scatter. Note that at $S = 0$, the first term is $(\Delta\rho V)^2$, where V is the volume of the macromolecule, and both the second and the third terms in the sum are 0 because the integral of $\delta(\mathbf{r})$ over the volume of the molecule is 0.

Suppose a series of samples were prepared that consisted of some macromolecule dissolved in solvents that differ in electron density, and macromolecular scattering profiles were measured for each such solution under otherwise identical conditions. Equation (8.4) tells us that the data obtained could be represented as a contrast-weighted sum of three parent scattering curves as follows:

$$I(S)_{\Delta\rho} = A(S)\Delta\rho^2 + B(S)\Delta\rho + C(S).$$

$A(S)$ is the scattering curve for $V(\mathbf{r})$, the molecular shape; $B(S)$ is the interference scattering curve; and $C(S)$ is the fluctuation scattering curve. Thus, it ought to be possible to recover $A(S)$, $B(S)$, and $C(S)$ from a set of scattering curves measured at different contrasts. In most contexts, $A(S)$ is the most interesting of these parent scattering curves because it reports on molecular shape.

8.9 SHAPE SCATTER DOMINATES MOST SOLUTION-SCATTERING CURVES

Most of the macromolecular solution-scattering data collected today is X-ray solution-scattering data, and with the aid of a few back-of-the-envelope calculations, it is easy to work out what scattering profiles of this sort will look like qualitatively. For example, hemoglobin is a protein that has a molecular weight of roughly 64,000, and like other proteins, its partial specific volume is $\sim 0.74 \times 10^{-3}$ m^3/kg. For these purposes, it is enough to think of it as being composed of roughly 5,000 carbon atoms and 5,000 hydrogen atoms; thus, a single hemoglobin molecule contains $\sim 35,000$ electrons. Because its volume is about 7.9×10^{-26} m^3, its average electron density is ~ 0.44 electrons/Å^3. The average electron density of pure water is ~ 0.33 electrons/Å^3. Hence, the forward scatter per hemoglobin molecule of an aqueous solution of hemoglobin will be proportional to $\sim 7.6 \times 10^7$

electrons squared. Using Wilson statistics (section 5.12), it can be shown that at high scattering angles, per molecule of hemoglobin, <I> will be roughly 400 times less than the forward scatter, that is, $\sim 1.8 \times 10^5$ electrons squared. Thus, at small S, both the shape component of the molecule's overall scattering curve, A(S), and its shape-fluctuation component, B(S), will be much stronger than its fluctuation component, C(S).

Even though it is usually only approximately correct over extended ranges of S, Guinier's law can be used to estimate how rapidly the intensity of the shape component of a macromolecule's scattering curve will decrease with increasing S. If hemoglobin were a sphere, its radius of gyration would be about 21 Å, and if Guinier's law exactly described A(S) for hemoglobin (which it would not if hemoglobin were a sphere!), the intensity of its shape scatter would be one-tenth of I(0) at $S \sim 0.02$ Å^{-1}, and it would be down by a factor of 100 at $S \sim 0.028$ Å^{-1}, where the contribution of fluctuation scatter would begin to count. Thus, if X-rays having wavelengths in the neighborhood of 1.0 Å were used to measure the solution-scattering curve of a molecule such as hemoglobin—as they commonly were in the era when laboratory X-ray generators were used for solution scattering experiments—0.028 Å^{-1} would correspond to a scattering angle of 1.6°. Thus, the most intense part of the solution-scattering profile of even quite small proteins will be found inside $2\theta = 5°$, which is why solution-scattering experiments are often referred to as **small angle scattering** experiments. For bigger macromolecules, the importance of shape scatter relative to the other two components of their solution-scattering profiles is even greater, but the region in reciprocal space within which shape scatter dominates is even smaller.

Solvent contrast also affects the diffraction data collected from macromolecular crystals, and it is instructive to consider what its impact might be. If hemoglobin were spherical, it would have a diameter of about 54 Å, and, therefore, it could be imagined as crystallizing in a cubic lattice, the unit cells of which have the dimensions: $a = b = c = 54$ Å. The (1, 0, 0) reflection from such a lattice would fall at $S = 0.0185$ Å^{-1}, where the shape scatter would be $\sim 1/7$ of I(0), the fluctuation scatter might be $\sim 1/300$ of I(0), and the fluctuation-shape correlation scatter might be $\sim 1/30$ of I(0). Two conclusions follow. First, the intensities of the lowest angle reflections in the diffraction patterns obtained from macromolecular crystals *are* affected by solvent contrast. Second, on average, the low-angle reflections in macromolecular diffraction patterns, which describe the shape of the macromolecule, are likely to be much stronger than the high-angle reflections that represent primarily intramolecular fluctuations in electron density.

It follows that solvent contrast must be taken into account whenever estimates for the intensities of the low-angle reflections in the diffraction patterns of macromolecules are computed using atomic resolution models for macromolecules, as they are during the refinement of macromolecular crystal structures. If they are not, the intensities predicted for these reflections will be much larger than they should be, on average, which could have baleful consequences for any structure refinements done that take those reflections into account. Experience shows that solvent contrast needs to be allowed for out to resolutions of roughly 10 Å.

8.10 MEASURED RADII OF GYRATION ARE CONTRAST-DEPENDENT

As Guiner's equation shows, the radii of gyration of macromolecules can be obtained by analyzing the shapes of solution-scattering curves in the region where S is close to 0. They can also be estimated by computing the second moment of $p(r)$, which might appear to be a completely different undertaking computationally, but it is invariably found that whenever the small angle data for a macromolecule are poorly measured, the second moment of the $p(r)$ will be poorly determined by the data also. Because this part of any macromolecule's solution-scattering curve is (usually) dominated by shape scatter, we might think that the radii of gyration measured for macromolecules would invariably be the same as the radii of gyration of uniform solids having the same shapes. This is not so, and a moment's thought reveals why. Although it is true that fluctuation and correlation contributions to the solution-scattering curve of a macromolecule are 0 at $S = 0$, for all finite values of S, they are not. Thus, the data used in Guinier plots necessarily include contributions from $B(S)$ and $C(S)$, and their importance relative to $A(S)$ depends on contrast.

Many years ago, H. Sturhmann pointed out that R_{obs}, the radius of gyration observed for a molecule in a solvent of some contrast, $\Delta\rho$, must show the following dependence on contrast:

$$R_{obs}^2 = R_\nu^2 + (\alpha/\Delta\rho) - (\beta/\Delta\rho^2),$$

where R_V is the radius of gyration of $V(\mathbf{r})$, and α and β are defined as follows:

$$\alpha = (I/V)\int \mathbf{r}^2\delta(\mathbf{r})dV$$

and:

$$\beta = (1/V^2)\int\int \mathbf{r}_i \cdot \mathbf{r}_j\delta(\mathbf{r}_i)\delta(\mathbf{r}_j)dV_i dV_j.$$

In both these expressions, V is the volume of the macromolecule.

α is the second moment of the fluctuation component of the electron density of a macromolecule, and it can be either positive or negative. For example, if the structure being studied consists of a protein shell surrounding a lipid-rich interior, α will be positive, because lipid has a lower electron density than protein does. Thus, if the radius of gyration of this assembly were measured under conditions where $\Delta\rho$ is small and positive, the measured radius gyration of the complex would be greater than R_s. For such molecules, a plot of R^2 versus $\Delta\rho^{-1}$, a **Stuhrmann plot**, will have a positive slope.

β characterizes a less intuitive property of $\delta(\mathbf{r})$. If $\delta(\mathbf{r})$ has an appreciable first moment, that is, if $\int \mathbf{r}\delta(\mathbf{r})dV \neq 0$, then the location of the molecule's center of scattering mass, the location where $\int[\langle\rho\rangle - \rho_{sol} + \delta(\mathbf{r})]\mathbf{r}dV = 0$, will change with contrast, and that change is bound to affect the radius of gyration we measure. β is always positive. A lipoprotein complex that is lipid-rich at one end and protein-rich at the other, for example, would have an appreciable β, and this would make its measured radius of gyration less than R_V whenever $|\Delta\rho|^{-1}$ is large. Thus, the Sturhmann plot for such a molecule will be curved downward.

8.11 CONTRAST VARIATION EXPERIMENTS ARE MORE EASILY DONE USING NEUTRON RADIATION THAN X-RAYS

As sections 8.8 and 8.10 show, the amount of information that can be recovered from solution-scattering data increases substantially when solution-scattering profiles are measured at several different contrasts. Thus, an important practical question arises: How can the contrast of a macromolecular solution be altered?

When X-rays are being used for solution-scattering measurements, the options available for adjusting contrast are limited. The electron densities of molecules are determined by their atomic compositions, and unacceptably massive chemical changes would be required to significantly alter them. There is more room to maneuver on the solvent side of the system, but still less than we might wish. The electron density of aqueous solvents can be adjusted upward by adding solutes to them. Sugars can be used, or high atomic number salts such as CsCl. However, the solute concentrations necessary to effect significant changes in solvent electron density are high, and biological macromolecules may aggregate, precipitate, or change their conformations when solute concentrations are that large. Even when the macromolecule of concern tolerates high concentrations of such solutes, the range of solvent electron densities that can be accessed may not be all that large due to solubility limitations.

Neutron radiation is far better suited for contrast variation experiments than X-radiation is. In section 3.14, it is pointed out that the neutron-scattering lengths of the different isotopes of a single element are not the same, and that the one of the biggest isotopic differences known is the one that distinguishes ^{1}H from ^{2}H (or D). For neutron beams having wavelengths in the angstrom range, the average scattering length density for thermal neutrons of ordinary, "light" water, H_2O, is less than that of any biopolymer, and the average scattering length density of heavy water, D_2O, is larger than that of any biopolymer. Thus, when neutron-scattering experiments are carried out on biological macromolecules, $\Delta\rho$ can be adjusted over a very wide range of both positive and negative values simply by using light water/heavy water mixtures as solvents. We can even study the scattering of macromolecules under conditions where $\Delta\rho = 0$, which is the condition called **contrast match**. When a macromolecular solution is contrast matched, only its fluctuation scatter, $C(S)$, will be observed. Because the $\Delta\rho$ changes that make this possible are obtained entirely by isotopic substitution, effects of its alteration on macromolecular structure are likely to be modest.

The average scattering length densities of macromolecules can be increased by replacing the ^{1}H's they normally contain with ^{2}H's. This kind of substitution, which can often be achieved biosynthetically at modest cost, and has minimal effects on conformation and activity, can dramatically increase the average scattering length density of these molecules. A plethora of neutron-scattering experiments have been reported over the years that depend on this fact.

Contrast variation data generated by neutron scattering are somewhat more complicated to analyze than the preceding discussion might suggest. All biological macromolecules contain hydrogen atoms that can exchange with protons in the solvent (i.e., water protons). Every hydrogen atom in a macromolecule that is not bonded to a carbon atom has the potential to do so, and the rate at which it does is strongly influenced by conformation. Thus, any biological macromolecule dissolved in heavy water rapidly becomes partially deuterated to a degree that can be determined accurately only by experiment, and its average scattering length density rises as a consequence. Exchange effects need to be taken into account in both the design and the execution of neutron-scattering experiments.

8.12 IT IS SURPRISINGLY DIFFICULT TO COMPUTE SOLUTION-SCATTERING CURVES

Three classes of experiments have tended to dominate in the solution-scattering community over the years. In the first type of experiment, which we might call the "Has it changed?" experiment, solution scattering is used to assay changes in macromolecular mass and/or conformation, often as a function of time following some perturbation in external conditions. The second type of experiment asks: "Is the structure of this molecule in solution the same as it is in crystals?" The third seeks to answer a more difficult question: "What is the shape of this molecule?" Experiments of the first type are so simple conceptually that they need no further discussion. Experiments of the second and third types deserve comment because a proper understanding of the relationship between the three-dimensional structures of macromolecules and their solution-scattering profiles is necessary for their successful execution.

Computation of the solution-scattering profile of a macromolecule of known three-dimensional structure from first principles is a prerequisite to carrying out experiments of the second type, and it is not as easy at it sounds. In the first place, no matter how the computation is approached, there are a lot of interatomic distances to consider. A macromolecule that consists of N atoms will have $(N/2)(N-1)$ such distances, and, thus, for hemoglobin, for example, about 5×10^7 distances would have to be determined to compute $p(r)$. In the second place, it is not obvious how the solvent should be treated, even if it is to be regarded as having a constant electron density/scattering length density. In that approximation, the function that needs to be computed is the rotational average of $[F(\mathbf{S}) - \Delta\rho S(\mathbf{S})]^2$, where $F(\mathbf{S})$ is the three-dimensional Fourier transform of the structure of the molecule in a vacuum, and $S(\mathbf{S})$ is the three-dimensional Fourier transform of $V(\mathbf{r})$, the function describing its three-dimensional shape.

You might think that once a macromolecule's crystal structure has been determined that $V(\mathbf{r})$ would be easy to compute, but this is not so. First, *all* the water and low molecular weight solutes that associate with a macromolecule in solution are components of that macromolecule for solution-scattering purposes, and their numbers, identities, and locations are generally unknown. (Note that the solvents in which macromolecules crystallize are usually not the solvents in which their solution behavior is studied.) Second, there are certain to be geometric, excluded volume effects at solvent/macromolecule interfaces that are hard to characterize, irrespective of solvent binding. Third, there are additional solvent effects to consider such as electrostriction, which is the compaction of the structure of water that occurs in the neighborhood of charged groups. The macromolecular component of the scattering produced by macromolecular solutions subsumes *all* such effects because the positions of these electron density perturbations correlate with the positions of the macromolecules responsible for them. Thus, it will only be by happy accident that the solution-scattering profile computed for some macromolecule on the basis of its crystal structure alone matches its solution-scattering profile, even if its structure in solution is exactly the same. (Problems of this type arise whenever attempts are made to rationalize bulk sample data [i.e., thermodynamic data], in molecular terms; they are not specific to solution scattering.)

In this context, it is useful to think about the impact the truncation of solution-scattering data sets has on the length distributions obtained by inverting them. Solution-scattering data sets commonly end at very low resolution, usually because the data have become too weak to measure. When the three-dimensional image of a molecule, $\rho(\mathbf{r})$, is computed using truncated three-dimensional diffraction data, $\rho(\mathbf{r})$ will be the convolution of the true electron density distribution of the molecule with the transform of the sphere in reciprocal space that describes that truncation. By atomic standards, the real-space images of the spheres that describe the truncation of data in solution-scattering experiments are very large. For example, if $S_{max} = 0.1$ Å^{-1}, the radius of the first 0 of its transform will be 7.15 Å. The convolution of the true electron density distribution of a molecule with a sphere transform that large will dramatically smooth the fluctuations in electron density both inside molecules and at molecule–solvent interfaces. The consequences of truncation for solution scattering are just the same. The $p(r)$ computed using truncated data will be smoother than the $p(r)$ computed using untruncated data, and not that much different from the $p(r)$ that would be obtained from an object of uniform density that has the shape resembling $V(\mathbf{r})$, but with less abrupt edges.

Does the smoothing just described cause serious harm? As already pointed out, the length distribution of hemoglobin describes ~50,000,000 interatomic distances, the longest of which is <75 Å. Thus, the density of interatomic distances in the length distribution of hemoglobin, i.e. the number of distances per unit length, is so high that there is no hope that its length distribution will include discrete peaks that correspond to distances between specific pairs of atoms, no matter how complete the data set used to compute it. Macromolecular length distributions will always be smoothly varying functions of distance that are dominated by molecular shape, just like the one shown in figure 8.2, and there is usually little point in collecting data past the point where $A(S)$ becomes small compared to $C(S)$.

All of these difficulties not withstanding, several useful algorithms have been developed for computing solution-scattering profiles for macromolecules of known structure. Many of them model solvent–molecule interfaces in sophisticated ways and use interesting algorithms that reduce both computational costs and the impact of computational artifacts, but it would be inappropriate to discuss the details here. Suffice it to say that the solution-scattering curves that have been computed for many biological macromolecules on the basis of their crystal structures match the corresponding experimental data to within experimental error over substantial ranges of S. Interestingly, the lowest angle regions of these computed scattering curves, where measurement error tends to be small, are often where the discrepancies between measurement and prediction are greatest, but this should not be surprising. The solvent effects discussed earlier modify the appearance of the periphery of molecules where $|\mathbf{r}|$ is large and, thus, might be expected to have a large impact on radius of gyration. Furthermore, the inner portions of a solution-scattering profile are extremely sensitive to aggregation phenomena. Nevertheless, computed radii of gyration commonly agree with measured radii of gyration to better than $\pm 5\%$. Thus, the solution structures and crystal structures of at least some biological macromolecules must be very similar. Conversely, it follows that it means something when measured solution-scattering curves and computed solution-scattering curves do not agree well.

8.13 USEFUL MODELS FOR THE SHAPES OF MACROMOLECULES CAN BE DERIVED FROM SOLUTION-SCATTERING CURVES

One of the most surprising developments in this field in the past 20 years has been the emergence of algorithms for extracting useful estimates of molecular shape from solution-scattering data. Ideally, the scattering profile analyzed should be $A(S)$, the shape scatter of a molecule, which can only be accessed by doing contrast variation experiments, but if the contrast is high, useful results can be obtained from a single data set. The reason this development is surprising is that formally, the number of independent parameters describing a macromolecular structure that can be extracted from its solution-scattering profile is so small that you would think that a range of molecular shapes would be compatible with any measured scattering curve.

Most of the algorithms for deriving molecular shapes from scattering profiles represent molecules as clusters of much smaller objects of uniform size, shape, and density, for example, spheres. Solutions-scattering profiles for such clusters can be computed using techniques that minimize the impact of irrelevant aspects of such clusters, such as their packing defects and the curvature of the surfaces of their components, thereby ensuring that the profiles obtained approximate those expected for objects of the same overall shape that have uniform internal density and a smooth exterior surface. The size and number of objects used to generate these models is set so that the volume of the cluster corresponds to the volume of the macromolecule responsible for the data under analysis.

An algorithm developed by D. Svergun, which is widely used, does a Metropolis Monte Carlo computation to find arrangements of sets of spheres of the appropriate total volume that: (1) have solution-scattering profiles similar to some measured scattering profile, and (2) have the least possible amount of surface area exposed to solvent. The results of numerous tests indicate that the molecular shapes recovered from solution-scattering data this way are surprisingly accurate, which implies that the requirement that surface area be minimized is a very powerful constraint. Because it is unclear why this should be so, the possibility remains that this algorithm

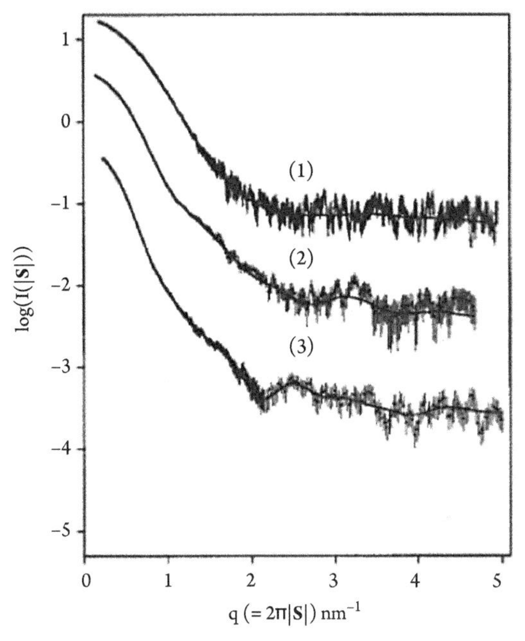

Figure 8.3 Typical solution-scattering curves. Solution-scattering profiles are shown for three proteins: (1) the monomeric form of yeast hexokinase, (2) the dimeric form of yeast hexokinase, and (3) yeast pyruvate dehydrogenase. Log of intensity is plotted in arbitrary units with adjacent curves offset by about 1 log to prevent overlap. The smooth curves in each profile are the scattering curves predicted for the protein in question based on its crystal structure. (Reprinted from Svergun, D. I., Petoukhov, M. V., and Koch, M. H. J. [2001], Determination of domain structure of proteins from X-ray solution scattering, Biophysical Journal 80, 2946–2953, with permission from the Biophysical Society.)

may occasionally fail. Figure 8.3 shows the scattering curves of three proteins, and figure 8.4 compares the shapes of those proteins inferred from their scattering curves with their known crystal structures. The agreement is quite good.

8.14 THE EXPERIMENTAL APPARATUS FOR SMALL ANGLE SCATTERING RESEMBLES THAT USED FOR COHERENT DIFFRACTIVE IMAGING

The issues that need to be addressed in the design of instruments for measuring solution-scattering data are similar to those that must be considered when designing instruments for measuring CDI data. In fact, the specifications that have to be met

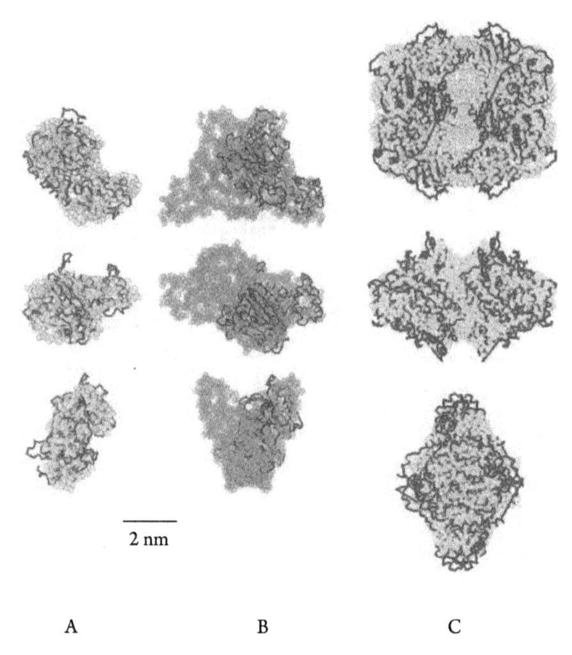

Figure 8.4 Protein shapes determined using small angle scattering data compared to protein crystal structures. The shapes (gray beads) are the molecular shapes recovered from small angle scattering data. The darker chains superimposed on the bead shapes represent the trajectories of the polypeptide chains of those same proteins determined crystallographically. Each superposition is shown in three orthogonal views. A: hexokinase monomer; B: the hexokinase dimer with the structure of a single monomer superimposed; C: pyruvate decarboxylase. (Reprinted from Svergun, D. I., Petoukhov, M. V., and Koch, M. H. J. [2001], Determination of domain structure of proteins from X-ray solution scattering, *Biophysical Journal* 80, 2946–2953, with permission from the Biophysical Society.)

by solution-scattering instruments are less stringent simply because lower transverse coherence lengths are acceptable in solution-scattering instruments. For most purposes, a solution-scattering profile that starts at a scattering angle corresponding to a Bragg spacing of $\sim 10^{-7}$ m will be more than adequate. CDI data need to begin at Bragg spacings that are 10 to 100 times larger.

No matter how well they are designed, both types of instruments share a shortcoming that has so far been ignored. Both are intended to measure scattering profiles that vary continuously with either $|\mathbf{S}|$ or $\mathbf{S}$, but both necessarily measure those profiles using radiation beams of finite cross-sectional size and finite beam divergence (see section 1.9). It follows that the data recorded on the detector, be it $I_{meas}(S)$ or $I_{meas}(\mathbf{S})$, will not be the data needed, $I(S)$ or $I(\mathbf{S})$, but instead will be the data required convoluted with the shape of the beam, as registered at the detector, $B(\mathbf{S})$. The effect beam shape has on data of this sort is called **slit-smearing**. From a practical point of view, a solution diffractometer, or an X-ray imaging diffractometer is operationally ideal if the peak of $B(\mathbf{S})$ is small compared to the length scale on which $I(S)$ changes appreciably on the detector surface because when that is so, $I_{meas}(S)$ (or $I_{meas}(\mathbf{S})$) will equal $I(S)$ (or $I(\mathbf{S})$) for all intents and purposes, and slit-smearing can be ignored.

The effects that slit-smearing have on continuous data sets can be corrected computationally after the fact. In the past, when the X-ray generators used for the collection of solution data were pathetically weak by modern standards, large, and often highly asymmetric, X-ray beams were used to collect solution data because this was the only way photon fluxes could be obtained that were high enough to make data collection practical. In that era many useful algorithms were devised for correcting data for slit-smearing effects. Nevertheless, within reason, every effort should be made to design experiments so that the data collected do not require correction, because at the very least, no matter how well it is done, correction causes or magnifies error.

Two further comments are in order. First, because most solution-scattering data sets end at 2θ less than $5°$, good solution-scattering data can often be collected using radiation beams that have remarkably modest coherence lengths; in other words, they need not be anywhere near as monochromatic as the beams used for the collection of single crystal data (see section 1.8). Second, the best X-ray solution-scattering diffractometers ever built are the ones now operating at synchrotron light sources around the world. The extremely bright, extremely well collimated X-ray beams that can be obtained from synchrotrons are ideal for the collection of solution data. The instruments that have been built to exploit these beams are equipped with area detectors, and routinely collect data sets of unprecedentedly high accuracy in unprecedentedly short times. The development of these instruments has revived an experimental technique that had been a scientific backwater for decades because of its inability to compete effectively with techniques like macromolecular crystallography and electron microscopy.

8.15 MOLECULAR WEIGHTS AND RADII OF GYRATION CAN BE MEASURED BY LIGHT SCATTERING

The theory of solution scattering developed in previous sections is valid independent of radiation type and wavelength. Only the expressions for form factors or electron densities need to be adjusted to make the equations for X-rays function for other types of radiation. Thus, it should come as no surprise that aqueous solutions of macromolecules scatter light more strongly than samples of pure solvent, and that information about macromolecular structure can be obtained by analyzing that incremental scatter.

Equation (8.4) can be used as a point of departure for discussing light scattering, and when doing so, it is important to remember that the wavelength of visible light is of the order of 5,000 Å, whereas the maximum linear dimensions of biological macromolecules are commonly 10 to 100 Å. It follows that as far as light is concerned, the atom-to-atom variations in electron density within macromolecules, which is so important in the context of X-ray scattering and diffraction, do not count. The length scale of those variations, which is of the order of 1 Å, is much too small. A biological macromolecule of modest molecular weight looks like a point in a light scattering experiment, and a very large biological macromolecule will appear to be only slightly more substantial than that. Thus, for light scattering purposes, only the first term in equation (8.4)—the volume term—needs to be taken into account, and because only its lowest S regions can be explored in a light scattering experiment, $I(S)$ can be adequately represented using Guinier's law:

$$I(S) \approx I(0) \exp\left[-(4\pi^2/3)Rg^2S^2\right].$$

For the record, note that if the radius of gyration of a protein were 50 Å and the wavelength of the light used to measure its scatter were 5,000 Å, then the intensity of the scatter at a scattering angle of 90° would be 0.997I(0), which means that it would be very hard to prove that radius of gyration of the molecule was not 0, (i.e., that the molecule was bigger than a point).

The only issue that remains is to determine what $(\Delta\rho V)^2$ corresponds to in the context of a light scattering experiment. The answer to this question is to be found in chapter 1, equation (1.11), which tells us that the property of substances that controls their light scattering is polarizability, α, and this means that I(0) will be proportional to $(\Delta\alpha V)^2$, where $\Delta\alpha$ is the difference in polarizability between the macromolecule and the same volume of solvent. Furthermore, in section 3.3, it is shown that for gases and dilute solutions, the square of the refractive index of a substance, n^2, is equal to $(1 + N\alpha)$, where N is the number of molecules per unit volume in a substance. Thus, $(\Delta\alpha V)$, which is the change in polarizability of a solution of the macromolecule in question, per macromolecule, should be proportional to $\left({n_{\text{sol}}}^2 - {n_o}^2\right)$, where n_o is the refractive index of the pure solvent. If the mass concentration of macromolecule in the solution is small, then:

$$\begin{aligned} n_{\text{sol}}^2 - n_o^2 &= [n_o + (\partial n/\partial c)c + \ldots]^2 - n_o^2 \\ &\approx 2n_o\,(\partial n/\partial c)\,c, \end{aligned}$$

where c is the weight concentration of the macromolecule in the solution in grams/liter, and $(\partial n/\partial c)$ is the refractive increment of the solution, which is a quantity that can be measured directly using a refractometer. Thus, if the concentration is 1 molecule/liter, then:

$$n_{\text{sol}}^2 = n_o^2 \approx 2n_o(\partial n/\partial c)(M/No),$$

where M is the molecular weight of the macromolecule, and N_o is Avogadro's number. It follows that $N(\Delta\alpha V)^2$, the macromolecular component of the forward scatter given by a sample that contains N molecules of some macromolecule, per liter, I(0), will be proportional to:

$$[2n_o(\partial n/\partial c)]^2\ (Mc/No).$$

When the leading constants in equation (1.11) are taken into account, we finds:

$$I(0) = (\pi^2/2\mu_o c)(1/\lambda^4)(\sin\ \varphi/R)E_o^2([2n_o\ (\partial n/\partial c)]^2\ (Mc/No)).$$

In this equation, λ is the wavelength of the light, R is the distance from the sample to the detector, and E_o is the electric field strength of the incident light, which is assumed to be polarized. Thus, as we would anticipate, the forward scatter in a light scattering experiment is directly related to the molecular weight of the macromolecule solutes responsible for it. Only if those solutes have very high molecular weights, and consequently very large linear dimensions, will the scatter they produce at high angles differ at all from the scatter they produce at low angles so that a radius of gyration estimate can be obtained from the data.

PROBLEMS

1. Show that $I(0) = \int_0^\infty p(r)dr$.
2. Show that $R_g{}^2 = (1/2)\left[\int_0^\infty r^2 p(r)dr \Big/ \int_0^\infty p(r)dr\right]$.
3. Using equation (8.3), it is trivial to show that:

$$SI(S) = (1/2\pi) = (1/2\pi)\int_0^\infty [p(r)/r]\sin(2\pi rS)dr.$$

   It is more challenging to demonstrate that:

$$p(r)/r = 8\pi\int_0^\infty [SI(S)]\sin(2\pi rS)dS.$$

   Prove that if the first equation is true, the second one must be also. [NB: The integral of $\delta(r)$ from $-\infty$ to $+\infty$ is 1.0, but its integral from 0 to $+\infty$ is 0.5.]
4. Derive the formula for the radius of gyration of a sphere.
5. The radius of gyration of a cylinder that is $2H$ long and R in diameter is: $(R^2/2 + H^2/3)^{1/2}$. What is the radius of gyration of sphere that has a radius of 50 Å? What would the radius of gyration be of an object of the same volume that is a cylinder whose length is 10 times its radius?
6. Most of the time structural biologists are interested in the wavelength of the X-rays they use, but when it comes time to generate those X-rays, it is often energies that count, not wavelengths. Using Planck's expression for photon energies, derive an equation that relates

photon energy in units of electron volts to wavelength in angstroms. What are the photon energies associated with photons that have wavelengths of 1 Å and 20 Å?

FURTHER READING

Feinup, J. R. (1982). Phase retrieval algorithms: a comparison. *Appl. Optics* 21, 2758–2769.

Glatter, O., and Kratky, O., eds. (1982). *Small Angle X-ray Scattering*. London: Academic Press.

Shapiro, D., Thibault, P., Beetz, T., Elser, V., Howells, M., Jacobsen, C., Kirz, J., et al. (2005). Biological imaging by soft X-ray diffraction microscopy. *Proc. Natl. Acad. Sci. USA* 102, 15343–15346.

Svergun, D. I. (1999). Restoring low resolution structures of biological macromolecules from solution scattering using simulated annealing. *Biophysical J.* 76, 2879–2886.

APPENDIX 8.1 DERIVATION OF THE DEBYE EQUATION

As we already know, the scattered intensity due to a single molecule that is held in a fixed orientation with respect to a beam of X-rays is given by the following formula:

$$\mathrm{I}(\mathbf{S}) = \sum_{\text{all}} \sum_{\text{atoms}} \mathrm{f}_i \mathrm{f}_j \exp\left[-2\pi i \left(\mathbf{r}_i - \mathbf{r}_j\right) \cdot \mathbf{S}\right].$$

Using the standard definition of the dot product, this equation can be rewritten as follows:

$$\mathrm{I}(\mathbf{S}) = \sum_{\text{all}} \sum_{\text{atoms}} \mathrm{f}_i \mathrm{f}_j \exp\left(-2\pi i \left|\mathbf{r}_i - \mathbf{r}_j\right| S \cos\theta_{ij}\right),$$

where θ_{ij} is the angle between $\mathbf{S}$ and $(\mathbf{r}_i - \mathbf{r}_j)$. It follows that the contribution made to the rotationally averaged value of $\mathrm{I}(|\mathbf{S}|)$ by any particular vector between atoms, $(\mathbf{r}_i - \mathbf{r}_j)$, will be the equation just given, averaged over all possible orientations of the molecule, with each orientation having the same weight as every other. Thus:

$$\langle \mathrm{I}(|\mathbf{S}|)\rangle = (1/4\pi) \sum_{\text{all}} \sum_{\text{atoms}} \mathrm{f}_i \mathrm{f}_j \int_0^{2\pi} \left[\int_0^{\pi} \exp\left(-2\pi i \left|\mathbf{r}_i - \mathbf{r}_j\right| S \cos\theta\right) \sin\theta \, d\theta\right] d\varphi,$$

where the factor $(1/4\pi)$ normalizes the double integral taken to evaluate that average (see figure 8.1.1). When these integrations are done, it is found that:

$$I(S) = \sum_{\text{all}} \sum_{\text{atoms}} f_i f_j \sin\left(-2\pi \left|\mathbf{r}_i - \mathbf{r}_j\right| S\right) / \left(-2\pi \left|\mathbf{r}_i - \mathbf{r}_j\right| S\right).$$

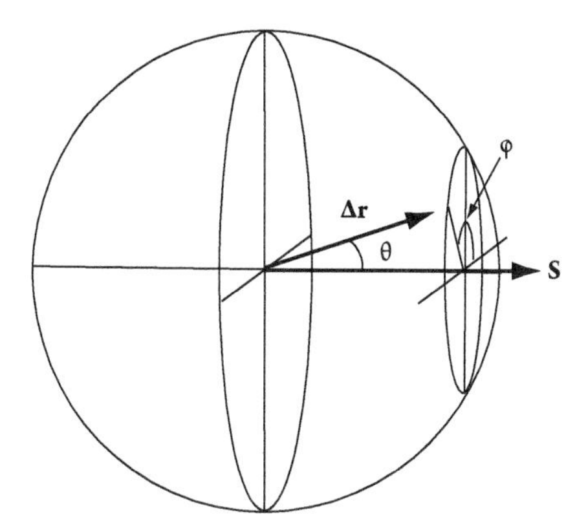

Figure 8.1.1 Geometry relevant to the integration required for rotational averaging. $\Delta\mathbf{r} = (\mathbf{r}_i - \mathbf{r}_j)$.

APPENDIX 8.2 SMALL ANGLE SCATTER AND THE RADIUS OF GYRATION

For very small values of S:

$$I(S) \approx I(0)\left\{I - \left(\left[4\pi^2/6I(0)\right] \sum_{\text{all}} \sum_{\text{atons}} f_i f_j \left|\mathbf{r}_i - \mathbf{r}_j\right|^2\right) s^2\right\}.$$

Because the distances between the atoms in a molecule do not depend on the location of the origin of the coordinate system used to specify their positions, it can be placed wherever convenient, and the obvious place to put it is at the center of scattering mass of the molecule of interest; in other words, at the location where $\sum f_i \mathbf{r}_i = 0$.

No matter where the origin is placed:

$$\left[4\pi^2/6I(0)\right] \sum_{\text{all}} \sum_{\text{atoms}} f_i f_j \left|\mathbf{r}_i - \mathbf{r}_j\right|^2$$
$$= \left[4\pi^2/6I(0)\right] \sum_{\text{all}} \sum_{\text{atoms}} f_i f_j \left(\mathbf{r}_i^2 + \mathbf{r}_j^2 - 2\mathbf{r}_i\mathbf{r}_j\right),$$

and the first two terms in the left-hand double sum both equal $\sum f_j \sum f_i \mathbf{r}_i^2$. However, if the origin is placed at the center of scattering mass, the third term in the left-hand double sum is 0, because it is a sum of terms each of which has the form $\mathbf{r}_j \cdot \sum f_i \mathbf{r}_i$, and the sum part of each term is 0. Thus:

$$[4\pi^2/6I(0)] \sum_{\text{all}} \sum_{\text{atoms}} f_i f_j \left|\mathbf{r}_i - \mathbf{r}_j\right|^2 = [4\pi^2/3I(0)] \sum f_j \sum f_i \mathbf{r}_i^2$$

$$= \left[4\pi^2/3\left(\sum f_j\right)^2\right] \sum f_j \sum f_i \mathbf{r}_i^2 = (4\pi^2/3)\left(\sum f_i \mathbf{r}_i^2 / \sum f_j\right)$$

Because the origin is the center of scattering mass, the expression $\left(\sum f_i \mathbf{r}_i^2 / \sum f_j\right)$ is the scattering factor–weighted (i.e., Z-weighted), average of the square of the distance between the atoms in a molecule and its center of scattering mass. Thus, it is the variance of the molecule's electron density distribution function.

PART FOUR

Optical Microscopy

9

Image Formation Using Lenses

The invention of the light microscope in the 17th century led to a revolution in biology, and the light microscope has been an essential part of the biologist's tool kit ever since. Microscopes produce images of structures, rather than images of the square of the Fourier transforms of structures, and thus the relationship between input and output is so obvious that even those who have no idea how microscopes work are able to use them to good effect.

By the early decades of the 20th century, the optical technology on which light microscopy is based had become so mature that most scientists stopped paying attention to it. However, in last few decades, many new methods have been devised for imaging biological materials with light microscopes. Some allow investigators to track the motions of specific macromolecules in living cells in real time. Others produce images that have resolutions much higher than were previously thought possible. Thus, surprisingly, optical microscopy is once again "in play" technically.

A sound appreciation of how microscopes work is a prerequisite to understanding these new developments, and it is to this issue that we now turn. Fortunately, because light microscopes and electron microscopes work in much the same way, many of the concepts we are about to explore will be useful in both contexts.

9.1 MAGNIFIED IMAGES OF SMALL OBJECTS CAN BE PRODUCED TWO DIFFERENT WAYS

Radiation can be used to generate magnified images of objects (i.e., do microscopy) in two fundamentally different ways: by **scanning** and by **direct imaging**. Scanning microscopy is the easier of the two alternatives to understand. A scanning microscope (figure 9.1, left) consists of a radiation source that produces a beam of light

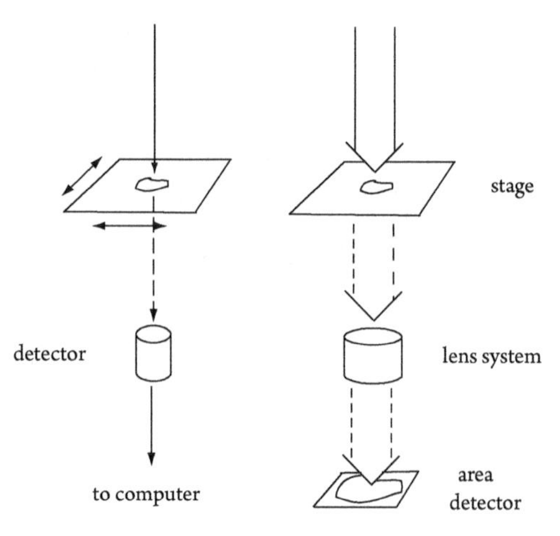

Figure 9.1 Schematic diagram comparing scanning microscopes (left) with conventional imaging microscopes (right).

(or some other radiation) whose diameter is small compared to the dimensions of the features of specimens a researcher wishes to image. The specimen is mounted on a stage that can be moved by accurately known amounts in a plane perpendicular to the beam. On the far side of the stage there is a detector, which is often aligned with the beam, but need not be, that measures the effect the specimen has on radiation passing through it. The specimen is scanned past the beam, and the output of the detector is measured as a function of the position of the sample. The result is usually presented on a computer screen as a two-dimensional array, each pixel of which displays a (perhaps) color-coded version of the output recorded when the sample was in the corresponding position. A scanning image may display the position dependence of the optical density of the specimen, or its color, or its light scattering, or its fluorescence, and so forth. The magnification of a scanned image is the ratio of the distance between adjacent pixels in the display and the distance between adjacent measurement points in the scan. Because specimen stages can be purchased that will position samples with subnanometer accuracy, the resolution of scanning microscopes can be very high. However, when visible light is the radiation used, image resolution is liable to be determined by diffraction effects, rather than the mechanical properties of the stage used, and, thus, not be much better than the wavelength of that light.

A direct imaging microscope (figure 9.1, right) has the same elements as a scanning microscope—a light source, a specimen stage, and a detector—but it is operated in a completely different way. First, the radiation source of an imaging microscope illuminates a large area of the specimen, not just a single pixel's worth. Second, the specimen stages of imaging microscopes are not moved as images are recorded; there is no scanning. Third, the detector system includes an (often) complex series of lenses that uses the radiation that passed through the specimen to form an image of the specimen that is projected onto an area detector, which may be the retina of a human eye, a piece of photographic film, or a CCD detector. What is seen at the detector is a magnified, two-dimensional, analog image of the illuminated area of the specimen. Thus, ignoring interference effects for the moment, an imaging microscope can be thought of as a massively parallel scanning microscope in that it simultaneously collects information about *all* points in a specimen, rather than gathering information about the specimen one point at a time.

The critical component of any imaging microscope is its lens system, and because light and electrons are the only kinds of radiation for which the lenses available today perform well enough to be useful for investigating biological ultrastructure, biologists need concern themselves only with only two kinds of direct imaging microscopes: optical microscopes and electron microscopes. Furthermore, once the way single focusing lenses operate is understood, it is easy to understand how the multilens systems in modern microscopes operate. We will deal with lenses that focus light first and will postpone our discussion of electron lenses until chapter 11. As we proceed, we will find that some kinds of scanning microscopes use focusing lenses.

9.2 THE DIRECTION PROPAGATION OF LIGHT CAN CHANGE AT INDEX OF REFRACTION BOUNDARIES

Since time out of mind it has been known that light **refracts**. In other words, its direction of propagation changes when it crosses boundaries between transparent substances that differ in refractive index, such as when it crosses air–glass or air–water interfaces. Optical lenses use refraction to focus (or defocus) light.

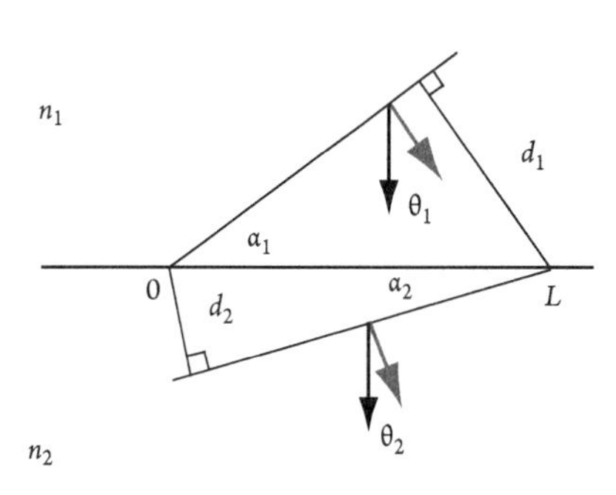

Figure 9.2 Geometric parameters used to derive Snell's law. The thick arrows indicate the direction of propagation of a planar wave front above and (at a later time) below the boundary between two transparent media. The refractive index of the upper medium, n_1, is less than the refractive index of the medium, n_2. The thin black arrows are both normal to the index of refraction boundary, and they point toward the medium that has the higher index of refraction. The system is viewed down the line, *O*, along which the front and the boundary intersect at time 0.

In chapter 3, it was shown that the speed of light in transparent substances is less than the speed of light in a vacuum, and that the index of refraction of a substance, n, is the speed of light in a vacuum divided by the speed of light in that substance. It is easy to understand why the change in speed of light that occurs when a beam of light crosses an index of refraction boundary would alter its direction of propagation. Suppose a planar wave front is moving toward the planar boundary that separates two transparent media that differ in refractive index. Figure 9.2 shows what that encounter will look like at the instant the incident wave front encounters the boundary, which is taken as $t = 0$, viewed down the line of intersection, which is designated "0". At $t = 0$, the distance between the boundary and the points on the wave front that will ultimately cross the boundary along the line parallel to 0 that is L away from it, d_1, will be:

$$d_1 = L \sin \alpha_1,$$

and the time that will elapse before that encounter occurs will be:

$$\Delta t = (n_1/c)L \sin\alpha_1,$$

because (c/n_1) is the speed of light in medium 1. By the time the front crosses the boundary at L, the part of the wave front that crossed the boundary at $t = 0$ will have traveled a distance of $[\Delta t(c/n_2)]$ into the medium. Thus:

$$d_2 = \Delta tc/n_2 = (n_1/n_2)L \sin \alpha_1 = L \sin \alpha_2,$$

which implies that:

$$n_1 \sin\alpha_1 = n_2 \sin\alpha_2.$$

Now if the normal to an index of refraction boundary is defined to be a vector that points into the medium that has the higher index of refraction, then $\alpha_1 = \theta_1$, and $\alpha_2 = \theta_2$, where θ_1 and θ_2 are the angles between that normal and the normal to the wave front in two media. (Both angles are measured clockwise from the direction defined by the boundary normal.) Thus:

$$n_1 \sin \theta_1 = n_2 \sin \theta_2. \tag{9.1}$$

Equation (9.1) is **Snell's law.**

9.3 IN GEOMETRICAL OPTICS, WAVES ARE REPLACED BY RAYS

The design of optical systems starts with Snell's law, and its application leads naturally to the time-honored approach to optical engineering called **geometrical optics**. In wave optics, the radiation entering an optical system is treated as a sum of coherent waves, each having its own frequency, amplitude, phase, and direction of propagation. In geometrical optics, each wave is represented as a **ray**, or possibly a bundle of rays, a ray being a thin pencil of light of the appropriate frequency that has no phase associated with it and that propagates in the direction perpendicular to the advancing wave front. Implicitly, rays have diameters at least as large as the wavelength of the radiation they represent but small compared to the dimensions of the optical system under consideration.

The goal in geometrical optics is to determine the way the trajectories of rays change as they pass through optical systems. This is done assuming that: (1) rays are straight lines in regions where the index of refraction is constant; (2) the directions of rays change as Snell's law requires when index of refraction boundaries are crossed; (3) the trajectories of the rays passing through any optical system are independent of each other; and (4) the distribution of electromagnetic energy emerging from an optical system is the sum of the energies carried by all the rays that entered it, duly propagated. No allowance is made for interference effects in geometrical optics; thus, the descriptions it provides of the spatial distributions of light energy emerging from optical systems are inaccurate at submicron length scales.

9.4 CURVED GLASS SURFACES CAN FOCUS LIGHT

The image-distorting effects of curved glass surfaces have been recognized for thousands of years. By the late 13th century, spectacles were being made in Europe, the lenses of which had spherical surfaces because it easy to make spherical glass surfaces by grinding and polishing glass blanks, and indeed, to this day, most lenses in most optical instruments have spherical surfaces.

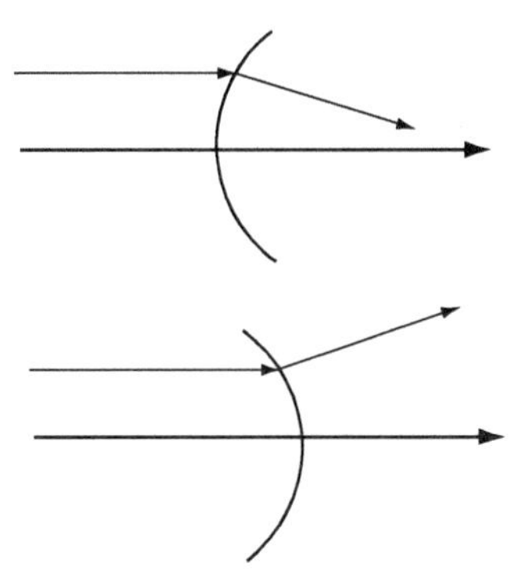

Figure 9.3 The effect of convex (top) and concave (bottom) glass surfaces on the trajectories of light rays originating on the left side of the diagram. The heavy arrow is the optical axis, and it is pointed in the direction that radiant energy is flowing. The medium to the left of both curved surfaces is air, and the medium to the right is glass.

Qualitatively, it is easy to understand why curved glass surfaces make light rays converge—**focus**—or diverge—**defocus**. Consider figure 9.3, which shows schematically what happens when light rays traveling from left to right cross an air–glass interface that is spherical in shape, but either convex (top) or concave (bottom). Most lenses are cylindrically symmetric about an axis, called the **optical axis**, which is normal to both its front and back surfaces. The center of the convex spherical surface in figure 9.3 (top) is located on its optical axis to the right of that surface. The center of the concave spherical surface (figure 9.3, bottom) is also located on the optical axis but to the left of the surface.

The trajectory of a light ray that is coincident with the optical axis of a curved glass surface does not change when it traverses the surface because the ray is either parallel or antiparallel to its normal at the point of entry (see equation (9.1)). However, most rays that cross such a surface off-axis or that are not parallel to the optical axis will change direction; that is, they will be refracted. Snell's law tells us that when a ray that has been propagating in a medium that has a low index of refraction enters a high refractive index medium, it will bend toward the normal to the boundary so that inside the high refractive index medium its trajectory is closer to parallel to that normal than it is in the low refractive index medium. Thus, off-axis rays passing from air into glass through a convex surface tend to **focus**, and if the surface is concave, they tend to **defocus**.

9.5 THE LENS LAW

For centuries, the designers of lens systems have been interested in working out what curved glass surfaces do to ray trajectories. It may come as a surprise, therefore, that those who do this kind of work still have nothing more powerful to guide them than Snell's law. There is no closed-form, mathematical expression that describes all the ways the trajectories of rays can change when they pass through even a single lens that has spherical surfaces. Lens designers still rely heavily on the approximate descriptions of lens systems performance they obtain using the brute-force technique called **ray tracing**, which involves working out what lenses do to ray trajectories, one ray at a time. In the old days, ray tracing was done by hand graphically. Today it is done using computer programs that make the process much faster and easier. Once the trajectories of sufficiently large numbers of different rays have been worked out for a lens system, quantitative conclusions can be drawn about its optical properties.

All of the preceding not withstanding, ray tracing can be dispensed for a single special case: When the surfaces of the lens of interest are spherical, the rays passing through it originate from points close to its optical axis, and the angle between those rays and the optical axis are small. Rays that meet those conditions are said to be **paraxial** (i.e., almost axial). The effects that lenses have on the trajectories of paraxial rays can be worked out mathematically, and the formulae that emerge provide a useful, first approximation description of the performance of any lens that has spherical surfaces.

Appendix 9.1 provides an analysis of what a lens with spherical surfaces does to the trajectories of a particular class of paraxial rays, namely those that are **meridional**. Meridional rays propagate entirely in the plane defined by the optical axis of the lens and their point of origin. Beyond the lens, y'', the distance from the optical axis of a paraxial, meridional ray that originates at a point that is y_o from the optical axis is:

$$y'' = \left(s_1 + s_2 - s_1 s_2/f\right)\theta_1 + \left(1 - s_2/f\right)y_o, \tag{9.2}$$

where f is a distance called the **focal length**, which is characteristic of the lens (see the following), θ_1 is the angle between that ray and the optical axis at its point of origin, s_1 is the distance along the optical axis from the point where the ray originates to the **principal point** of the front surface of the lens, and s_2 is the distance along the optical axis from the back principal point of the lens to the point where the ray is observed. (The equations that describe locations of the principal points of a lens are provided in appendix 9.1.)

Comment: When numbers are plugged into the expressions for the locations of the principal points of a lens, we find that for many lenses it is impossible to measure s_1 and/or s_2 directly because its principal points are inside the body of the lens. Why not measure those distances from the front and back surface of the lens instead, which are always accessible? The practice of referencing s_1 and s_2 to their principal points has several important advantages. First, as appendix 9.1 shows, the formula that describes the trajectories of paraxial rays is radically simplified when distances along the optical axis are measured from principal points instead of from lens surfaces. Second, but not so obviously, when this convention for measuring distances is adhered to, the focal length of a lens is the same for rays entering it from the right as its focal length for rays entering it from the left, which would not otherwise always be the *case. Finally, it turns out that the effect that a multielement lens assembly has on the trajectories of paraxial rays can be described using equations (9.2) and (9.3), which are single-lens equations, provided f and the locations of the principal points are appropriately specified.*

The focal length of a lens with spherical surfaces is given by the following expression:

$$f = \left[(n-1)(1/R_1 - 1/R_2) + D(n-1)^2/nR_1R_2\right]^{-1}, \qquad (9.3)$$

which depends *only* on the shape of the lens and the refractive index of the glass of which it is made (see figure 9.4). By convention, the R_I surface of a lens is the one on the left side of diagrams like figure 9.4, which is the side from which light approaches the lens, and if that surface is convex, viewed from the left, R_I is *positive*. The R_2 surface of the lens is the one on the right, and if it is concave to light approaching from the left, R_2 is *negative*. D is the thickness of the lens, which is the distance between its two surfaces measured on the optical axis. (NB: The locations of the principal points of a lens are likewise determined entirely by its shape and refractive index.)

It is easy to measure focal lengths. If a lens is illuminated with a bundle of rays all of which are parallel to its optical axis, θ_1 will be 0, and thus for those rays, downstream of the lens, equation (9.2) becomes: $y'' = [1 - s_2/f]y_o$, which means that y'' will be 0 for all those rays at the point where $s_2 = f$, regardless of y_o. In other words, they will come to a focus at that point on the optical axis. Conversely, on the far side of a lens having a focal length of f, the trajectories of all the (divergent) paraxial rays that emanate from a point source on its optical axis located at $(0, -f)$ will be parallel to each other and to the optical axis.

What about the (paraxial) rays originating at a point on the optical axis that is further from the lens than the distance f? For those rays, $y_o = 0$, and in that case, equation (9.2) tells us that all the rays in question, regardless of their angle with respect to the optical axis as they left their point of origin, will converge on the optical axis at a distance s_2 downstream of the lens, where s_2 satisfies the equation:

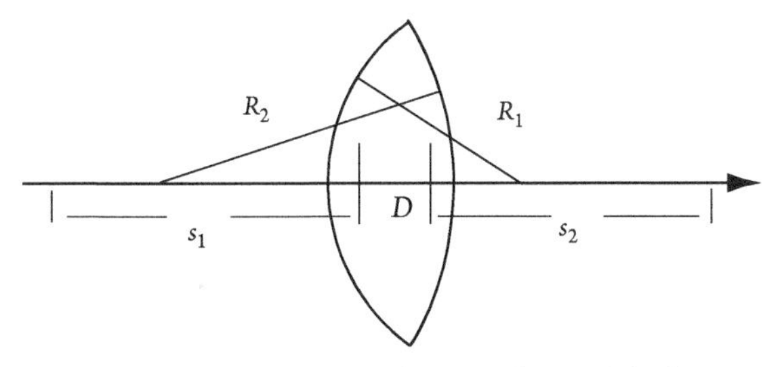

Figure 9.4 Lens geometry. The optical axis of the lens shown is the heavy arrow. The distances s_1 and s_2 are both measured from their respective principal points, which are the intersections of the two gray lines inside the lens with the optical axis, and D is the thickness of the lens, measured along its optical axis (see appendix 9.1).

$$s_1 + s_2 - s_1 s_2/f = 0.$$

This equation is commonly written as follows:

$$1/f = 1/s_1 + 1/s_2, \tag{9.4}$$

and it is called the **lens law**. If the lens it characterizes is thin, that is, if D is small compared to R_1 and R_2, the lens law can be written in a form that may be more familiar:

$$(n - 1)(1/R_1 - 1/R_2) = 1/s_1 + 1/s_2.$$

(See equation (9.3).) The lens law is the starting point for all discussions of lens performance.

Any lens for which f is positive, which it is for lenses like the one shown in figure 9.4, will focus paraxial rays originating at a source that is further upstream from the lens than its focal length because if s_1 is larger than f, then $[1/f - 1/s_1]$, i.e. ${s_2}^{-1}$, will also be positive. (NB: If s_1 is altered, s_2 will be altered also.)

9.6 THE LENS LAW IS VALID FOR PARAXIAL SKEW RAYS

It is now time to address an aspect of optics that is all too easily overlooked when optical phenomena are discussed in the framework provided by two-dimensional diagrams such as figure 9.4, as they commonly are. Optical systems operate in three dimensions, and three-dimensional systems must be analyzed in three dimensions.

It is standard practice in optics to make the optical axis of a lens system the z-axis the three-dimensional coordinate system used to describe its performance. The line perpendicular to the optical axis that goes through the (off-axis) origin of the set of rays under consideration is usually taken to be the y-axis of that coordinate system, and the intersection of the y-axis with the optical axis is the origin of the axis system. The x-axis is the line perpendicular to both the x-axis and the z-axis that passes through that origin, and its direction is usually chosen so that the right-hand rule is obeyed.

The trajectory of any ray that originates at an off-axis point can be projected onto the (y, z) plane of the axis system just described, which is referred as the **meridional plane**, and also onto its (x, z) plane, which is termed the **sagittal plane**. Clearly, if both the meridional and sagittal projections of some ray are known, its trajectory in three dimensions is known also. A ray that emanates from an off-axis point in a direction that takes it out of the meridional plane is a **skew ray**, and by definition, its projection onto the sagittal plane cannot coincide with the optical axis (figure 9.5). In fact, whatever the projection of a skew ray onto the meridional plane may be, its sagittal projection will always reveal that it is diverging from the optical axis as it approaches the lens.

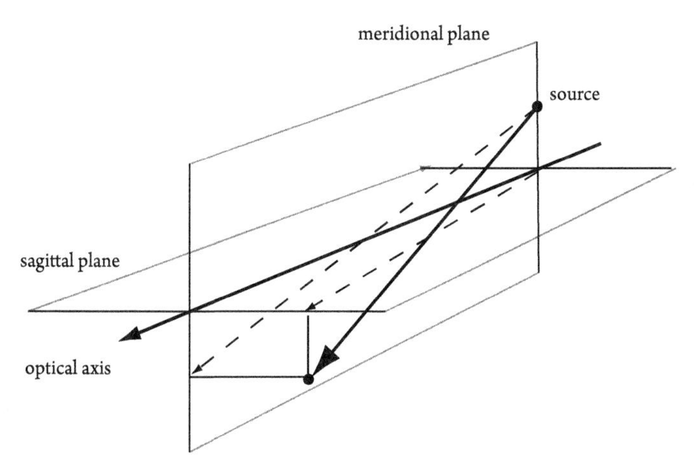

Figure 9.5 The trajectory of a skew ray coming from an off-axis point source (solid arrow) described in terms of its projections (dotted arrows) onto its meridional and sagittal planes.

In fact, the analysis of ray trajectories provided in appendix 9.1 is valid *only* for meridional rays. (NB: All the rays emanating from an on-axis point are meridional.) It should also be obvious that a similar derivation that encompassed both meridional and skew paraxial rays would be much messier mathematically. The reason this more general problem is not treated in appendix 9.1 is that equations (9.2) and (9.3) are just as valid for skew paraxial rays as they are for meridional paraxial rays. Therefore, no additional insights would have emerged had we made that extra effort.

9.7 FOCUSING LENSES PRODUCE IMAGES

There is another special case addressable using equation (9.2) that is of great interest. Consider an off-axis point source that is located in a plane perpendicular to the optical axis of a lens that has a positive focal length, and that this plane intersects the optical axis upstream of the lens at a distance $s_1 > f$. (We will call that plane the **object plane.**) Suppose that we examine the intersections of the (paraxial) rays emanating from that point with a plane perpendicular the optical axis that intersects it at a distance s_2 "downstream" from the lens, and that s_2 is chosen so that the lens law is obeyed. (We will call that plane the **image plane** for reasons that will shortly become apparent.) If these conditions are met, then equation (9.2) reduces to:

$$y'' = [1 - s_2/f]\, y_o,$$

which indicates that *all* the paraxial rays originating at $(0,\ y_o,\ -s_1)$, regardless of angle, will focus at a single point in the image plane, namely the one at $(0,\ y'',\ s_2)$. Because the lens law is valid for all paraxial rays, this expression implies that there will be a one-to-one correspondence between points in the object plane and points in the image plane. The point $(x,\ y,\ -s_1)$ in the object plane will correspond to the point $(x'',\ y'',\ s_2)$ in the image plane, where x'' bears the same relationship to x as

y'' does to y. Furthermore, the intensity and color of the light associated with each point in the image plane will be determined by the intensity and color of the light originating at the corresponding point in the object plane. In short, the distribution of light in the image plane will be an image of whatever is in the object plane.

Using the lens law (equation (9.4)), it is easy to show that $[1 - s_2/f] = -(s_2/s_1)$. Because both s_1 and s_2 are positive quantities under the circumstances specified, the negative sign of this ratio indicates that if the coordinates of a point in the object plane are (x, y), its coordinates in the image plane will be $[-(s_2/s_1)x, -(s_2/s_1)y]$, which is a formal way of saying that the images generated by single, focusing lenses that are operated this way are *inverted* relative to the objects they portray. Finally, the quantity $|(s_2/s_1)|$ is the scale factor that converts distances in the object plane into distances in the image plane. It is called the **magnification** of the image, M. If s_1 is close to f, but greater than f, as assumed here, then s_2 will be larger than s_1, and M will exceed 1. Thus, a single focusing lens used in the manner described is a microscope. [It is perfectly possible to position an object relative to a focusing lens so that $M < 1$.]

If s_1 is less than f, then, interestingly, s_2, the z-coordinate of the image plane will be a negative number. What this means physically is that the "spray" of rays originating from each point in the object plane is so divergent that the lens cannot refract them enough to focus them, in other words, the lens is not "strong" enough. However, to the right of the lens each such spray of rays will be *less* divergent than it was when it left the object plane. Thus, an observer looking at the light coming through the lens will see an image of the object in an image plane that appears to be on the left-hand side of the lens (i.e., on the same side of the lens as the object itself), but that looks larger than the object actually is. Thus, magnification is still achieved, and the magnification factor is still $|s_2/s_1|$, but the image is *not* inverted: The point at (x, y) in the object will correspond to a point at (Mx, My) in the image. The lenses in magnifying glasses are used in this manner; the objects to be examined are placed inside of f.

It should be noted that magnifying glass images and microscope images differ profoundly in another way. When a focusing lens is operated in the microscope mode, an image will register on a piece of photographic film placed in its image plane, or will be seen projected on a screen placed in the same location. Images of this sort are **real images**. On the other hand, if the same lens is used as a magnifying glass, nothing will register on a piece of film placed in the image plane because none of the light that passes from the object through the lens ever reaches that plane. What the lens is doing is making it *appear* that there is light energy coming from the image plane, not making light energy actually come from there. Images of this sort are **virtual images**.

Van Leeuwenhock's microscopes consisted of a single, focusing lens. The inverted images they produced were projected onto the retina of the user's eye. The first lens in any microscope is its **objective lens**, and the optical properties of that lens largely determine not only the magnification of the images a microscope produces, but, for reasons that will be explained shortly, they also largely determine their quality. Hooke's microscopes, which were built at about the same time, were more sophisticated in that they were **compound microscopes**, as all modern microscopes are.

In addition to having objective lenses, they had **ocular lenses**—an eyepiece lens positioned so that the images produced by their objective lenses could be additionally magnified. An ocular is basically a magnifying glass. Van Leeuwenhock's microscopes were superior to Hooke's because their objective lenses were better.

9.8 LENS PERFORMANCE IS LIMITED BY ABERRATION

A lens is **ideal** if the trajectories of *all* the rays passing through it conform to the lens law, irrespective of their inclinations with respect to the optical axis, or the distances between the optical axis and the points where they enter the lens. In short, if a lens is ideal, all the rays that pass through it will behave the way paraxial rays do when they transit a real lens that has spherical surfaces. Thus, if $\mathrm{i}(x, y)$ describes the spatial distribution of radiant energy just as the incident beam of radiation emerges from some specimen, the image produced of that specimen by an ideal lens would be $\mathrm{i}(-Mx, -My)$. However, just as there is no Santa Claus, there are no ideal lenses, and the image of $\mathrm{i}(x, y)$ is never precisely $\mathrm{i}(-Mx, -My)$. The deviations from paraxial behavior displayed by real lenses are called **aberrations**. Aberration so seriously limits the quality of the images generated by single, spherical lenses that modern microscopes would be sorry devices indeed if assemblies of lenses, i.e., multielement lenses, could not be manufactured that are remarkably close to ideal in their performance.

The easiest kind of aberration to understand is **chromatic aberration**. It arises because the refractive indices of transparent substances vary with wavelength; n is always larger for blue light than for red. Thus, the focal length of a single element lens will not be the same for red light as it is for blue (see equation (9.3)); hence, no such lens can produce an image of a colored object that is focused at all wavelengths. The effects of chromatic aberration are particularly obvious in the images generated by single lenses at the boundaries between regions of different color, where colored fringes will often be (annoyingly) evident. (Chromatic aberration is described quantitatively in section 11.11.) The impact of chromatic aberration on microscope images can be reduced dramatically by replacing the single lenses we might have otherwise have used as objective lenses with appropriately designed multielement lenses of the same focal length, the components of which are made of glasses that differ in refractive index. Lens assemblies of this kind are called **achromats** if their focal lengths are identical for light of two different wavelengths, and **apochromats**, if the correction for chromatic aberration is perfect at three wavelengths.

When objects are imaged using monochromatic light, other kinds of aberration become manifest, the so-called **monochromatic aberrations**, the most important of which are: **spherical aberration**, **coma**, **astigmatism**, **curvature of field**, and **distortion**. They cause points in the object plane to be imaged as disks, ellipsoids, or even more irregular shapes, and they distort the overall shapes of the objects being imaged.

A version of the lens law that encompasses monochromatic aberrations can be obtained by generalizing the theory outlined in appendix 9.1 so that it will deal with both skew and meridional rays, and then increasing the number of terms used in the power series approximations for the trigonometric functions that appear in the resulting equations, so that they will deal with rays that are not paraxial. For example, $\sin(\theta)$ could be approximated as $[\theta - (1/3!)\theta^3]$ instead of θ, and $\cos(\theta)$ could be represented as $[1 - (1/2!)\theta^2]$ instead of 1. The larger the number of terms included in these trigonometric power series, the wider the range of ray angles for which the resulting equations will be usefully accurate, but the more horrifying the resulting algebra. The monochromatic aberrations named earlier are the ones that can be described using a theory that takes account of *only* the second terms in these power series expansions. They are commonly referred to as the **primary** or **Seidel** aberrations, and they are the aberrations that most seriously limit the performance of real lenses.

9.9 SPHERICAL ABERRATION IS THE MOST IMPORTANT OF THE SEIDEL ABERRATIONS

If a lens displays spherical aberration, as all lenses with spherical surfaces do, points in the object plane will be represented as disks in its image plane. This happens because the focal length of a lens that has spherical aberration depends on the distance between its optical axis and the point where a ray enters it; in other words, it depends on y (see appendix 9.1), which, the focal length of an ideal lens does not. The reason spherical aberration is more objectionable than the other four Seidel aberrations is that it alters the appearance of the on-axis points, which the other four do not. Only chromatic aberration has this same malignant property.

It is comparatively easy to derive an expression that describes what spherical aberration does to images (see appendix 9.2), and the reason for doing so here is not because spherical aberration seriously affects the images produced by modern optical microscopes, but because it has a huge impact on the images produced by electron microscopes, and it is easier to understand in the context of glass lenses and light than it is in the context of electron beams and electron lenses.

We derived the lens law by working out what spherical lenses do to ray trajectories, and we could produce a theory that deals with aberrations by generalizing the arguments made in appendix 9.1. However, aberrations are more conveniently approached by thinking about what they do to the shapes of wave fronts. As figure 9.6 shows, to the left of a focusing lens, the wave fronts originating at any point on its optical axis that is further from the lens than its focal length will be spherical surfaces that expand in radius as they approach the lens. If an image of the origin point is to appear to the right of the lens, the shapes of those wave fronts must be altered by their passage through the lens so that they become spherical surfaces

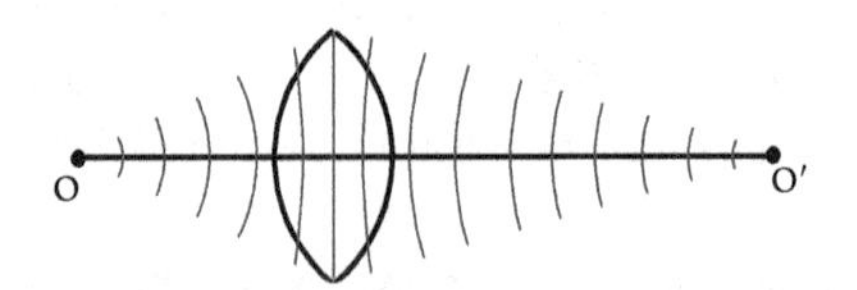

Figure 9.6 The effect of focusing lenses on divergent wave fronts.

centered on some point on the optical axis that lies to the right of the lens, which is where the image of the origin point will appear. This happens because the speed of light is lower in glass than it is in air, and focusing lenses are thickest on-axis and get progressively thinner as the distance from the optical axis increases. Thus, the peripheral parts of a wave front passing through a focusing lens "get ahead" of its on-axis parts, and if the effect is big enough, a divergent, concave wave front will be converted into a convergent, convex wave front.

A *wave front* is a surface in space on which the phase of the electromagnetic radiation being considered is constant. It follows that if the different regions of a particular wave front that started at o are to focus at o', the lengths of the paths traversed by all the rays that would have to be used to represent that wave front must be the same, measured not in meters, but in *wavelengths*. Only if this is so will constructive interference occur at the focal point so that an image forms there. In section 3.3, we discovered that the wavelength of light is its wavelength *in vacuo* divided by the refractive index of the medium through which it is propagating. Thus, measured in number of wavelengths, the length of the trajectory a ray takes as it travels between two points is proportional to $\sum n_i l_i$, where the sum is over all the line segments in the trajectory, l_i is the geometric length of the ith segment, and n_i is the index of refraction of the medium the ray traverses as it transits the ith segment. The refractive index–weighted length of a ray's trajectory is its **optical path length**.

An equation is developed in appendix 9.2 for the difference in optical path length between a ray traveling from $(0,\ 0,\ -s_1)$ to $(0,\ 0,\ +s_2)$ through a lens with spherical surfaces on axis, and a ray traveling between the same two points via an off-axis trajectory. If the angle between the optical axis and the off-axis ray is θ at $(0,\ 0,\ -s_1)$, then for a thin lens, to fourth order in θ, the difference in the optical path lengths of the two rays, ΔOP, will be:

$$\Delta OP = \left(\theta^2/2\right) f^2\,[-1/f + 1/s_1 + 1/s_2] - \left(\theta^4/4\right) C_S, \qquad \mathbf{(9.5)}$$

where C_S is a quantity having the dimensions of length that is called the spherical aberration coefficient of the lens. Its value is determined (largely) by lens parameters (see appendix 9.2).

Three points need to be made about equation (9.5). First, if θ is small enough so that θ^4 terms can be ignored, that is, if the off-axis rays are paraxial, ΔOP will be 0 when the term in square brackets is 0 (i.e., when the lens law is obeyed). Thus, equation (9.5), which was derived using optical path length arguments, is consistent with the lens law, which we derived using Snell's law. Second, if the value of θ associated with some particular ray is outside the paraxial range, ΔOP will not be 0 at the paraxial image point. Thus, that ray will not contribute to the intensity observed at the paraxial focal spot. Third, by setting ΔOP and $1/s_1$ both to 0, making the substitution: $s_2 = \mathrm{f}(\theta) = [f + \Delta(\theta)]$, and solving equation (9.5) for $\Delta(\theta)$, the distance between the paraxial focal point and the focal point for some particular set of nonparaxial rays, we can show that:

$$\Delta\,(\theta) = -C_S\left(\theta^2/2\right).$$

Because C_S is always positive for a focusing lens, the off-axis focal length of a focusing lens that has spherical aberration is *always* smaller than its paraxial focal length.

9.10 THERE ARE FOUR OTHER SEIDEL ABERRATIONS

For our purposes, qualitative descriptions of the four remaining Seidel aberrations will suffice. The impact they have on the appearance of the on-axis parts of images is zero, but it increases as the distance from the optical axis increases.

Coma makes the images of off-axis points in the object plane look like comets—a bright head and a diffuse tail. Coma is seen in the images formed by spherical lenses because the point in space where rays from an off-axis point in the object plane are brought to a focus depends on where those rays went through the lens.

Astigmatism makes the images of off-axis points look ellipsoidal. The reason this occurs is that for spherical lenses, the focal length for the meridional rays originating from an off-axis point in the object plane is not the same as the focal lengths for the skew rays coming from that same point.

Comment: This kind of astigmatism is not the astigmatism optometrists talk about, which is caused by differences in the curvature of the lens of an eye in different directions. The astigmatism we are talking about here is characteristic of spherical lenses that have the same curvature in all directions.

Field curvature is a consequence of the fact that the proper surface on which to record the image produced by a spherical lens is a sphere, not a plane. Another way of putting it is that for real lenses, as opposed to the ideal lenses, the image surface is well approximated as a plane only close to the optical axis. Thus, if an image is recorded on a planar surface with the center of the image plane in perfect focus, the focus will degrade as the distance from the optical axis increases; the exterior portions of the image will be blurred due to poor focus.

Distortion results from a change in the magnification of an image that increases (or decreases) with the distance from the optical axis. It also is a product of position-dependent variations in the focal length of lenses.

The larger the working radius of a lens, and the wider the range of angles between the rays that enter it and its optical axis, the more severe the impact of its aberrations on image quality, which explains why the aberrations of the objective lens of a microscope are so critical for its performance. The range of angles between the optical axis and the rays originating in the object plane of a microscope that pass through its objective lens is *much* larger than the angular range spanned by the same rays when they reach the image plane, because the object plane of a microscope is always much closer to its objective lens than the image plane is. Thus, the ocular lenses of microscopes, which are used to magnify the image formed by its objective lens, deal with bundles of rays that are much closer to paraxial than those the objective lens of that microscope must image; consequently, their aberrations have a comparatively small impact on overall image quality.

9.11 PARALLEL BUNDLES OF RAYS FOCUS IN THE BACK FOCAL PLANE OF IDEAL LENSES

In the 19 century, as the art of lens making advanced, it became more and more interesting to ask what a point in the object plane would look like if it were imaged by an ideal lens. Would the image of a point be a point, as geometrical optics predicts? This question was answered definitively by Ernst Abbé in 1873, and the answer is "no". His analysis is so fundamental to everything that has happened in optics since 1873 that we need to understand his reasoning.

The starting point of Abbé's analysis is a generalization of the idea elaborated in section 9.5, namely that all (paraxial) rays entering some lens parallel to its optical axis will focus on-axis at a point on its optical axis that is f beyond the lens. In fact, all bundles of parallel (paraxial) rays focus in the **back focal planes** of a focusing lens, no matter what the angle they make with its optical axis on its upstream side. (The back focal plane of a lens is the plane perpendicular to its optical axis that intersects the optical axis at a distance of f beyond the lens.)

This fact is easily demonstrated using equation (9.2). When s_2 is made equal to f, then equation (9.2) becomes:

$$y'' = f\theta_1. \quad \mathbf{(9.6)}$$

The angle a ray makes with the optical axis as it leaves its point of origin in the object plane is the sole determinant of where it will cross the back focal plane; the displacement of its origin from the optical axis makes no difference. Figure 9.7 is a ray-tracing diagram showing how ideal lenses work that makes the same point. The dotted lines in that figure are rays leaving two different points in the image plane on parallel trajectories. On their way to their respective image points, they clearly intersect in the back focal plane.

It is possible to record the distribution of light energy in the back focal plane of a lens by placing a piece of photographic film there. Rays that passed through the system parallel to the optical axis will produce the on-axis signal in that picture. Rays that are not parallel to the optical axis will account for the signals registered

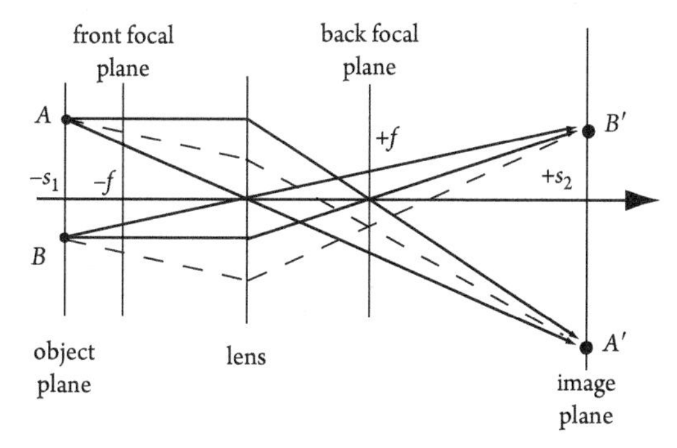

Figure 9.7 Parallel families of rays focus in the back focal plane.

at off-axis points, and the larger the angle between such a ray and the optical axis, the further off-axis its contribution will be. Thus, these geometric considerations suggest that each point in a back focal plane image corresponds to what observers would have seen in the absence of the lens if they had viewed the specimen from the corresponding angle at a distance so far from the specimen that it looked like a point. If that is so, the image in the back focal plane ought to correspond to the Fraunhofer diffraction pattern of the specimen. Is this true?

9.12 THE LIGHT WAVE IN THE BACK FOCAL PLANE IS THE FOURIER TRANSFORM OF THE LIGHT WAVE AT THE OBJECT PLANE

In section 9.9, we argued that at the instant a plane wave front perpendicular to the optical axis of some focusing lens clears the back surface of that focusing lens, or, more accurately, clears its back principal plane, it will have been transformed into a spherical surface of radius f that is centered on the point where the optical axis intersects the back focal plane. Only if this is so will the front converge on that point as it progresses, as experiment tells us it does.

For a thin lens, the distance from the center of the back focal plane to a point on the right surface of the lens that is located a distance r from the optical axis is $(f^2 + r^2)^{1/2}$. If the part of the front that traversed the lens at r is to converge on the focal point in phase with the part of the front that traversed the lens on its optical axis, the net effect of the lens must be to *advance* the phase of the peripheral part of the front relative to the central part of the front by an amount equal to $(2\pi/\lambda)\Delta$, where:

$$\Delta = \left(f^2 + r^2\right)^{1/2} - f.$$

If r is small compared to f (the usual paraxial approximation), then:

$$\Delta = f(1 - r^2/2f^2) - f \approx r^2/2f.$$

Another way of putting it is that by comparison with a ray passing through a lens on-axis, a ray that passes through the periphery of a lens behaves as though the length of its path to wherever it is headed on the right side of the lens is *shorter* than it really is geometrically by an amount equal to Δ. (The fact that this alteration in relative path length results from the retardation of the on-axis part of the front relative to its peripheral parts rather than by advancement of its peripheral parts with respect to its on-axis parts is immaterial. It is only the sign of the difference that counts.)

Knowing how lenses alter the effective lengths of the trajectories of rays passing through them at different distances from their optical axes, we can work out what the total amplitude and phase will be of the light that is brought to a focus at a specific point in the back focal plane, namely the point $(\xi,\ \eta, +f)$. The trajectory length of the ray passing through that point in the back focal plane that originates at the intersection of the object plane and the optical axis (i.e, at $(0,\ 0,\ -s_1)$) will be taken as the standard of comparison. If we designate the point where that ray crosses the lens $(x', y', 0)$, then the geometric length of its trajectory from its origin to the back focal, d_o, is:

$$d_o = \left(x'^2 + y'^2 + s_1^2\right)^{1/2} + \left[\left(x' - \xi\right)^2 + (y' - \eta)^2 + f^2\right]^{1/2}.$$

Now consider a ray parallel to the reference ray that originates in the object plane at the point $(x, y, -s_1)$. It will reach the lens at $(x' + x, y' + y, 0)$, and travel from there to the back focal plane, which it will cross at (ξ, η, f) because parallel rays focus in the back focal plane. The geometric distance the second ray travels from its point of origin to the back focal plane, d_1, will be:

$$d_1 = (x'^2 + y'^2 + s_1^2)^{1/2} + \left[(x' + x - \xi)^2 + (y' + y - \eta)^2 + f^2\right]^{1/2}.$$

The difference in *effective optical* distance traveled by the two rays as they pass from the object plane to the back focal plane, $\Delta\Delta$, will thus be:

$$\Delta\Delta = d_1 - d_o - (1/2f)\left[\left(x' + x\right)^2 + (y' + y)^2 - \left(x'^2 + y'^2\right)\right].$$

If s_1 and f are both large compared to x, y, x', y', ξ, and η, $\Delta\Delta$ can be evaluated using the approximation $(1 + x)^{1/2} \approx 1 + x/2$. When this is done, it is found that:

$$\Delta\Delta \approx (-1/f)(\xi x + \eta y).$$

Now suppose that the object plane is illuminated by a plane wave that is coherent over its entire extent and oriented so that it's normal is parallel to the optical axis. As the wave emerges from the object plane, its amplitude and phase can be described as $\mathrm{f}(x, y) \exp[i\alpha(x, y)]$, where $\mathrm{f}(x, y)$ is the amplitude of the wave at (x, y), and $\alpha(x, y)$ is the phase change that occurred as the wave traversed the sample at (x, y). (For obvious reasons, functions like this are called **object transmission functions**. We will discuss their properties further in section 10.2.) Note that if you were able to visualize the light emerging from the sample right at the object plane, the image your eye would see would be proportional to $\mathrm{f}(x, y)\mathrm{f}^*(x, y)$. At (ξ, η, f), the effect of the wave will be:

$$\mathrm{F}(\xi, \eta) = \int\int \mathrm{f}(x, y) \exp\left[i\alpha(x, y)\right] \exp\left[(-2\pi i/f\lambda)(\xi x + \eta y)\right] dxdy, \quad \mathbf{(9.7)}$$

where the integration is over the entire object plane. Now, $\tan(2\theta) = (x_2 + y_2)^{1/2}/s_1$, and at small angles, $\tan\ 2\theta \approx \sin\ 2\theta \approx 2\ \sin\ \theta \approx 2\theta$. Furthermore, from equation (9.6), we know that $(\xi^2 + \eta^2)^{1/2}/f = 2\theta$. Thus, $\xi/(f\lambda)$ and $\eta/(f\lambda)$ correspond to the x and y components of **S**, the scattering vector, and if we describe positions in the object plane in terms of a vector **r**, we can rewrite equation (9.7) as follows:

$$F(\mathbf{S}) = \int f(\mathbf{r}) \exp\left[i\alpha(\mathbf{r})\right] \exp(-2\pi\mathbf{r} \cdot \mathbf{S})d\mathbf{r}.$$

Thus, just as the geometry of lens systems suggested, the distribution of light energy in the back focal plane of a focusing lens is directly related to the Fourier transform of the object transmission function.

Comment: In the derivation of the preceding equation, the phase of $F(\mathbf{S})$ *is determined relative to the phase of the radiation reaching that point in the back focal plane that originated at (0,0) in the object plane. Because the length of that ray varies with scattering angle, the phase of* $F(\mathbf{S})$ *also varies systematically with* $|\mathbf{S}|$*. If the preceding derivation is redone using the on-axis distance between the object plane and the image plane as the distance/phase standard for all points in the back focal plane, we discovers that:*

$$F'(\mathbf{S}) = \exp\left(-\pi i\lambda f^2 S^2[1/f - 1/s_1 + f/(Ms_1)^2]\right)F(\mathbf{S}),$$

where $F'(\mathbf{S})$ *is proportional to the field in the back focal plane, and* $F(\mathbf{S})$*, is proportional to* $FT(f(\mathbf{r}))$*. Clearly, if* $\mathbf{S}$ *is small (i.e., in the paraxial limit), the phase correction described by the exponential term will be small, and, and furthermore, if* s_1 *is not much larger than f (i.e., if the lens in question is being used as a microscope, M), the magnification, will be large, and the difference in phase between* $F'(\mathbf{S})$ *and* $F(\mathbf{S})$ *will be small even for quite large values of S. For both reasons, we will ignore this systematic phase error in our analysis of image formation in microscopes. Finally, it is important to realize that this phase correction does not affect the distribution of radiant energy in the back focal plane because* $F'(\mathbf{S})[F'(\mathbf{S})]^* = F(\mathbf{S})[F(\mathbf{S})]^*$.

9.13 AS LIGHT TRAVELS FROM THE BACK FOCAL PLANE TO THE IMAGE PLANE IT GETS FOURIER TRANSFORMED AGAIN

Huygen's principle tells us that every point in any plane through which light passes can be considered a source of light in its own right, and that the light seen further on in an optical system can be computed by summing the contributions made by all these point sources. Thus, now that we know what the light wave looks like in the back focal plane, we can use Huygen's principle to compute what the wave will look like when it reaches the image plane.

At the point (X, Y) in the image plane, the light wave will be the sum of the contributions from all the points in the back focal plane, each with its phase adjusted to take into account the distance between (ξ, η, f), and (X, Y, s_2). For this part of the computation, light originating at $(0, 0, f)$ is used as the phase standard, the series approximation for the square roots of sums of quantities of dissimilar magnitude is taken advantage of, and the phase "error" commented on previously is ignored, which is reasonable in the paraxial limit. Thus, the expression that emerges will be valid if X and Y are small compared to $(s_2 - f)$, which it always is for microscopes. It will be assumed further that in general, η^2 and ξ^2 terms can be ignored. When this is done, it is found that the phase correction that must be applied to the radiation from (ξ, η, f) that reaches (X, Y, s_2) will be:

$$-\left[2\pi/\left(s_2 - f\right)\right](x\xi + Y\eta).$$

It follows that in the image plane, the light wave will be:

$$\begin{aligned} \mathrm{f}'(X, Y) = & \iint \left\{ \iint \mathrm{f}(x, y)) \exp\left[i\alpha(x, y)\right] \right. \\ & \times \exp\left[-(2\pi i/f\lambda)(x\xi + y\eta)\right] dx\, dy \Big\} \\ & \times \exp\left\{-\left[2\pi i/\lambda(s_2 - f)\right](X\xi + Y\eta)\right\} d\xi\, d\eta \\ = & \iint \mathrm{f}(x, y) \exp\left[i\alpha(x, y)\right] dxdy \\ & \times \iint \exp\left[-(2\pi i/\lambda)(\xi\left[x/f + X/(s_2 - f)\right]\right. \\ & + \eta\left[y/f + Y/(s_2 - f)\right])\Big] d\xi\, d\eta \end{aligned}$$

The second double integral equals 0 unless:

$$x/f + X/(s_2 - f) = 0, \text{ and } y/f + Y/(s_2 - f).$$

Using the lens law, it is easy to shown that this means that the integral will be nonzero only when:

$$X = -(s_2/s_1)x, \text{ and } Y = -(s_2/s_1)y.$$

Thus, the second double integral is a Dirac delta function that can be written: $\delta[X + (s_2/s_1)x,\ Y + (s_2/s_1)y]$, and thus:

$$\mathrm{f}'(X, Y) = \mathrm{f}\left[-(s_2/s_1)x, -(s_2/s_1)y\right],$$

and, hence, the image observed in the image plane will be $\mathrm{f}'(X, Y)\mathrm{f}'^{*}(X, Y)$. Thus, diffraction theory confirms that the wave at the image plane is a magnified, inverted version of the wave that left the object plane, which is consistent with what we had concluded from geometrical considerations.

9.14 THE IMAGES OF POINTS HAVE THE FUNCTIONAL FORM $\mathrm{J}_1(X)/X$

As we know from earlier discussions, the Fourier transform of a point has uniform amplitude from $|\mathbf{S}| = 0$ to $|\mathbf{S}| = \infty$. Thus, the light scattered from a point source in the object plane of a lens should illuminate its back focal plane uniformly. That said, when it comes time to consider what happens as the light scattered by a point makes its way from the back focal plane to the image plane, it is important to remember that the area illuminated in the back focal plane is finite in extent. Every lens has a diameter, and only rays that pass through a lens can illuminate its back focal plane. Thus, the illuminated area in the back focal plane will (usually) be circular, and it is easy to show that its radius will be the radius of the lens times $[(s_2 - f)/s_2]$. In many optical devices, the radius of the illuminated area of the back focal plane that contributes to image formation is further reduced by a circular aperture that is deliberately included in the optical track. That limiting radius, however it is set, determines the maximum value of $|\mathbf{S}|$, $|\mathbf{S}|_{max}$, that can contribute to the final image. Thus, at the image plane, the wave produced when a point is imaged by a lens–aperture system will be the Fourier transform of a function in reciprocal space that is 1.0 everywhere inside a radius of $|\mathbf{S}|_{max}$, and 0 everywhere else. If the specimen of interest is more complicated than a single point, at the image plane, the wave will be the convolution of the Fourier transform of the specimen with the Fourier transform of a single point.

The Fourier transform of a two-dimensional object the density distribution of which is a circle of radius a that has an amplitude of 1 within its boundary and is 0 everywhere else is the two-dimensional equivalent of the Fourier transform of a one-dimensional square wave. Because this two-dimensional object has cylindrical symmetry, its transform will also, and we anticipate that like $\sin x/x$, it will have its maximum value at $\mathbf{S} = 0$, decline to 0 at a radius in reciprocal space in the neighborhood of $(1/2a)$, and then oscillate with an ever-decreasing amplitude. As appendix 9.3 shows in theory, and figure 9.8 confirms in practice, this is true. Normalized for the area of the circle, the Fourier transform of a circle of radius a, $F(R, \Theta)$, is:

$$F(R, \Theta) = 2J_1(2\pi aR)/(2\pi aR), \qquad \textbf{(9.8)}$$

where J_1 is the cylindrical Bessel function of order 1, and qualitatively, $J_1(x)/x$ resembles $\sin x/x$, as you can easily confirm by evaluating equation (9.8) numerically.

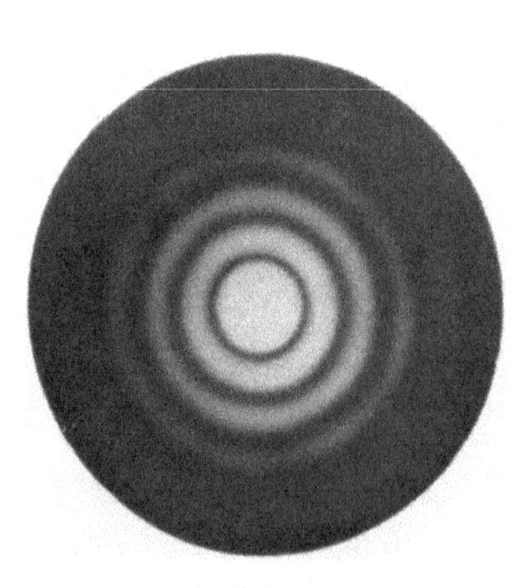

Figure 9.8 The image of a point produced using an optical instrument of finite aperture. Note that this photograph has been grossly overexposed to make its peripheral rings visible. (Reprinted with permission from Born, M., and Wolfe, E., [1999] *Principles of Optics*, 7th edition, New York: Cambridge University Press.)

Figure 9.8, which corresponds to $F(R, \Theta)^2$, is the image of a point produced by an optical instrument of finite aperture, and its bright, central maximum is called **Airy's disk**. For the record, the first 0 of $J_1(x)/x$, which corresponds to the innermost dark ring in figure 9.8, occurs at $x = 3.832$; the first 0 of $\sin x/x$ occurs at $x = 3.141$.

9.15 RAYLEIGH'S CRITERION IS USED TO ESTIMATE THE RESOLUTION OF MICROSCOPES

As we pointed out in section 7.3, the resolution of an optical device is the separation that two points have in object space in order that the image of the two of them be unmistakably different from the image of a single point. Now that we know what the image of a point is, we have the information we need to address this question for the images formed by microscopes. In the 19th century, Lord Rayleigh suggested that it would be reasonable to consider two points in some image resolved if the distance between them is larger than the separation at which the peak of the $J(x)_1/x$ function that corresponds to the image of one of them coincides with the first 0 of the $J_1(x)/x$ function that is the image of the other. (We will revisit Rayeigh's criterion again in chapter 10.) Because the first 0 of $J_1(x)$ falls at $x = \sim 1.22\pi (= 3.832\ldots)$, Rayleigh's criterion is met if the separation between the two points, ΔR, satisfies the equation:

$$|\mathbf{S}|_{max} \Delta R = 0.61.$$

Suppose that the radius of the circular region in the back focal plane that contributes to images in some microscope is ρ_{max}. If that is so, then the maximum value of the scattering angle of the scattered radiation the system images, $2\theta_{max}$, will be $\tan^{-1}(\rho_{max}/f)$. Furthermore, by definition, $|\mathbf{S}|_{max} = (2/\lambda) \sin \theta = (2/\lambda) \sin[\tan^{-1}(\rho_{max}/f)/2]$. Thus:

$$\Delta R = 0.61\lambda/(2 \sin[\tan^{-1}(\rho_{max}/f)/2]).$$

Microscopes are normally operated with the object plane only slightly further away from the objective lens than its focal length. This means that the radius of the operative part of the back focal plane is almost the same as the radius of the part of the objective lens that contributes to image formation, a. It follows that we can write:

$$\Delta R = 0.61\lambda/(2 \sin[\tan^{-1}(a/f)/2]).$$

Now $\tan^{-1}(a/f)$ is half the angle subtended by the lens when the lens is viewed from the on-axis point in the object plane, and it is equal to $2\theta_{max}$. Thus:

$$\Delta R = 0.61\lambda/[2 \sin(2\theta_{max}/2)].$$

In order to convert this expression for resolution into the one encountered in most textbooks, we need to ask how close $[2 \sin(2\theta_{max}/2)]$ is to $\sin(2\theta_{max})$ numerically. The answer is that at 30°, the difference between the two is about 3%, and at 45°, the difference is only about 8%. When that angle is 45°, the specimen is already very close to the front surface of the objective lens, and given how rough and ready the resolution estimation business is to begin with, who is going to quarrel about 10%? So, because in the optical world, unlike the crystallographic world, $2\theta_{max}$ is usually designated θ, we write:

$$\Delta R = 0.61\lambda / \sin\theta.$$

There is one more subtlety to deal with. The velocity changes observed when light crosses refractive index boundaries correspond to changes in wavelength, as we already know. As is pointed out in section 3.3, in a medium of refractive index n, the wavelength of light having a wavelength of λ in vacuo is λ/n. It follows that for any given lens geometry and frequency of light, $|\mathbf{S}|_{max}$ will be larger than it would have been in vacuo if the medium through which light travels between the sample and the objective lens has an index of refraction greater than 1.

For this reason, light microscopists intent on obtaining high-resolution images commonly fill the gap between the coverslip, which sits on top of their specimens, and the front surface of the objective lens of their microscope with a transparent fluid that has a high-refractive index. These fluids are called **immersion oils**, and the refractive indices of the immersion oils in common use are similar to those of ordinary glass; they are of the order of 1.5. This has the effect of dividing the wavelength term in the resolution equation by n, where n is the refractive index of whatever is in the gap. Thus, at last, we arrive at the standard expression for optical resolution:

$$\Delta R = 0.61\lambda / n \sin\theta. \quad (9.9)$$

The quantity $(n \sin\theta)$ is the **numerical aperture** of an objective lens.

When the gap between the specimen and the objective lens is filled with immersion oil, the refraction that would otherwise occur as light enters the front surface of an objective lens is greatly reduced, or eliminated entirely, which means that the focal lengths of objective lenses used for oil immersion microscopy are determined primarily by the curvatures of their back surfaces, and they are designed with this in mind. Thus, the wise microscopist does not use oil immersion lenses to image objects without also using immersion oil because the optical properties of such lenses are bound to be suboptimal if an abnormally large amount of refraction occurs at their front surfaces.

Comment: The specimens biologists examine typically consist of a drop of water that includes the object of interest on a glass slide that is covered by a thin piece of glass called a "coverslip" that both flattens the top surface of the drop and reduces the rate of evaporation. If the gap between the objective lens and the coverslip is filled with air, the refraction that occurs at the coverslip/air interface will increase the angular divergence of the rays leaving any given point in the specimen, and that increase will reduce the resolution of the resulting image. This effect is minimized when oil immersion techniques are used.

9.16 MICROSCOPES DISPLAY DEPTH OF FIELD AND DEPTH OF FOCUS

So far, we have attended to only two properties of the images produced by microscopes: magnification and resolution. High resolution correlates with high magnification because high resolution can only be achieved using lenses that have high numerical apertures, and a lens will have a high numerical aperture only if its focal length is small compared to its diameter. When specimens are placed close to the focal planes of such lenses, as they must be if the resolving power of such a lens is to be taken advantage of, high magnification images will invariably result. That said, it is easy to generate images that have high magnifications but low resolutions using lenses that have low numerical apertures. Microscopic images have two other properties that depend on numerical aperture: **depth of field** and **depth of focus**.

In much of the discussion so far, we have implicitly assumed that the specimens to be examined are so thin that they can be considered as two-dimensional objects, and, thus, if the plane of such a specimen is coincident with the object plane of a microscope, the entire specimen will be in perfect focus in the image plane. The depth of field of a microscope specifies how thick a specimen can be and still be in satisfactory focus through its entire thickness.

Depth of field is easily estimated for microscopes if we start with the premise that the only reason the images of points are not points is that the apertures of objective lenses are finite. Consider a point in the object plane of the objective lens of a microscope that is on its optical axis, and thus perfectly focused in its image plane. To a useful first approximation, we know that the image of that point will be a disk the radius of which is about the same as the resolution of the microscope, suitably magnified. If the rays contributing to the image of that point were projected onto a plane that is parallel to the object plane, but slightly displaced from it, the image they would create would still be a disk. However, the further that second plane was from the object plane, the larger the radius of that disk would be because the rays contributing to that image diverge on either side of the object plane. The radius of that disk, r, is related to the distance between the object plane and the plane of observation, Δx, as follows:

$$r = \Delta x \tan \theta,$$

where θ is the angle between the optical axis and a ray from the center of the object plane to the edge of the objective lens. If the radius of the disk in the second plane were only half the resolution of the microscope, it would be hard for the observer to distinguish the image of that (out of focus) disk, which would be the convolution of that disk with the appropriate $J_1(x)/x$ function, from the image of a perfect point, which is $J_1(x)/x$ convoluted with a delta function. For most purposes, that slightly out of focus disk would be just as well focused as the image of a perfect point. If we take that degree of image degradation as the limit of what is acceptable, then the depth of field of an image, D, is twice the distance from the object plane at which the diameter of the projected image equals the resolution of the microscope. Thus:

$$D = 2\Delta x_{\max} = 2\Delta R/2 \tan\theta = 0.61\lambda/n \sin\theta \tan\theta.$$

Clearly, as θ goes toward $\pi/2$, which is to say as the numerical aperture increases, D goes to 0. Thus, when a specimen is examined that is thicker than D, its image will be an in-focus, projection image of whatever is in the slab of thickness D that is centered on the object plane, superimposed on out of focus images of whatever is in the adjoining slabs, and the further those slabs are from the object plane, the more out of focus they will be.

Comment: For future reference, you should note the similarity between the equation for depth of field just provided and the expression developed in section 8.2 for the relationship between sample thickness and the resolution at which the projection theorem starts to fail as a description of images:

$$|\mathbf{S}|_{\max} = (2/t\lambda)^{1/2}.$$

If you equate t, the sample thickness, with D, and $|\mathbf{S}|_{max}$ with $\sin\theta/\lambda$, you discover that at small scattering angles, the two expressions are almost the same. The messages conveyed by this fact are: (1) that the image of a thin specimen ceases to correspond well to a projection of its density distribution function onto the image plane when its thickness begins to exceed the depth of field of the device used to create it, which is not too surprising, and (2) that the thickness of specimens that is compatible with the projection theorem is inversely related to resolution squared; a doubling of the resolution reduces the allowable thickness by a factor of 4. This observation will have consequences when it comes time to consider the way in which high-resolution structures are recovered from electron microscopic images (chapter 13).

Anyone who has ever used a light microscope is familiar with the effect that moving the specimen stage up and down has on the images of thick objects. You get the sensation that you are traveling up and down through the specimen, which is true, optically speaking. Save for the confusion produced by the images of out-of-focus slabs, you are looking at the specimen a slice at a time. The effect is called **optical sectioning**, and, quite obviously, there is information about structure of thick specimens in the direction perpendicular to the optical axis contained in the way their images change as the distance between the specimen and the objective lens is altered. The higher the numerical aperture of the objective lens, the thinner the slab will be that is in focus at any position of the specimen, and the more obvious this sectioning effect will be. The depths of field of high-resolution objective lenses can approach their lateral resolutions. For low resolution (i.e., low numerical aperture lenses), the depth of field will be tens of microns.

Depth of focus is the image plane equivalent of depth of field. The question that is answered about an imaging system by determining its depth of focus is how far from the true image plane of the system a piece of film, or some other light detector, can be placed, and still get a well-focused image of whatever is in the object plane. Using the same arguments that are used to estimate depth of field, we are led to the conclusion that depth of focus, D_f, should be:

$$D_f = 0.61\lambda M/n \sin\theta \tan\theta', \tag{9.10}$$

where M is the magnification of the image, and θ' is one-half the angle subtended by the lens (or the limiting aperture) when seen from the point where the optical axis crosses the image plane. However, because $\tan\theta' \approx (\tan\theta)/M$, it follows that:

$$D_f = M^2 D. \tag{9.11}$$

For the objective lenses that are supplied with modern microscopes, depth of focus does not vary a lot with numerical aperture (or magnification), and it is of the order of millimeters.

PROBLEMS

1. In section 9.12, the following equation appears:

$$\Delta\Delta = -(1/f)(x\xi + y\eta).$$

   Do the algebra necessary to demonstrate the validity of this equation.
2. Van Leeuwenhock's microscopes had only a single lens, an objective lens, which was a spherical glass bead. Suppose you wanted to build a microscope of this sort that would produce an image that has a magnification of 50 times at a distance from the back principal point of the lens, s_2, of 15 cm.
   a. At what distance from the front principal point of the lens would objects have to be viewed to obtain this result?
   b. Using the expressions provided for the lens law and the focal length, determine what the radius of the glass bead would have to be, assuming that the glass available had an index of refraction of 1.5.
   c. Using the expressions provided for the locations of the principal points, determine what the distance would have to be from the front surface of the bead to the object being imaged in order to obtain this result.
3. Suppose you wanted to make a microscope based on a spherical objective lens that produced an image that has a magnification of 50 times at a distance from the back principal point of the lens, s_2, of 15 cm, but that was to operate using an immersion oil that has an index of refraction of 1.5 to fill the space between the front surface of the lens and the object plane.
   a. Derive an expression for the lens law that is valid for this set of circumstances, again assuming that the glass to be used for the lens has an index of refraction of 1.5. Hint: Rework the proof in appendix 9.1, noting that only the encounter with the R_2 surface of the lens will result in refraction.
   b. What would the radius of the glass bead have to be now in order to obtain the desired focal properties?

c. Would it make any difference if the front surface of the lens were flat instead of spherical?

4. Suppose that the lens you "designed" in problem 3 was capable of bringing to a focus all the light entering it from a point on its optical axis that is 0.3 cm from its front principal point. Furthermore, assume that the lens itself constituted the limiting aperture of the Van Leeuwenhock microscope you made with that lens. What would the resolution of the microscope be when operating with light having a wavelength of 500 nm?
5. Show that the radius of the disk that is the image of a point produced by a lens that has spherical aberration is proportional to the cube of the radius of the aperture of that lens.
6. Some lenses are plano-convex, which is to say that one of their surfaces is convex and the other a plane.

   a. When it comes to spherical aberration, does it make any difference which way around such a lens is placed in an image-forming device?
   b. If so, will it be better to orient it so that its R_1 surface is the flat surface, or will it work better the other way around?

FURTHER READING

In the 1960s, the textbooks used in first-year college physics courses invariably included chapters on optics, but they seldom do today. Anyone of those older physic texts would be a good place to find discussions of the basics of optics.

Max Born and Emil Wolf's *Principles of Optics*, 7th edition (New York: Cambridge University Press, 1999.) covers the all the material in this chapter with magisterial authority, but might be found by some to be less than entirely user-friendly.

Two texts that might be more comforting are:

Jenkins, F. A., and White, H. E.(1976). *Fundamentals of Optics*, 4th ed., New York: McGraw-Hill.

Klein, M. V., and Furtak, T. E. (1986). *Optics*, 2nd ed., New York: John Wiley & Sons.

APPENDIX 9.1 DERIVATION OF THE LENS LAW

Consider a glass lens having an index of refraction of n that has two spherical surfaces, one of radius R_1 and the other of radius R_2. If the lens is cylindrically symmetric, which most are, it will have a unique optical axis that passes through the lens normal to both of its surfaces (figure 9.1.1). The thickness of a lens, if it is appreciable, can be specified as D, the distance between its two surfaces along its optical axis. (The convention for naming lens surfaces and determining the signs of radii is described in section 9.5.)

Suppose a point source of light is placed on the optical axis of a lens at a distance s_1 to the left of its left surface, and for the moment, assume that s_1 is large compared to both $|R_1|$ and $|R_2|$. Where will a ray leaving that point at an angle of θ_1 relative to the optical axis intersect with the optical axis on the right side of the lens, if indeed

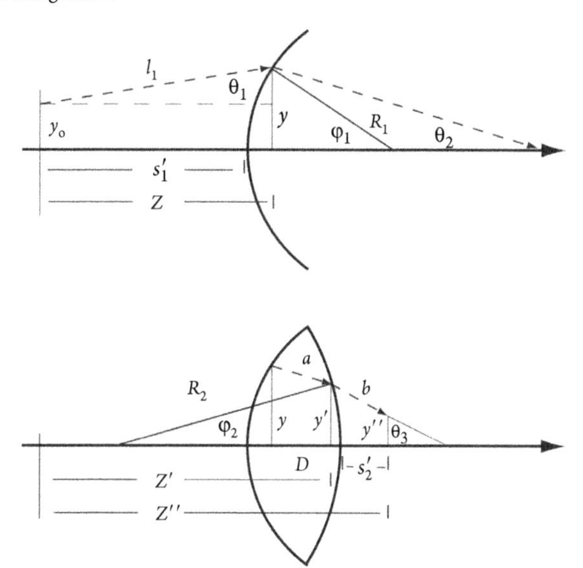

Figure 9.1.1 The trajectory of light through a biconvex lens. The top panel defines the parameters that have to be taken into account to describe the refraction that occurs at the front surface of the lens. The bottom panel does the same for the parameters that have to be taken into account in dealing with the refraction that occurs at the right surface of the lens. The heavy arrow is the optical axis, and the dotted arrows show the trajectory of the ray that originates at $(y_o, -s_1')$, continues along the trajectory marked *a* inside the lens, and then exits from the lens along the trajectory marked *b*.

it does so at all? This problem is solved in two steps. First, the direction the ray will take after it has been refracted by its encounter with the left face of the lens at (y, z) is determined (figure 9.1.1, top). Second, using the result of the first step as input, the direction the ray takes after it has crossed the right-hand boundary of the lens at (y', z') is worked out (figure 9.1.1, bottom).

Figure (9.1.1, top) identifies the geometric quantities relevant to the encounter of a ray with the left surface of the lens, and to simplify matters, we assume that the lens is surrounded by air ($n = 1.0$). We can easily verify that:

$$\sin(\theta_1 + \varphi_1) = n \sin(\varphi_1 - \theta_2) \qquad \text{(Snell's law)} \tag{9.1.1}$$

$$y = l_1 \sin \theta_1 + y_o = R_1 \sin \varphi_1. \tag{9.1.2}$$

However, $z = s_1' + R_1(1 - \cos \varphi_1)$, and $l_1/z = \cos \theta_1$. Therefore,

$$l_1 = (1/\cos \theta_1)\left[s_1' + R_1(1 - \cos \varphi_1)\right]. \tag{9.1.3}$$

If the point where the ray reaches the right surface of the lens is designated (y', z'), then the following must be true:

$$y' = y - a \sin\theta_2 = -R_2 \sin\varphi_2, \tag{9.1.4}$$

where a is the distance between (y, z) and (y', z'). Furthermore:

$$z' = z - a \cos\theta_2 = s_1' + D + R_2(1 - \cos\varphi_2). \tag{9.1.5}$$

At the left boundary of the lens, a second refraction event occurs:

$$n \sin(\varphi_2 + \theta_2) = \sin(\varphi_2 + \theta_3), \tag{9.1.6}$$

and at (y'', z'') beyond the lens:

$$y'' = y' - b \sin\theta_3, \tag{9.1.7}$$

and:

$$z'' = z' - b \cos\theta_3, \tag{9.1.8}$$

where b is the distance between (y', z') and (y'', z'').

In the paraxial limit, the sine and cosine terms in these equations are well approximated by the first terms in their power series expansions. Thus, all the sine terms can be replaced by the value of the argument itself (in radians), and cosine terms can be replaced by either by 1.0, if the angle is close to 0, or by -1.0, if the angle is close to 180°. When this is done, we find:

$$\theta_1 + \varphi_1 = n(\varphi_1 - \theta_2), \tag{9.1.1'}$$

$$y = l_1\theta_1 + y_o = R_1\varphi_1, \tag{9.1.2'}$$

$$l_1 = s_1' = z, \tag{9.1.3'}$$

$$y' = y - a\theta_2 = -R_2\varphi_2, \tag{9.1.4'}$$

$$z' = z + a = s_1' + D, \tag{9.1.5'}$$

$$n(\varphi_2 + \theta_2) = \varphi_2 + \theta_3, \tag{9.1.6'}$$

$$y'' = y' - b\theta_3, \tag{9.1.7'}$$

$$z'' = z' + b. \tag{9.1.8'}$$

Comparison of equations (9.1.3′), and (9.1.5′) reveals that $a = D$.

The goal is to find an expression for y'', the distance of the ray from the axis, that depends only on s_1', s_2', the distance from the back face of the lens to the point where y'' is measured, y_o, θ_1, D, R_1 and R_2. By combining equations (9.1.7′) and (9.1.4′), we find that:

$$y'' = y - D\theta_2 - b\theta_3.$$

From equation (9.1.6′), we know that $\theta_3 = (n-1)\varphi_2 + n\theta_2$, and $b = s_2{}'$, and thus:

$$y'' = y - \left(D + ns_2{}'\right)\theta_2 - s_2{}'(n-1)\varphi_2.$$

Using equation (9.1.4′), we can write:

$$\begin{aligned} y'' &= y - \left(D + ns_2{}'\right)\theta_2 + (s_2{}'/R_2)(n-1)(y - D\theta_2) \\ &= \left[1 + (s_2{}'/R_2)(n-1)\right]y - \left[D + ns_2{}' + (Ds_2{}'/R_2)(n-1)\right]\theta_2. \end{aligned}$$

From equation (9.1.1′), we know that $\theta_2 = [(n-1)/n]\varphi_1 - (1/n)\theta_1$, and from equation (9.1.2′) that that $\varphi_1 = (1/R_1)(s_1{}'\theta_1 + y_o)$. Thus:

$$\begin{aligned} \theta_2 &= [(n-1)/n]\,\varphi_1 - (1/n)\theta_1 \\ &= (1/n)\left[(n-1)(s_1{}'/R_1) - 1\right]\theta_1 + [(n-1)/nR_1]\,y_o. \end{aligned}$$

Further, $y = s_1{}'\theta_1 + y_o$. Hence, after gathering terms, we obtain the equation desired:

$$\begin{aligned} y'' = {} & [s_1{}' + s_1{}'s_2{}'(n-1)/R_2 + D/n + s_2{}' \\ & + Ds_2{}'(n-1)/nR_2 - D(n-1)s_1{}'/nR_1 \\ & - (n-1)s_1{}'s_2{}'/R_1 - D(n-1)^2 s_1{}'s_2{}'/nR_1R_2]\theta_1 \\ & + \left[1 - s_1{}'\left\{(n-1)/R_1 - (n-1)/R_2 + D(n-1)^2/nR_1R_2\right\}\right. \\ & \left. - D(n-1)/nR_1\right]y_o. \end{aligned}$$

This ghastly looking expression can be written in a much simpler form if the conventions for measuring distances along the optical axis are redefined. Inspection of the y_o term shows how we might proceed. The quantity in curly braces in the y_o term has the dimensions of distance^{-1}, and for reasons that will become clear later, we will call that distance f:

$$f = \left[(n-1)\,(1/R_1 - 1/R_2) + D\,(n-1)^2\,/nR_1R_2\right]^{-1}. \qquad \textbf{(9.1.9)}$$

Making this substitution, the y_o term becomes:

$$\left[1 - s_2{}'/f - D(n-1)/nR_1\right]y_o,$$

and if we replace s_2', with a new variable that we will call s_2:

$$s_2 \equiv s_2{}' + (n-1)Df/nR_1,$$

we discover that the y_o term can be made even simpler:

$$[1 - s_2/f]\, y_o.$$

Thus, in effect, instead of measuring distances along the optical axis to the left of a lens from an origin that is the back surface of the lens, we are suggesting that they should be measured from an origin *inside* the lens that is located at $(n - 1)Df/nR_1$. The fact that this rather odd suggestion further simplifies the algebra of the y_o term suggests that it might be interesting to see what happens if we define a new variable, s_1, as follows:

$$s_1 \equiv s_1' - (n - 1)Df/nR_2,$$

and see what happens to the θ_1 part of the y'' equation when s_1' and s_2' are replaced by s_1 and s_2. The results are gratifying:

$$y'' = [s_1 + s_2 - s_1 s_2/f]\,\theta_1 + [1 - s_2/f]\, y_o. \qquad \textbf{(9.1.10)}$$

The locations from which s_1 and s_2 are measured are referred to as the **principal points** of a lens. As D goes to 0, the differences between s_1 and s_1', and between s_2 and s_2' go to 0 also.

APPENDIX 9.2 OPTICAL PATH LENGTH AND SPHERICAL ABERRATION

Using the geometric parameters defined in figure 9.1.1, it is easy to write expressions for the on-axis optical path length between two points located at s_1' and s_2' on the optical axis, OP_d, and the optical path length for any off-axis trajectory that connects the same two points, OP_i:

$$OP_d = s_1{}' + s_2{}' + nD$$

$$OP_i = \left[\left(s_1{}' + z_1\right) + y^2\right]^{1/2} + \left[\left(s_2{}' + z_2\right)^2 + y'^2\right]^{1/2} + n\,(D - z_1 - z_2)\,(1 + \theta_2^2/2).$$

In the second equation, $z_1 \equiv z - s_1{}'$, and $z_2 \equiv s_1 + D - z'$. The equation for OP_i will be usefully accurate provided θ_2 is small, as it is rays originating at points on or close to the optical axis. ΔOP, the difference between these two optical path lengths is:

$$\Delta OP = OP_i - OP_d.$$

The two square root terms in the expression for OP_i can be evaluated as power series in $s_1{}'$ and $s_2{}'$, assuming that y and y' are small compared to $s_1{}'$ or $s_2{}'$, and by recognizing that if that is so, then z_1 and z_2 will be small also. Taking the first three terms of both power series, subtracting terms where possible, and taking advantage of some trigonometric relationships, we finds that:

$$\Delta OP = \left(y^2 + z_1^2\right)/2s_1' - \left(2z_1s_1' + z_1^2 + y'^2\right)/8s_1'^3 + \left(y'^2 + z_2^2\right)/2s_2' \\ - \left(2z_2s_2' + z_2^2 + y'^2\right)/8s_2'^3 + (1-n)(z_1+z_2) + n(y-y')\theta_2/z.$$

At this juncture, the geometry of the lens must be considered, and when that is done, it is found that to fourth order in y and y':

$$z_1 = R_1(1 - \cos\varphi_1) \approx y^2/2R_1 + y^4/8R_1^3,$$

and:

$$z_2 = -y'^2/2R_2 - y'^4/8R_2^3.$$

(Remember the sign convention for lens radii!) Substituting these expressions for z_1 and z_2 into the equation for ΔOP, we find that to fourth order in y and y':

$$\Delta OP = \left(y^2/2\right)\left[(1-n)/R_1 + 1/s_1'\right] + \left(y'^2/2\right)\left[(n-1)/R_2 + 1/s_2'\right] \\ - \left(y^4/8\right)\left[\left(1/s_1'^2\right)\left(2/R_1 + 1/s_1'\right) + (n-1)/R_1^3\right] \\ - \left(y'^4/8\right)\left[\left(1/s_2'^2\right)\left(1/s_2' - 2/R_2\right) - (n-1)/R_2^3\right] + \left(y - y'\right)\left(n\theta_2/2\right).$$

The next step is to determine the relationship between y and θ_2, which can be done using Snell's law (see appendix 9.1). Snell's law requires that:

$$\sin(\theta_1 + \varphi_1) = n\,sin(\varphi_1 - \theta_2),$$

which is equivalent to:

$$\sin\theta_1\cos\varphi_1 + \cos\theta_1\sin\varphi_1 = n\sin\varphi_1\cos\theta_2 - n\cos\varphi_1\sin\theta_2.$$

Because $\sin\varphi_1 = y/R_1$, $\cos\varphi_1 \approx 1 - y^2/2R_1^2 - y^4/8R_1^4$. In addition, $\sin\theta_1 = y\left[\left(s_1' + z_1\right)^2 + y^2\right]^{1/2}$, and $\cos\theta_1 = \left(s_1' + z_1\right)\left[\left(s_1' + z_1\right)^2 + y^2\right]^{1/2}$. Substituting these expressions into the preceding expression, and making the usual approximations, we can show that to third order in y:

$$\theta_2 = \left(y/n\right)\left[(n-1)/R_1 - 1/s_1'\right] \\ + \left(y^3/2n\right)\left[\left(1/s_1'^2\right)\left(2/R_1 + 1/s_1'\right) + (n-1)/R_1^3\right],$$

which for notational convenience, we will write as:

$$\theta_2 = \alpha y + \beta y^3.$$

We also need to understand the relationship between y and y'. Now:

$$y' = y - [D - R_1 (1 - \cos\varphi_1) + R_2 (1 - \cos\varphi_2)] \tan\theta_2,$$

but because θ_2 is generally small, this the relationship is well approximated by the following expression:

$$y' = y - \left(D - y^2/2R_1 + y'^2/2R_2\right) \tan\theta_2,$$

a quadratic equation that is easily solved. The next steps are: (1) to expand the square root term in the solution of this equation as a power series, truncating after the third order term in $\tan\theta_2$, (2) to replace $\tan\theta_2$ everywhere it appears in the resulting expression by the appropriate power series in θ_2, retaining the first few terms, and, finally, (3) to replace θ_2 everywhere it appears in the equation so generated using the expression for θ_2 as a function of y given previously. To third order in y:

$$\begin{aligned} y' = y(1 - D\alpha) + y^3 \big[&(1/2)(1/R_1 - 1/R_2)\,\alpha - D\left(\beta + \alpha^3/3\right) \\ &+ \left(D\alpha^2/2R_2\right)(1 - D\alpha)\big], \end{aligned}$$

which (again) for notational convenience, we will write in the following, abbreviated form:

$$y' = y(1 - D\alpha) + Ey^3.$$

Furthermore, to fourth order in y:

$$n\left(y - y'\right)(\theta_2/2) = (n/2)\left(D\alpha^2 y^2 + \alpha(\beta D - E)\,y^4\right).$$

Substituting these expressions into the equation for ΔOP, and gathering terms, we find:

$$\begin{aligned} \Delta OP = \left(y^2/2\right)&\left\{(1-n)/R_1 + 1/s_1{}' + \left(1 - D\alpha^2\right)\left[(n-1)/R_2 + 1/s_2{}'\right] + nD\alpha^2\right\} \\ &+ y^4\left\{(1-\alpha D)\,E\left[(n-1)/R_2 + 1/s_2{}'\right] + nD\alpha\left(\beta + \alpha^3/3\right)/2 - nE\alpha/2\right\} \\ &- \left(y^4/8\right)\left\{2n\beta + \left(1/s_2'^2\right)\left(1/s_2' - 2/R_2\right) - \left[(n-1)/R_2^3\right](1 - \alpha D)^4\right\}. \end{aligned}$$

Luckily, everything we need to know about spherical aberration can be extracted from the thin lens version of this equation. In the thin lens limit, $D = 0$, $s_1 = s_1{}'$, $s_2 = s_2{}'$, $\alpha = (n-1)/nR_1 - 1/ns_1$, $\beta = (1/2n)[(1/s_1{}^2)(2/R_1 + 1/s_1) + (n-1)/R_1{}^3]$, and $E = (\alpha/2)(1/R_1 - 1/R_2)$. When these substitutions are made, a much simpler equation emerges:

$$\Delta OP \approx \left(y^2/2\right)[-1/f + 1/s_1 + 1/s_2]$$

$$+ \left(y^4/2n\right)(1/R_1 - 1/R_2)\left[(n-1)/R_1 - 1/s_1\right]$$
$$\times \left\{-1/f + 1/s_1 + 1/s_2 + (1/2)\left[(n-1)/R_1 - 1/s_1\right]\right\}$$
$$- \left(y^4/8\right)\left[\left(1/s_1^2\right)(2/R_1 + 1/s_1) + (n-1)/R_1^3 + \left(1/s_2^2\right)\right.$$
$$\left.(1/s_2 - 2/R_2) - (n-1)/R_2^3\right].$$

If the system being examined is a microscope, then s_1 will be only slightly greater than f, and s_2 will be $\sim Mf$, where M is >10. Under these conditions, close to the paraxial focal plane, the second term in the expression for ΔOP will be (approximately):

$$\left(y^4/4\right)\left[n(n-1)f\right]^{-1}\left[(n-1)/R_1 - 1/f\right]^2.$$

However, $[(n-1)/R_1 - 1/f] = (n-1)/R_2$; hence, that second term can also be written:

$$\left(y^4/4\right)\left[(n-1)/nfR_2^2\right].$$

When these assumptions are valid, the third term becomes:

$$-\left(y^4/4\right)\left\{\left[(n-1)/2\right]\left(1/R_1^3 - 1/R_2^3\right) + \left(1/2f^2\right)\left(2/R_1 + 1/f\right)\right\}.$$

Finally, rather than using y as the variable to describe the deviation of some ray from the axial direction, we commonly use θ, the scattering angle. For small θ, which is all we will be interested in going forward, $y \approx f\theta$. Thus:

$$\Delta OP \approx (\theta^2/2)f^2\left[-1/f + 1/s_1 + 1/s_2\right]$$
$$-\left(\theta^4/4\right)f^4\left\{\left[(n-1)/2\right]\left(1/R_1^3 - 1/R_2^3\right)\right.$$
$$\left.+\left(1/2f^2\right)\left(2/R_1 + 1/f\right) - (n-1)/nfR_s^2\right\}$$

Now if we define the spherical aberration coefficient, C_s, as follows:

$$C_s \equiv f^4\left[((n-1)/2)\left(1/R_1^3 - 1/R_2^3\right) + \left(1/2f^2\right)\left(2/R_1 + 1/f\right) - (n-1)/nfR_s^2\right],$$

we can then write:

$$\Delta OP \approx \left(\theta^2/2\right)f^2\left[-1/f + 1/s_1 + 1/s_2\right] - \left(\theta^4/4\right)C_s. \qquad \textbf{(9.1.1)}$$

APPENDIX 9.3 THE FOURIER TRANSFORM OF A CIRCULAR APERTURE

Whenever we seek to describe the properties of a system that has circular symmetry, the most economical way to do it is using a polar coordinate system. In such a system, the real-space coordinates x and y are replaced by r and θ using the following relationships:

$$x = r\cos\theta; y = r\sin\theta, \text{ and } r = (x^2 + y^2)^{1/2}.$$

Polar coordinates should be used in reciprocal space also. Thus:

$$X = R\cos\Theta; Y = \sin\Theta, \text{ and } R = (X^2 + Y^2)^{1/2}.$$

The Fourier transform of a function, f(x, y), is:

$$\mathrm{F}(X,Y) = \int\int \mathrm{f}(x,y)\exp\left[-2\pi irR(xX + yY)\right]\, dxdy.$$

Replacing x, y, X, and Y with their polar equivalents, we find:

$$\mathrm{F}(R,\Theta) = \int\int \mathrm{f}(x,y)\exp\left[-2\pi irR(\cos\theta\ \cos\Theta + \sin\theta\ sin\Theta)\right] rd\theta\, dr.$$

Using a little elementary trigonometry, we can write this expression as follows:

$$\mathrm{F}(R,\Theta) = \int\int \mathrm{f}(x,y)\exp\left[-2\pi irR\ \cos(\theta - \Theta)\right] rd\theta\, dr.$$

For a circular aperture of radius a, this integral simplifies to:

$$\int_0^a r\, dr \int_0^{2\pi} \exp\left[-2\pi irR\ \cos(\theta - \Theta)\right] d\theta.$$

The system under consideration has circular symmetry, so the value of this integral will not depend on Θ. Thus, nothing will be lost if Θ is set to 0. Hence:

$$\mathrm{F}(R,\Theta) = \mathrm{F}(R,0) = \int_0^a r\, dr \int_0^{2\pi} \exp(-2\pi irR\ \cos\theta)d\theta.$$

The integral in θ is challenging, and the simplest way to evaluate it is to consult any mathematical tables that happen to be lying around in hopes that some overachiever solved it in the past. In this instance, we are delighted to discover that:

$$\mathrm{J_n}(x) = \left(i^{-n}/2\pi\right)\int_0^{2\pi} \exp\left(ix\ \cos\theta\right)\exp\left(in\theta\right) d\theta,$$

where $\mathrm{J_n}(x)$ is the cylindrical Bessel function of order n. Bessel functions oscillate with decreasing amplitude as x increases; hence, they have the character we expect of the solution to this particular problem. If n is taken to be 0, then the form of this function is exactly the one needed. Thus:

$$F(R,\ 0) = 2\pi \int_0^a J_o(-2\pi rR)\, rdr = 2\pi \int_0^a J_o(2\pi rR)\, rdr.$$

(The last equality holds because J_0 is symmetric about 0.) This is progress, but we are still left with a nasty looking definite integral to evaluate. Happily, the tables provide a solution to this problem too:

$$d\left[x^{n+1} J_{n+1}(x)\right] = x^{n+1} J_n(x)\, dx.$$

It follows that, normalized for the area of aperture:

$$F(R,\ 0) = 2\left(J_1(2\pi aR) / (2\pi aR)\right).$$

In short, the form of $F(R,\ 0)$ is $J_1(x)/x$, and for the record, when $J_1(x)$ is expressed as a power series, the first term is $(x/2)$. Thus, the value of $2J_1(x)/x$ at $x = 0$ is 1.0, just like $\sin x/x$.

10

The Light Microscope

As van Leeuvenhock knew, a single focusing lens is all you need to make a microscope. Nevertheless, all modern microscopes are compound microscopes, like Hooke's instruments, which have eyepieces to magnify the images produced by their objective lenses. Moreover, modern objective lenses are invariably multielement assemblies that focus light with much lower levels of aberration than can be achieved using single element lenses, and their oculars are usually multielement lenses as well.

In this chapter, we discuss a few of the astonishingly large number of ways that microscopes are used to obtain information about biological structure today. The goal is to provide you with an appreciation of the richness of this area of experimental biophysics and with confidence in your capacity to understand how these different kinds of microscopy work.

10.1 INCOHERENT LIGHT IS THE ILLUMINATION OF CHOICE FOR ORDINARY MICROSCOPY

In chapter 9, much is made of interference effects in light microscopy, and thus it may come as a surprise to learn that biological microscopy is commonly done using light sources that generate incoherent light, or **Koehler illumination** (figure 10.1). The light source for a Koehler illumination system is often an incandescent lamp, the heated filament of which generates light thermally, and hence incoherently. That light is passed through an opalescent diffusing screen to further degrade its transverse coherence before being focused by a lens, the auxiliary lens, on the point where the optical axis of the microscope intersects the front focal plane of a second focusing lens, called the **condenser**. The specimen is illuminated by the rays that

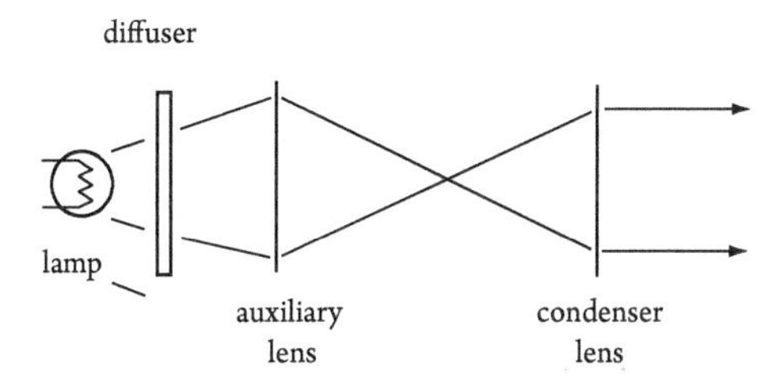

Figure 10.1 Schematic diagram of a Koehler light source.

emerge from the condenser, all of which are parallel to the optical axis, but in the plane of the specimen, the transverse coherence length is small.

The reason incoherent light is often preferred for microscopy is that, all else being equal, images produced using incoherent illumination are better resolved than those produced using coherent illumination. No matter how a specimen is illuminated, the light it scatters can be described as the appropriately summed scatter from an array of pointlike volume elements. The radiation scattered by a single point is coherent with itself and will progress through the lenses of a microscope as Abbé's theory prescribes. Its image will have the form $[J_1(x)/x]^2$. If a specimen consisting of two such points is illuminated incoherently, the resulting image will have the form $([J_1(x)/x]^2 + [J_1(x - \Delta)/(x - \Delta)]^2)$, where Δ is the distance between the images of the two points. By contrast, if the same specimen is imaged coherently, its image will be $[(J_1(x)/x) + (J_1(x - \Delta)/(x - \Delta)]^2)$. Because $[J_1(x)/x]^2$ is a more sharply peaked function than $[J_1(x)/x]$ is, and because interference effects tend to fill the "gaps" between adjacent $[J_1(x)/x]$ functions in coherent images, incoherent images are better resolved than coherent images.

If a pair of adjacent points are imaged incoherently, there will be a noticeable trough between their two maxima if the distance between them meets or exceeds Rayleigh's criterion, That is, if their separation is $0.61\lambda/n \sin \theta$, or greater. However, if they are imaged coherently, there will be no dip between maxima unless their separation exceeds $\sim 0.9\lambda/n \sin \theta$, which is consistent with the conclusion we reached about crystallographic resolution in section 7.3. Another way of thinking about it is that when microscopes are operated using incoherent illumination, the images they produce correspond as closely to the predictions of geometrical optics as the laws of physics permit; each ray produces its own point image, and the overall image is a sum of independent point images. Thus, for many applications, incoherent illumination is a good thing, and the more perfectly incoherent the better.

10.2 INCOHERENT IMAGE FORMATION IS EASILY DESCRIBED IN FOURIER TERMS

The concepts developed to describe coherent imaging can provide a Fourier description of incoherent imaging. As we discussed in section 9.12, when coherent light passes through a specimen, the electromagnetic wave that emerges is the product of the original wave and an object transmission function, $o(x, y)$ that reflects the reductions in wave amplitude and phase changes that occurred during the passage of the wave through the specimen. The amplitude part of $o(x, y)$, $f(x, y)$ is a positive number ≤ 1, and its phase component can be conveniently written as $\exp[i\alpha(x, y)]$, where $\alpha(x, y)$ is the corresponding phase change. Thus, if only a single point of some specimen were to be illuminated, say the point at (x, y), then the object transmission function would be:

$$o(x', y') = f(x', y') \exp[i\alpha(x', y')]\delta(x' - x, y' - y).$$

Now $\alpha(x, y)$ equals $(2\pi/\lambda) \int n(x, y, z)dz$, where the z-axis is the optical axis, and $n(x, y, z)$ is the refractive of the specimen at (x, y, z), and the integration runs over the thickness of the specimen. In practice, the increase in optical path length, $\int n(x, y, z)dz$, observed when biological samples are examined in the light microscope is quite large, but most of it is due to the passage of light through the glass slide on which the sample is mounted and the coverslip that lies on top of it. Those contributions to the total phase change have nothing to do with the structure of the sample and do not vary with position. It follows that what counts for the images these instruments produce is not the total phase change that results as light passes through a mounted specimen, but rather the (often) much smaller, point-to-point variations in $\alpha(x, y)$ due to the structure of the specimen:

$$(2\pi/\lambda)\left[\int n\left(x, y, z\right) dz - \Delta n_o\right],$$

where Δn_o is the contribution made to the total index of refraction observed at (x, y) by the slide–coverslip assembly, which is assumed to be constant. That difference is what "$\alpha(x, y)$" will refer to hereafter. These comparatively small point-to-point variations in phase caused by the passage of light through the object of interest report on the distribution of nonaqueous material in the sample. Generally speaking, the higher the mass density of stuff in any volume element in the object, the higher its refractive index. (NB: The integrations specified lead to an accurate description of images only if the specimens are thin.)

The Fraunhofer diffraction pattern of the point specimen just described is proportional to the Fourier transform of its object transmission function:

$$\mathrm{FT}(f\left(x', y'\right) \exp\left[i\alpha\left(x', y'\right)\right] \delta\left(x' - x, y' - y\right)),$$

and, as we already know, that transform describes the impact that the object transmission function has on light passing through the back focal plane of the microscope being used to image the specimen. The amplitude of $\mathrm{FT}[\delta(x' - x, y' - y)]$ is a constant everywhere in the back focal plane, and it is multiplied both by $f(x, y) \exp[i\alpha(x, y)]$, which does not vary over the back focal plane, and by $\exp[-2\pi i(\xi x + \eta y)]$, which does, ξ and η being spatial coordinates in the back focal plane. The $\exp[-2\pi i(\xi x + \eta y)]$ term "encodes" information about the position of the illuminated point in the object plane.

Due to the finite apertures of microscopes, only a portion of the transform of a specimen can be accessed in the back focal plane. Thus, the image observed when light from the specimen reaches the image plane depends on the product of the specimen's Fourier transform and a **contrast transfer function**, which represents all the things a microscope does to light passing through it. Contrast transfer functions always include terms that represent the limiting aperture, $A(\xi, \eta)$, and they may also include terms that describe the phase distortions produced by lens aberrations, $\exp[i\omega(\xi, \eta)]$. Thus, in general, the contrast transfer function can be written $A(\xi, \eta)\exp[i\omega(\xi, \eta)]$, and the disturbance in the back focal plane that ultimately gets imaged is:

$$\mathrm{FT}(f(x', y')\exp[i\alpha(x', y')]\,\delta(x' - x, y' - y)) \cdot A(\xi, \eta)\exp[i\omega(\xi, \eta)].$$

It follows from what has just been said that the contribution the point specimen makes to the light wave that reaches the image plane will be:

$$\mathrm{FT}[\mathrm{FT}(f(x', y')\exp[i\alpha(x', y')]\delta(x' - x, y' - y))] * \mathrm{FT}(A(\xi, \eta)\exp[i\omega(\xi, \eta)]).$$

Ignoring magnification, and identifying $\mathrm{FT}(A(\xi, \eta)\exp[i\omega(\xi, \eta)])$ as the **point image function** of the microscope, $P(X, Y)$, where X and Y are the coordinates of the point in the image plane, the image, $I(X, Y)$, which is proportional to the square of the Fourier transform of the function in the back focal plane, will be:

$$I(X, Y) = \left[f(x', y')\exp[i\alpha(x', y')]\,\delta(x' + x, y' + y) * P(X, Y)\right]^2. \quad \mathbf{(10.1)}$$

It should be emphasized that equation (10.1) is correct for *both* incoherent and coherent illumination because the light scattered by a single point is coherent with itself, as we noted earlier. Now:

$$\begin{aligned} &f(x', y')\exp[i\alpha(x', y')]\delta(x' + x, y' + y) * P(X, Y) \\ &\qquad = f(-x, -y)\exp[i\alpha(-x, -y)]\,P(x - X, y - Y). \end{aligned}$$

Hence:

$$\begin{aligned} I(X, Y) &= \{f(-x, -y)\exp[i\alpha(-x, -y)]\,P(x - X, y - Y)\}^2 \\ &= [f(-x, -y)]^2\,P(x - X, y - Y)\,P * (x - X, y - Y). \end{aligned}$$

$P(X, Y)P^*(X, Y)$, which is the image of a single point in the microscope in question, is called a **point spread function**. It is a characteristic of the microscope being used, and we will write it subsequently as $S(X, Y)$. If there are no phase shifts in the contrast transfer function of a microscope and its limiting aperture is circular, then $S(X, Y)$ will be an Airy disk pattern. Because an incoherent image is a sum of appropriately weighted and appropriately located point spread functions, it is easy to write an expression for the incoherent image of a specimen that consists of many points:

$$\mathrm{I}(X,\ Y) = \left[\mathrm{f}\left(-x,\ -y\right)\right]^2 * \mathrm{S}(X,\ Y), \tag{10.2}$$

which is **not** the same as its coherent counterpart:

$$\mathrm{I}(X,\ Y) = \left\{\mathrm{f}\left(-x,\ -y\right)\exp\left[i\alpha\left(-x,\ -y\right)\right] * \mathrm{P}(X,\ Y)\right\}^2. \tag{10.3}$$

If a piece of photographic film having an infinitely fine-grained emulsion were placed on the objective lens side of an illuminated microscope specimen, the image recorded on it would be proportional to $[\mathrm{f}(x,\ y)]^2$, which is the function the microscopist (usually) wants to visualize in magnified form. Let us call it $\mathrm{i}(x,\ y)$. Because equation (10.2) correctly describes incoherent images, the image that would be recorded if a piece of photographic film were placed in the back focal plane of a microscope operating with incoherent light must be proportional to:

$$\mathrm{FT}\left[\mathrm{i}\left(x,\ y\right)\right] \cdot \mathrm{FT}^{-1}\left[\mathrm{S}(X,\ Y)\right]. \tag{10.4}$$

Functions such as $\mathrm{i}(x,\ y)$ and $\mathrm{S}(X,\ Y)$ are proportional to the square of the electric field of the light illuminating a sample and, therefore, proportional to radiant energy. Thus, the message conveyed by equations (10.2) and (10.4) is that incoherent imaging is best understood by tracking the flow of light *energy* through microscopes (i.e., rays). Coherent image formation, on the other hand, must be approached by asking how the *amplitudes*—the electric fields of light waves—evolve and interfere as light moves through microscopes (equation (10.3)). Coherent images correspond to the squares of the amplitude distributions in the image plane.

10.3 THE FOURIER TRANSFORM OF THE SQUARE OF ANY FUNCTION IS THE AUTOCORRELATION FUNCTION OF ITS FOURIER TRANSFORM

We already know a lot about the Fourier transforms of functions such as $\mathrm{i}(x,\ y)$ and $\mathrm{S}(X,\ Y)$ because $\mathrm{i}(x,\ y)$ and $\mathrm{S}(X,\ Y)$ are both the products of a function multiplied by its complex conjugate. Thus, like Patterson functions, both transforms are autocorrelation functions (see section 6.6). $\mathrm{FT}[\mathrm{i}(x,\ y)]$ is the autocorrelation function of $\mathrm{f}(x,\ y)\exp[i\alpha(x,\ y)]$, whereas $\mathrm{FT}^{-1}[\mathrm{S}(x,\ y)]$ is the autocorrelation function of $\mathrm{P}(X,\ Y)$. Figure 10.2 compares the simplest of all one-dimensional aperture functions, the square wave, with those of its Fourier transform, which is a one-dimensional $\mathrm{P}(X,\ Y)$, the square of the Fourier transform of the square wave, which is a one-dimensional $\mathrm{S}(X,\ Y)$, and the Fourier transform of $\mathrm{S}(X,\ Y)$, which is the autocorrelation function of the Fourier transform of the square wave.

Note that the Fourier transforms of autocorrelation functions have a property that we have not commented on before, and which may seem counterintuitive. The range of frequencies spanned by an autocorrelation function is always twice that of its parent function. The reason is that when $\cos\,\theta$ is squared, all its troughs become

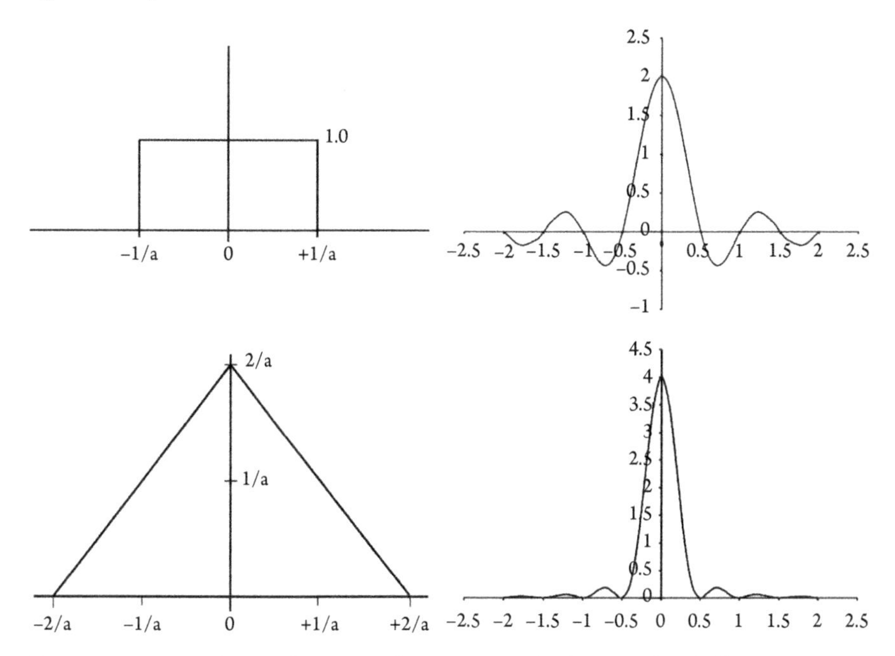

Figure 10.2 A simple, one-dimensional contrast transfer function (top left) compared to the corresponding point image function (top right) and point spread function (bottom right), and to its own autocorrelation function (bottom left). The contrast transfer function is a square wave of amplitude 1 that extends from $-1/a$ to $+1/a$ in reciprocal space. The amplitude of its point spread function, which is its Fourier transform, is measured in units of $(1/a)$, and its x-axis has units of a. The point spread function is the contrast transfer function squared, and its amplitude is measured in units of $(1/a^2)$. The point spread function and the autocorrelation function of the aperture function are Fourier mates.

positive peaks, and thus the frequency of the oscillations associated with $\cos^2\theta$ is twice that of $\cos\theta$, as can easily be verified algebraically. The same is true of $\sin\theta$.

The theories for coherent and incoherent image formation represent the two extremes of a reality that is almost always somewhere in between. It is difficult to make an illumination system that generates light that is totally incoherent down to the wavelength length scale. There is bound to be some contribution of coherent imaging in microscopes that use Koehler illumination. It is easier to approach the perfectly coherent extreme by using lasers as light sources, but it is still hard to attain absolute perfection. You will appreciate that the theory of image formation in instruments operating with partially coherent light could pose challenges, and we will leave it at that.

10.4 LIGHT ABSORPTION ACCOUNTS FOR MUCH OF THE CONTRAST IN ORDINARY MICROSCOPE IMAGES

In order for an image to be of any use at all, the amount of light falling on the image plane must vary from point to point. That variation is the **contrast** of an image, and clearly an image that has no contrast is worthless, no matter what its nominal resolution. For that reason, it is as important to understand the sources of the contrast seen in microscopic images as it is to understand how the light responsible for them propagated through that microscope, which is all that we have discussed so far.

Light absorption is an important source of contrast in many microscopic images. Some biological specimens include chromophores that absorb visible light, and if they are localized in a specimen, and white light is used to illuminate it, its image will vary in color appropriately from one location to the next. That said, most biological molecules do not absorb electromagnetic radiation in the visible range; to first approximation, cytoplasm is transparent goo. For that reason, biological specimens are often **stained**—treated with dyes—before being imaged in the light microscope. Over the centuries, a large number of stains have been discovered that are useful for microscopy, some of which bind specifically to certain types of biological molecules. Hence, not only can staining improve the contrast in images of cellular material generally by making them colored, it can also be used to localize specific kinds of molecules (e.g., protein, DNA, RNA) in specimens. Pathologists and histologists routinely stain the specimens they examine, often after having stabilized them chemically (i.e., after **fixing** them).

10.5 FLUORESCENCE PLAYS AN INCREASINGLY IMPORTANT ROLE IN BIOLOGICAL MICROSCOPY

A few of the chromophores found naturally in biological materials fluoresce in the visible range. Chlorophyll, for example, the pigment that makes green plants green, absorbs blue light and fluoresces strongly in the red. It is easy to modify conventional light microscopes so that they will selectively image the fluorescent molecules in a specimen.

A chromophore that fluoresces appreciably is a **fluorophore**, and every fluorophore absorbs light at wavelengths shorter than the wavelengths at which it emits light. Thus, a microscope that is to image a particular fluorophore must be equipped with a light source that generates light in the wavelength range that the fluorophore to be imaged absorbs, and it must be constructed so that as little of the light emanating from that source will reach the image plane as possible, but so that all the light emitted by the fluorophore will. Colored filters can be used for this purpose. A colored filter can be inserted into the optical track of the microscope between the specimen and the light source that selectively absorbs light in the wavelength range that the fluorophore fluoresces, but does not absorb the light the fluorophore absorbs. A second colored filter can then be placed between the specimen and the image plane that absorbs all the light that gets past the first filter, but does not absorb any of the light the first filter absorbs. The images produced by a microscope that is equipped this way will be dark ("black") at every point that corresponds to a location in the specimen where there are no fluorophores, and bright ("colored") at points that correspond to locations where there are fluorophores.

Fluorescence microscopy has two interesting characteristics. First, the light used to excite the fluorescence in a sample does not have to enter it along the optical axis of the microscope that is to image it because fluorophores emit light in all directions,

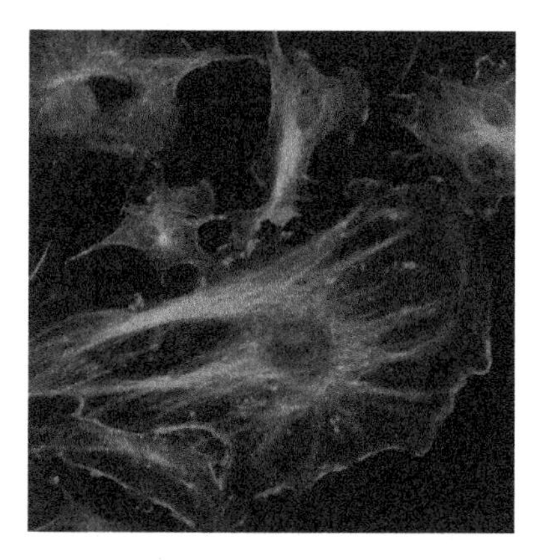

Figure 10.3 Tissue culture cells imaged by fluorescence microscopy. Image ["FluorescentCells.jpg"] obtained from the ImageJ archive maintained by the Research Services Branch of National Institutes of Health: http:/rsb.info.nih.gov/ij/images/. Readers interested in seeing this image in all its glory can download it from NIH or from www.oup.com/us/visualizingtheinvisible. In that image, nuclei are stained blue. Microtubules are stained green. Actin is stained red.

more or less. Second, once a fluorophore absorbs a photon, its emission of a longer wavelength photon, which is to say its fluorescence, which returns it to the ground state, is a stochastic process that (usually) has a half-life of nanoseconds. Thus, the light emitted by the different fluorophores in a sample will have phases that are totally unrelated, no matter what the coherence properties of the light used to excite them, and this means that fluorescence images are *perfectly* incoherent.

Fluorescence microscopy has been practiced for decades, but until recently it was not widely used because only a modest number of problems can be addressed using the fluorophores found naturally in biological specimens, and because the technology for labeling (i.e., staining) nonfluorescent biomaterials with fluorescent dyes was primitive. Times have changed! Many methods now exist for rendering specific biological molecules fluorescent. For example, antibodies raised against a particular protein can be made fluorescent by treating them in vitro with reactive, low molecular weight compounds that include fluorophoric groups, and antibodies that have been derivatized this way can be used as protein-specific, fluorescent stains. As figure 10.3 shows, the images obtained from fluorescently labeled samples can be as spectacular visually, as they are informative biologically. Images like this one have stimulated a generation of chemists and biochemists to synthesized new fluorophores and to develop new ways of coupling them specifically to biological macromolecules.

Comment: All molecules, including those that fluoresce, become more reactive chemically when they absorb light, and when photo-activated fluorophores react with substances such as oxygen, the products are usually not fluorescent. Thus, the longer the time a fluorescently labeled specimen is exposed to light, the weaker the fluorescence it emits. This phenomenon is called ***photobleaching****. The half-lives of fluorophores, measured in numbers of photons fluoresced per molecule, vary a lot from one fluorophore to the next, and they often depend on environmental factors, such as the local concentration of molecular oxygen. Because photobleaching limits our ability to accurately determine the locations of fluorescently labeled macromolecules in microscope images, photobleaching is a bad thing in most contexts.*

It is also possible make proteins fluoresce genetically. The jellyfish *Aequoria victoria* produces a small protein called green fluorescent protein (GFP), which, as its name suggests, absorbs blue light and emits a green fluorescence. Techniques that are now standard in molecular biology allow you to construct genes for chimeric proteins that have a GFP domain added to either the N-terminal or the C-terminal ends of any proteins you want, insert them into living cells, and get them expressed. Remarkably, not only are the these chimeric proteins fluorescent, they are often still able to perform the same functions as their nonchimeric counterparts. Fluorescence microscopy can be used to determine the locations of proteins that have been labeled this way in cells, tissues, organisms. Since the discovery of GFP, GFP mutants have been discovered that are even more useful for microscopy than the native protein. The GFP variants available today include several that have emission spectra distinctly different from wild-type GFP, and thus, it is now possible to track the locations of many different proteins simultaneously in the same cell or organism on the basis of color differences.

As the power of the methods available for labeling cellular components fluorescently has increased, so too has the power of the methods available for doing fluorescence microscopy. If ever there was an area of cell biology that has a bright future, it is this one, both literally and figuratively.

10.6 LIGHT SCATTERING CONTRIBUTES TO IMAGE CONTRAST

Light scattering generates contrast in images, although its contribution is often less obvious than the contribution made by absorption (or fluorescence). Macromolecular solutions scatter light (see chapter 8), and the efficiency with which they do so depends on three factors: (1) the difference in refractive index between the solvent, which is usually water, and the macromolecular material dissolved in it; (2) the concentration of the macromolecular material in the solution; and (3) the molecular weights of those macromolecules. Because macromolecular material is not uniformly distributed in cells and tissues, the fraction of the light incident on a biological specimen that is scattered by it varies with location.

The reason scattering causes image contrast is that, viewed from the sample, the solid angle subtended by the objective lens of a microscope is not 4π steradians, and radiation scattered at angles so high that it does not enter the objective lens of a microscope cannot be focused by it. Thus, the more strongly some region in a specimen scatters light, the smaller the amount of light energy available to contribute to its image. Hence, the regions in an image that correspond to locations in a specimen that scatter strongly will appear darker than the regions that correspond to positions in the specimen that scatter light weakly. As you would expect, the smaller the aperture of the microscope (i.e., the smaller its numerical aperture), the more important the contribution of made by this kind of contrast, **aperture contrast**, to overall image contrast. However, the price paid for increasing the aperture contrast in images this way is severe: reduced resolution.

10.7 DARK FIELD ILLUMINATION CAN BE USED TO IMAGE OBJECTS THAT SCATTER LIGHT

Most biological specimens neither absorb light efficiently, nor scatter it strongly. When they are examined in the microscope, most of the light incident on them passes through unaltered in any way, and that unaltered incident light contributes a bright background to their images. Because contrast-generating processes commonly produce variations in image brightness that are small by comparison, the signal-to-noise ratio in conventional microscope images of unstained biological specimens is often very low. In this connection, it is worth emphasizing that one of the many attractions of fluorescence microscopy is that fluorescence images have black backgrounds, and thus their signal-to-noise ratios are about as high as they could possibly be.

By appropriate engineering of illumination systems, the bright background that unscattered light normally contributes to biological images can be eliminated from them altogether (figure 10.4). A point source of light is placed on the optical axis of the condenser lens of the microscope at a location beyond its front focal plane, and an opaque sheet is positioned between that source and the condenser that has an annular hole in it. If the annulus in the opaque sheet is centered on the optical axis, the only light rays reaching the condenser will be those that diverge from the optical axis over a narrow range of angles. The condenser will focus that cone-shaped set of rays at the center of its own image plane, which is where the specimen must be mounted when this kind of microscopy is done. On the far side of the specimen, the rays that do not change direction as they pass through the specimen will again diverge from the optical axis, and if angles and distances are chosen correctly, they will not enter the objective lens. Under these circumstances, the only light entering the objective lens will be light that was scattered by the specimen. Because points scatter light uniformly in all directions, the image of the specimen that emerges will be the inverse of the image that would have been produced using conventional illumination if all the contrast in that image was aperture contrast. Regions of the sample that scatter strongly will appear to be bright objects set in a dark background. Thus, this technique is called **dark field** microscopy. It generates images that have high signal-to-noise ratios, but only those components of a specimen that scatter strongly (or fluoresce) are visualized.

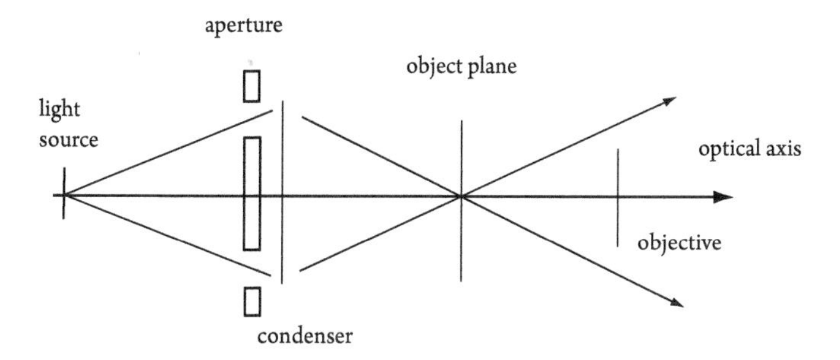

Figure 10.4 Schematic diagram of the illumination system for dark field microscopy.

10.8 PHASE OBJECTS CAN BE VISUALIZED USING PHASE CONTRAST MICROSCOPY

As we have already pointed out, many biological specimens barely absorb any visible light, and thus have object transmission functions that are effectively $\exp[i\alpha(x, y)]$. Objects of this sort are called **phase objects**, and theory predicts that (to first order) aperture contrast should be the only kind of contrast in their images. In 1935, Zernicke devised a novel method for doing light microscopy that greatly increases the contrast of the images of phase objects. It is called **phase contrast** microscopy, and it has been widely used ever since.

Phase contrast microscopy is easy to understand. We start by assuming that $\alpha(x, y)$ is small, an assumption called the **weak phase approximation** that is usually well justified. If $\alpha(x, y)$ is small, then the object transmission function of a phase object, $\exp[i\alpha(x, y)]$, can be approximated using the first two terms of its power series expansion: $[1 + i\alpha(x, y) + \ldots]$, and if such a specimen is illuminated with coherent light, which it must be if this technique is to work at all, then the image in the back focal plane will be:

$$\mathrm{FT}\left(\left[1 + i\alpha\left(x, y\right)\right]\right) \cdot \mathrm{A}\left(\xi, \eta\right) \exp\left[i\omega\left(\xi, \eta\right)\right]$$

which is the same as:

$$\left(\delta\left(\xi, \eta\right) + i\mathrm{FT}\left[\alpha\left(x, y\right)\right]\right) \cdot \mathrm{A}\left(\xi, \eta\right) \exp\left[i\omega\left(\xi, \eta\right)\right].$$

In the back focal plane of a phase contrast microscope, there is a glass element inserted called a **phase plate**, the design of which depends on the design of the rest of the microscope. The phase plate appropriate for the microscope being discussed here would be a flat glass plate that has parallel surfaces, and that at its center, has a very small, cylindrical protrusion that is just thick enough so that the phase angle of light passing through that part of the phase plate is retarded by an angle of $\pi/2$ relative to light passing through the rest of the plate. If that protrusion is centered on the optical axis of the instrument, then in the back focal plane beyond the phase plate the image function will be:

$$\left(i\delta\left(\xi, \eta\right) + i\mathrm{FT}\left[\alpha\left(x, y\right)\right]\right) \cdot \mathrm{A}\left(\xi, \eta\right) \exp\left[i\omega\left(\xi, \eta\right)\right].$$

In the image plane, ignoring magnification factors, the specimen wave will be:

$$i\left[1 + \alpha\left(-x, -y\right)\right] * \mathrm{P}\left(X, Y\right)$$

and the image, will be:

$$\begin{aligned} \mathrm{I}\left(X, Y\right) &= \left(i\left[1 + \alpha\left(-x, -y\right)\right] * \mathrm{P}\left(X, Y\right)\right)\left(-i\left[1 + \alpha\left(-x, -y\right)\right] * \mathrm{P}\left(X, Y\right)\right) \\ &= 1 + \alpha\left(-x, -y\right)\left[\mathrm{P}\left(X, Y\right) + \mathrm{P}*\left(X, Y\right)\right] + \text{higher order terms} \\ &\approx 1 + 2\alpha\left(-x, -y\right) * \mathrm{Re}\left[\mathrm{P}\left(X, Y\right)\right], \end{aligned}$$

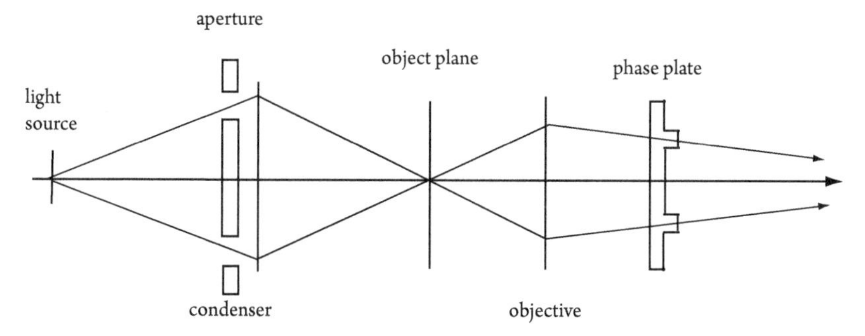

Figure 10.5 Schematic diagram of a phase contrast microscope showing its illumination system and phase plate. Note that the phase plate is in the back focal plane of the objective lens.

and thus the contrast in the image will be proportional to $\alpha(-x, -y)$, as desired.

There are a few subtleties that should be pointed out. First, $\alpha(x, y)$ is real, by definition. Second, the convolution of $P(X, Y)$ with 1.0 is simply the integral of $P(X, Y)dXdY$ over all X and Y; it is a constant. Third, by hypothesis, $\alpha(x, y)$ is small so that terms proportional to $[\alpha(x, y)]^2$ can be ignored. Fourth, the expression inside the square brackets is twice the real part of $P(X, Y)$, the point image function. Thus, if $P(X, Y)$ is close to entirely real, which it will be if the objective lens being used has low aberration, the point spread function of a phase microscope will be $P(X, Y)$, not $S(X, Y)$. Thus, the images of phase objects produced by microscopes operating in the phase-contrast mode are not as well resolved as the conventional, bright field images produced by the same microscopes.

The phase plates in the phase microscopes available commercially do not modulate the phase of light passing through the back focal plane in quite the way we just described. Instead, the illumination systems in these microscopes resemble those used for dark field microscopy, except that they are designed to allow the unscattered portion of the (coherent) illumination that they deliver to specimens to enter the objective lens. The image produced by that unscattered light in the back focal plane is thus an annulus, and for that reason the phase plates in these microscopes include have annuli of the appropriate diameter and width that are raised (or lowered) relative to the rest of the plate, as desired (figure 10.5).

Phase contrast can contribute to the images of phase objects formed by microscopes that lack phase plates provided the light used to illuminate samples has an appreciable transverse coherence length. Both spherical aberration and defocus produce phase shifts in the back focal plane that depend on the distance between the point where radiation crosses the back focal plane and the optical axis, in other words, $\rho = (\xi^2 + \eta^2)$, which is all that is needed. Thus, if the specimen is thicker than the depth of field, phase effects will enhance the contrast of the images of the parts of the specimen that are out of focus. This kind of phase contrast, which plays a comparatively minor role in optical microscopy, is very important in electron microscopy, as we shall see (section 12.8).

10.9 CONFOCAL MICROSCOPY

In recent decades, an imaging technique called **confocal microscopy**, or **confocal scanning microscopy**, has grown in popularity, and it is often used to image fluorescently labeled samples. Confocal microscopy is a form of scanning microscopy, as the long form of its name implies; thus, a confocal microscope measures the optical properties of specimens, one volume element at a time. For that reason, they generate images much more slowly than conventional, nonscanning microscopes of otherwise similar optical capability.

Confocal microscopes are designed to illuminate as small a volume in a specimen as is physically possible, and then to the extent possible, record only the light originating in that volume. There are a lot of ways this goal can be achieved optically, but for the sake of discussion, we will consider the system outlined schematically in figure 10.6. A high numerical aperture objective lens is placed at a distance from some specimen that would be appropriate if that lens were being used to image it at high resolution. On axis, in what would normally be the image plane of that lens, an opaque plate with a pinhole in it is placed with its pinhole on axis. By illuminating that plate from behind, a demagnified image of the pinhole can be made to appear in what would normally be the object plane of the objective lens, which has a light distribution similar to $S(MX, MY)$, where M is a number much less than 1.0. The intensity of the light delivered to the specimen by this means not only falls off rapidly as the distance from the optical axis increases in the object plane, it also falls off rapidly both above and below the object plane because the divergence of the rays that are brought to focus in the object plane is large. The response of the illuminated volume to the light falling on it is observed using a second objective lens that is on the far side of the specimen, on axis, and positioned so that the illuminated volume in the sample is in focus for that lens. (The double focusing that is characteristic of microscopes of this sort is why they are called confocal.) A magnified image of the illuminated volume will form in the image plane of the second lens. An opaque plate, similar to the plate on the source side of the microscope, is positioned in the image plane, with its pinhole on the optical axis of the second lens. A detector on the far side of that

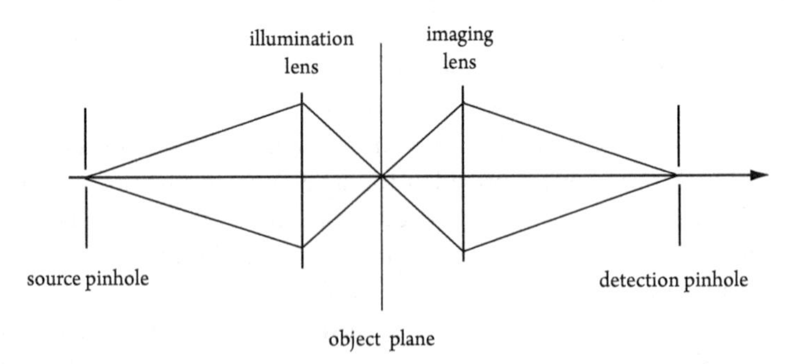

Figure 10.6 Schematic diagram of a two-lens, confocal microscope. Light passes through the system from left to right. Immediately beyond the detection pinhole there is a detector that measures the amount of light passing through the system.

pin hole records the amount of light coming through it, and an image is created by displaying the variation in the output of the detector as the specimen is translated relative to the illuminating spot in two, or, often, three dimensions. Confocal microscopes built for fluorescence imaging commonly have only a single objective lens, which is used both to deliver light to the specimen and to image its fluorescence.

To understand why confocal images are interesting, you must think carefully about what they should look like. In the first place, as we pointed out earlier, the image of a thick specimen in a conventional microscope is an in-focus image of a thin slab of the specimen, superimposed on images of all the slabs above and below the focal plane that are out of focus by amounts that increase with distance from the focal plane. Because all of the slabs in a thick specimen contribute light to any image produced by a conventional microscope, no matter where the focal plane happens to be in the specimen, the out of focus "clutter" in such images tends to be high.

Confocal images are superior in this regard. Confocal instruments illuminate cone-shaped volumes in the sample that converge on the focal point in the object plane on the light source side, and diverge from it on the image-plane side. All the off-axis sample points inside both cones are illuminated, but they all engender off-axis images in the image plane that will contribute to the amount of light entering the image-plane pinhole only to the extent that their point spread functions overlap that pinhole. On-axis points in the specimen that are above or below the focal plane will be out of focus in the image plane, and even though their images will be centered on the detector pinhole, a comparatively small fraction of the light energy they contain will register at the detector because the images of out-of-focus points are larger than the images of in-focus points. Clearly, the volume element in the specimen that is best illuminated will be the one at the focal point of the first lens, and that volume element will also be the one that is most effective in delivering light to the detector. Thus, in confocal images, there should be less cross talk between adjacent points within the image plane than there is in conventional images, and it should also be possible to obtain cleaner images of single planes in the specimen.

10.10 THE POINT SPREAD FUNCTION OF A CONFOCAL MICROSCOPE IS THE PRODUCT OF TWO OBJECTIVE LENS POINT SPREAD FUNCTIONS

These concepts can be expressed in mathematical form. If the point spread function of the lens used to illuminate samples in some specimen is cylindrically symmetric, its point spread function will be a function of distance from the optical axis, r, and distance from the focal plane, $z: S_o(r,\ z)$. If the pinhole is effectively a Dirac delta function, the signal generated by the volume element in the specimen that is located at $(x,\ y,\ z)$, relative to the focal point of the illuminating objective lens, will be proportional to $S_o(r,\ z)$, where $r = (x^2 + y^2)^{1/2}$. Relative to the focal point of the second objective lens in the image plane, the center of the signal arising at $(x,\ y,\ z)$ will be found at $(-Mx,\ -My,\ -M^2z)$, where M is the magnification of the image created by the second objective lens. (See question 1 at the end of this chapter for an explanation of why the z coordinate is $-M^2z$, rather than $-Mz$.) Furthermore, the fraction of the light energy originating at $(x,\ y,\ z)$ that enters the image-plane pinhole will be proportional to $S_i(Mr,\ -M^2z)$, where S_i is the point spread function of the objective lens used to form the final image. Thus, the contribution the volume element at $(x,\ y,\ z)$ in the specimen makes to the signal measured at the detector will be proportional to:

$$S_o\,(r,z)\,S_i\left(Mr,\ -M^2z\right).$$

[If the same lens is used both to illuminate samples and to image the radiation coming from them, then this expression becomes: $S(r,z)S(Mr,\ -M^2z)$.]

Because the optical sectioning properties of confocal microscopes are important for the way they are used, we now need to consider a topic that we have so far ignored, namely what point spread functions look like in three dimensions, $S(X,\ Y,\ Z)$. For microscopes having aberration low enough so that only the aperture contribution to their contrast transfer functions need be taken into account and for circular apertures, the appropriate expressions were derived decades ago, but they are quite messy (see Born and Wolf, 1999). Nevertheless, we already know what $S(X,\ Y,\ Z)$ looks like in the image plane:

$$S\,(X,\ Y,\ 0) = \left\{2J_1\left[4\pi R\ \sin\left(\theta_{\max}/2\right)/\lambda\right]/\left[4\pi R\ \sin\left(\theta_{\max}/2\right)/\lambda\right]\right\}^2,$$

where $R = (X^2 + Y^2)^{1/2}$, and $\theta_{\max}$ is the angle subtended by the radius of the lens measured from its focal point (see appendix 9.2). [Implicit in this equation is the notion that the lens being used is ideal, and that the refractive index of the medium between the lens and the image plane is 1.0.] Furthermore, it is not too difficult (see appendix 10.1) to arrive at an expression that is valid for the distribution of energy along the optical axis:

$$S\,(0,\ 0,\ Z) = \left\{\sin\left[2\pi\ (Z/\lambda)\sin^2\left(\theta_{\max}/2\right)\right]/\left[2\pi\ (Z/\lambda)\sin^2\left(\theta_{\max}/2\right)\right]\right\}^2.$$

For our purposes, it will be enough to approximate $S(X,\ Y,\ Z)$, the full three-dimensional function, as the product of those two functions:

$$S\,(X,\ Y,\ Z) \approx S\,(X,\ Y,\ 0)\cdot S\,(0,\ 0,\ Z)\,.$$

Although this expression is not accurate, the qualitative message it conveys is correct: namely that the intensity of three-dimensional point spread functions dies very rapidly with distance from its center at $X = Y = Z = 0$ in all directions. Thus, in the neighborhood of the focal point:

$$\begin{aligned} S\,(X,\ Y,\ Z) \approx {} & \left\{2J_1\left[4\pi R\ \sin\left(\theta_{\max}/2\right)/\lambda\right]/\left[4\pi R\ \sin\left(\theta_{\max}/2\right)/\lambda\right]\right\}^2 \\ & \times\left\{\sin\left[2\pi\ (Z/\lambda)\sin^2\left(\theta_{\max}/2\right)\right]/\left[2\pi\ (Z/\lambda)\sin^2\left(\theta_{\max}/2\right)\right]\right\}^2. \end{aligned}$$

It follows that if the confocal microscope being used has only a single objective lens, or if its two objective lenses are identical, then $C(r, z)$, the contribution made by the volume element at (r, z) in the image plane to the signal registered for the volume element at $(0, 0)$, will be:

$$C(r, z) \approx \{2J_1(4\pi r \sin(\theta_{max}/2)/\lambda)/[4\pi r \sin(\theta_{max}/2)/\lambda]\}^2 \quad (10.5)$$
$$\times \{\sin(2\pi(z/\lambda)\sin^2(\theta_{max}/2))/[2\pi(z/\lambda)\sin^2(\theta_{max}/2)]\}^2$$
$$\times \{2J_1(4\pi Mr \sin(\theta'_{max}/2)/\lambda)/[4\pi Mr \sin(\theta'_{max}/2)/\lambda]\}^2$$
$$\times \{\sin(2\pi(M^2 z/\lambda)\sin^2(\theta'_{max}/2))/[2\pi(M^2 z/\lambda)\sin^2(\theta'_{max}/2)]\}^2.$$

It is important to remember that θ_{max}, the angle subtended by the illuminating lens measured from the point where the optical axis intersects the object plane, is much larger than θ'_{max}, the angle subtended by the imaging lens measured from the point where the optical axis intersects the image plane. Even so, for any given r and z, the numerical values of the first two terms in equation (10.5) are not terribly different from the values of the second two terms, no matter what the numerical apertures of the lenses being used, provided the two lenses have the same specifications. Thus, in most discussions of confocal microscopes, the point spread function of the instrument as a whole, S_{conf}, is taken to be the square of the point spread function of its objective lens(es):

$$S_{conf}(x, y, z) \approx S^2_{obj}(x, y, z) = C(r, z).$$

The reason the lateral resolution of a confocal microscope should be better than that of a conventional microscope is because $[J_1(x)/x]^4$ is more sharply peaked than $[J_1(x)/x]^2$, and similarly, the superiority of the optical sectioning properties of confocal microscopes results from the fact that $(\sin x/x)^4$ is more sharply peaked than $(\sin x/x)^2$.

In practice, the point spread functions of confocal microscopes are not as good as this analysis suggests they should be. For one thing, in order for any photons to get to the focal spot and then to the detector, the two pinholes in the microscope must have finite diameters. Thus, the point spread function of a real confocal microscope will be the point spread function just given, convoluted with the aperture functions of the two pinholes. The degradation in performance that results makes the in-plane resolution of confocal microscopes about the same as that of a conventional bright field microscope equipped with the same objective lens.

It is instructive to compare the location of the first 0 of the axial component of $S(x, y, z)$ with the depths of field of microscope images, which are discussed in section 9.16. That first 0 occurs at the point on the optical axis where $[(2\pi/\lambda)\sin^2(\theta_{max}/2)z]$ equals π, and thus at:

$$z_o = \lambda/2 \sin^2(\theta_{max}/2).$$

For objective lenses with high numerical apertures, that z_o is $\sim 4D$, where D is the depth of field, which confirms that the slab of a specimen that is within $\pm D$ of the focal plane will certainly be safely in focus. Using the same criterion to estimate resolution in z as is used for estimating resolution in the focal plane, two points in a specimen that have the same (x, y) coordinates but differ in z will be resolved in a series of images taken at different levels of the focal plane if the axial point spread function of one of them superimposes on the first 0 of the axial point spread function of the other. That distance is z_o, and for high-resolution microscopes, z_o is around 1 to 2 μ. The in-plane resolution of such microscopes is a 0.3 to 0.4 μ.

10.11 THREE-DIMENSIONAL IMAGES CAN BE RECOVERED FROM TWO-DIMENSIONAL MICROSCOPIC IMAGES

Biological microscopists are often satisfied by the two-dimensional images their instruments naturally provide, but it has long been obvious that three-dimensional information about biological structures can be recovered from (two-dimensional) micrographs. Here, we will briefly address two ways this can be accomplished: one of them simple and straightforward, and the other far more complicated and far from straightforward.

The simplest way to determine the three-dimensional structure of an object with a microscope is to cut it up into thin slices (i.e., sections), image each section separately, and then stack the images of sections on top of each other in the appropriate order. This brute-force approach has been used for generations, and it is still the only way to obtain reliable images from specimens that are millimeters thick because even if specimens that thick were not opaque, which they often are, light passing through them is liable to be scattered many times. When multiple scatter dominates, the first Born approximation can no longer be relied on, and the distribution of light in the back focal plane will no longer be related to $\mathrm{FT}[\rho(x, y, z)]$ in a simple way, as we already know.

The renderings of three-dimensional structures that emerge from direct sectioning approaches approximate $\rho(x, y, z)$ as a sum of projections, that is, as:

$$\sum_{0}^{N} \int_{n\Delta z}^{(n+1)\Delta z} \rho\left(x, y, z\right) dz,$$

were Δz is the thickness of the sections imaged, and $(N + 1)\Delta z$ is the thickness of the object being imaged. In the limit as Δz goes to 0, this sum exactly equals $\rho(x, y, z)$, but as long as Δz is greater than the in-plane resolution of the microscope being used for imaging, the resolution of the approximation to $\rho(x, y, z)$ that emerges is likely to be inferior in the z direction.

The instruments used to cut biological materials into thin slabs suitable for microscopy are called **microtomes**. Light microscopy requires sections $10^{-5}\,\mathrm{m}(10\text{-}\mu)$ thick, or less, and the microtomes used to section samples for electron microscopy can produce sections a hundred times thinner than that.

Microtomes, whatever the type, are refined versions of the devices butchers use to slice baloney. Their capabilities vary enormously depending on the kinds of specimens they are meant to process, the thickness of the slices they make, and other factors Biological materials do not lend themselves to being cut into ultrathin slices because they are soft and deformable. Consequently, before sectioning is done, biological specimens are often "stiffened" by being frozen, or, alternatively, by being fixed (i.e., chemically stabilized), and embedded in some kind of hard supporting medium. For example, it is sometimes possible to replace the water in biological specimens with liquid acrylic resins that can be made to solidify in place, effectively turning the specimen into a piece of epoxy. This sounds like a horrible thing to do with a delicate sample whose microscopic structure you wish to preserve, but it is commonly far less damaging than you might imagine.

For those interested in the theory of image formation, the computational approaches that have been developed for recovering three-dimensional information from micrographs are more interesting than those that depend on cutting up samples into slices. They are more interesting for another reason that is entirely practical. In the former case, the sample being examined might just possibly emerge from the imaging process intact, or if it is a live cell, it might even survive. In the latter case, neither is possible.

All computational approaches to three-dimensional microscopy exploit the optical sectioning phenomenon that is manifest whenever thick specimens are examined in the microscope, and the stage on which they are mounted is moved up and down with respect to the objective lens. The appearance of the specimen changes as the stage moves up and down, and it is a simple matter to obtain a series of micrographs of some specimen, each taken at an accurately known (relative) elevation of the stage. As we already know, it is hard to interpret a single image of a thick specimen because it will always be the superposition of an in-focus image of the parts of the structure that were in the focal plane with out-of-focus images of adjacent planes. A priori, there is no way to distinguish the in-focus features from the out-of-focus features in a single image.

All of the computational approaches that have been proposed for removing out-of-focus noise from images start with the proposition that if the true, in-focus three-dimensional image of some object is $\mathrm{i}(x, y, z)$, then its image in a microscope, $\mathrm{i}'(x, y, z)$, must be:

$$\mathrm{i}'\left(x, y, z\right) = \mathrm{i}\left(x, y, z\right) * \mathrm{f}\left(x, y, z\right), \tag{10.6}$$

where $\mathrm{f}(x, y, z)$ is the weighting function that describes the cross talk between adjacent points in the image. For points that are in plane, $\mathrm{f}(x, y, z)$ is the $[\mathrm{J}_1(x)/x]^2$ point spread function we talked about previously. For points that are out of plane, the $\mathrm{f}(x, y, z)$ corresponds to the image of a point that is not in focus, which is *not* the same as the $\sin(x)/x$ part of equation (10.5), but is also known from theory. [NB: $\mathrm{f}(x, y, z)$ can be measured directly, if need be.]

The convolution theorem suggests that there might be a simple way to recover $\mathrm{i}(x, y, z)$ from a set of measured images, $\mathrm{i}'(x, y, z)$. If equation (10.6) is true, then:

$$\mathrm{FT}\left[i'\left(x, y, z\right)\right] = \mathrm{FT}\left[\mathrm{i}\left(x, y, z\right)\right] \cdot \mathrm{FT}\left[\mathrm{f}\left(x, y, z\right)\right].$$

Thus, the convolution operation expressed in equation (10.6) might be reversed, or the data **deconvoluted**, as follows:

$$\mathrm{i}\left(x, y, z\right) = \mathrm{FT}^{-1}\left\{\mathrm{FT}\left[\mathrm{i}\left(x, y, z\right)\right]/\mathrm{FT}\left[\mathrm{f}\left(x, y, z\right)\right]\right\}.$$

This is not the only context in the physical sciences where it would be desirable to effect a deconvolution, in other words, to reverse the damage done to data by instrument transfer functions. However, the simple equation just written conceals a host of notoriously difficult practical problems. The reason deconvolutions are problematic is that at some locations in reciprocal space $\mathrm{FT}[\mathrm{f}(x, y, z)]$ is bound to be either very small, or even dead 0. In all such locations, theory predicts that $\mathrm{FT}[\mathrm{i}'(x, y, z)]$ *must* have an amplitude that is correspondingly small. It follows that in those parts of reciprocal space, the deconvolution equation calls for the division of one small number by another, or, worse, the division of 0 by 0, which is something you were told never to attempt in fourth grade. The fact is that the images we might use for this purpose, like all real data, include noise, and the process just described is guaranteed to magnify the impact of image noise on the rendering of $\mathrm{i}(x, y, z)$ that emerges because the signal-to-noise ratio in the transforms of these images will be poorest where $\mathrm{FT}[\mathrm{f}(x, y, z)]$ is small, or 0. Mathematicians describe problems like these as being "ill-posed".

The fact of the matter is that because $\mathrm{f}(x, y, z)$ is not a Dirac delta function, some of the information that $\mathrm{i}(x, y, z)$ contains gets lost when specimens are imaged in microscopes. In the end, you cannot recover information from experimental data that it does not contain to begin with; you cannot make a silk purse out of a sow's ear. In microscopes, the loss of information is most serious in the z-direction, which is to say in the direction of the optical axis. This is why microscope images tend to resemble projections in z, and why resolution in z is invariably inferior to resolution in (x, y). All of this not withstanding, several algorithms have been developed for processing optical image data that produce approximations to $\mathrm{i}(x, y, z)$ that are substantially more accurate than the images fed into them, $\mathrm{i}'(x, y, z)$.

10.12 LIGHT MICROSCOPES CAN RESOLVE POINTS THAT ARE CLOSER TOGETHER THAN THE RAYLEIGH LIMIT

For most of the 20th century, most scientists were persuaded that no image-forming device could have a resolution better than the Rayleigh limit, their quarrels over the details of how that limit should be estimated not withstanding. This is no longer the case. Over the past two decades, many techniques have been developed that make it possible to obtain images from light microscopes that have resolutions as much as 10-fold better than the Rayleigh limit. It is far too soon to tell which of them, if any, will prove to have the combination of resolution gain, wide applicability, and user-friendliness that physical techniques must possess to win a wide following in the biological community.

Here, we will discuss two of these approaches, and like almost all the rest of them, they are forms of fluorescence microscopy, and it is no accident that this is so. The only way a small object, or small volume element of a larger object can be imaged is if its interaction with the working radiation of the instrument being used to image it produces a signal large enough so it can be differentiated from the signals being produced by all the other stuff in the specimen of which it is a part. Everything else being equal, the smaller a volume, the less radiation it will scatter or absorb, and the smaller its effect on optical path lengths. Thus, poor signal-to-noise ratios can make the effective resolution of the images of biological materials substantially lower than the theoretical resolution of the instruments used to create them. The reason fluorescence microscopy is so attractive in this context is that a single fluorophore molecule can emit enough photons to form an image that is bright enough to be detected against the dark-field background fluorescence microscopes provide. Because the typical fluorophore has a molecular weight of a few hundred and linear dimensions 1% the wavelength of visible light, the image of a single fluorophore in a microscope will be indistinguishable from the point spread function of that microscope, which has physical dimensions far larger than those of the (magnified) image of that fluorophore. How might this set of facts be exploited to generate images that have ultrahigh resolutions?

The first of the two ultrahigh-resolution techniques we will describe is called **stimulated emission depletion microscopy**, or **STED** microscopy, and it was invented by S. W. Hell and his coworkers in Germany. It depends on the kinds of systematic manipulations of the electronic states of fluorophores that in recent decades have become routine in chemical physics laboratories around the world because of advances in laser technology. The phenomenon on which STED is based, stimulated emission, has been understood in its essence for over a century; the rate at which molecules in excited electronic states return to the ground state *increases* in the presence of light having a frequency equal to the energy of the transition in question divided by Planck's constant.

A STED microscope is a confocal microscope that uses two pulsed lasers to illuminate specimens. The first operates at a wavelength in the range the fluorophor of interest absorbs, and the second functions in the range characteristic of its emission. A STED light pulse sequence starts with a short pulse of light from the first laser that illuminates a volume in the sample that has the shape of the microscope's point spread function. A few tenths of a nanosecond later, before the fluorophores excited by the first pulse have had time to fluoresce appreciably, the second laser delivers a light pulse to the sample. Like the light pulse from the first laser, the second pulse emerges from its laser as a coherent, plane wave of circular cross section. However, unlike the first pulse, the second pulse passes through a device called a *spatial light modulator* before reaching the microscope. The modulator modifies the phase of the light from the second laser so that if we were to use a polar coordinate system centered on the axis of its beam as the means for describing the locations of points within that beam, we would find that after the modulator, the phase at the point (r, θ) is *equal* to θ, regardless of r. (In this coordinate system, r is the distance from the axis of the beam, and θ is the angle with respect to some arbitrarily chosen reference direction.) Before the modulator, the phase of the light in any wavefront of

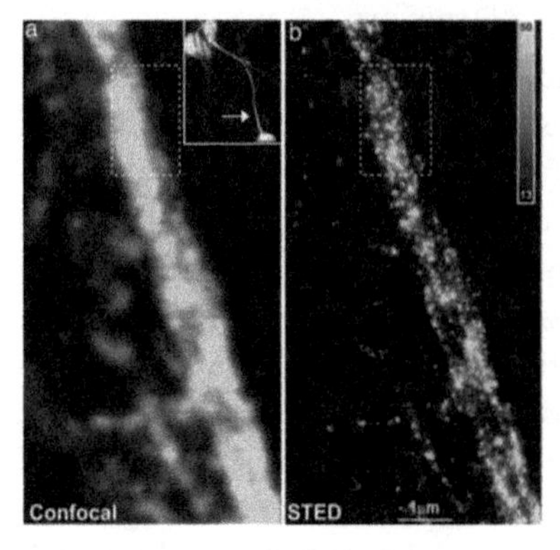

Figure 10.7 Confocal and STED microscopies compared. The specimen is a neuroblastoma cell stained with fluorescently labeled antibodies to reveal the location of 200 kD neurofilament protein it contains. The insert in image (a) is a low-resolution image that shows the region of the cell imaged at high resolution. The rest of image (a) is a conventional, high-resolution, confocal image of that bit of axon. Image (b) shows what the same bit of axon looks like when imaged using STED. The point-to-point resolution in the right image is about 45 nm. (Reproduced with permission from Donnert, G., Keller, J., Medda, R., Andrei, M. A., Rizzoli, S. O., Luhrmann, R., Jahn, R., Eggeling, C., and Hell, S. W. [2006], Macromolecular-scale resolution in biological fluorescence microscopy, *Proc. Natl. Acad. Sci. U.S.A.* 103, 11440–11445. Copyright (2006) National Academy of Sciences, U.S.A.)

the second beam is independent of both r and θ. Thus, the modulator has the same effect on the beam as a phase plate would that had the form of a helical ramp of the appropriate pitch and width.

The phase-modulated light from the second laser is brought to a focus in the sample by the same objective lens used to focus the first light pulse, but its point spread function is *not* the same as the point spread function of the first light pulse. As we already know, the reason point spread functions have a central maxima is that, normally, the amplitudes of the light in all the rays focused there interfere constructively; they add. For the second pulse, at the focal point there is perfect *destructive* interference between the rays coming together there because the ray originating at (r, θ) will interfere destructively with the ray originating at $(r, \theta + \pi)$. Thus, although the peripheral parts of the spread function of the second pulse will resemble the point spread function of the first pulse, its central point will be a minimum, not a maximum, and its central maximum will look like a doughnut.

If the intensity and duration of the second pulse is set correctly, most of the fluorescent molecules at the periphery of the volume illuminated by the first pulse will return to the ground state during the second pulse due to stimulated emission, leaving only the molecules at the center of the point spread function of the first laser still able to fluoresce. The resulting reduction in the effective size of the illuminated volume translates directly into a reduction in the size of the region in the image that is associated with the point in the specimen on which the illuminating objective lens is focused. Thus, both the effective transverse and longitudinal widths of the point spread function that corresponds to the expression in the first curly brackets in equation (10.5) will be reduced, and the more pointlike it becomes, the more pointlike $C(r, z)$ will be as a whole.

With STED, the in-plane resolution of images can be improved by as much as a factor of 10 compared to the Rayleigh limit (see figure 10.7). The improvement achieved depends on the number of photons delivered to the sample in the second pulse because as that number increases, the size of the volume that is effectively illuminated drops. However, the smaller that volume, the smaller the number of photons available to image it. In principle, that loss in signal can be offset by increasing the number of times the fluorescence of any given point in the specimen is excited, but the larger the number of times each volume is excited, the longer it will take to obtain a complete image. However, there are limits to the resolution that can be achieved this way because the amount of photobleaching that occurs within any given volume of the specimen will increase with the number light pulse sequence, not with the number of useful photons that each pulse sequence produces.

The second technique for obtaining hyper-resolved optical images that we will consider is called photo activated localization microscopy, or PALM. PALM microscopy exploits an unusual property of GFP (and other fluorescent proteins) that is called photoactivation. When a naïve GFP molecule is exposed to short wavelength visible light, its absorption spectrum shifts dramatically toward the red, and the efficiency with which it will subsequently emit green fluorescence in response to excitation at longer wavelengths increases dramatically. This effect is magnified in some GFP mutants, which can show a $\sim 1,000$-fold increase in the fluorescence they emit in response to 480-nm light following brief exposure to 400-nm light.

The objective in PALM microscopy is to image single, GFP-labeled protein molecules in specimens that may contain tens of thousands of them. This is achieved by irradiating a naïve specimen with 400-nm light at low intensity. The duration and intensity of this activating pulse is adjusted so that only a handful of the GFP domains in the sample get activated. Their locations in the specimen are determined by irradiating the sample with 480-nm light and imaging the fluorescence that emerges. The detection phase of the experiment is allowed to continue until the activated GFP domains disappear due to photobleaching. At that point, the image is stored in memory, and the activation and detection process is repeated so that another set of GFP domains can be imaged. The process is repeated until either the experimenter runs out of patience or there are no more GFP domains left to activate in the sample, which can take many hours to occur.

Comment: Light absorption is quantized; at the molecular level, it is all or nothing. As light intensities fall, the probability that a given molecule will absorb a photon drops, but if it absorbs any light at all, it will absorb an entire photon and respond to that absorption event the same way it would have if the light intensity had been higher.

Each image produced for a specimen is processed assuming that each of the comparatively small number of fluorescent spots it includes represents the radiation emitted by a single GFP domain; if the number of activated GFP domains is small, the probability of overlap in the plane of the image will be small. The location of the center of gravity of each such spot is determined by fitting its image to the point spread function of the microscope. A digital version of the image is created in which each spot in the original image is replaced by a point located at its center of gravity, or possibly a disk centered on that location, the diameter of which is determined by the magnitude of the error associated with the determination of its position. The final image of the specimen is created by superimposing all of the resolution-enhanced images taken of it.

Three comments are in order at this point. First, GFPs are not the only photoactivatable fluorescent proteins available today. Second, the first PALM experiments reported used a special illumination technique called total internal reflection fluorescence, or **TIRF**, to ensure that only fluorophores in the uppermost layers of the specimens being examined would be imaged. The shallowness of the specimen volumes excited by the incident radiation simplified the imaging problem that had to be solved by reducing the total number of fluorophores contributing to these first images. However, PALM can be done without TIRF. Third, resolutions below 10 nm have been reported for PALM images, which is even better than the highest resolutions that have been reported for STED images.

Comment: Because TIRF illumination is used for many kinds of single-molecule experiments, as well as for microscopy, it should be explained here. Suppose that a light ray traveling through a high index of refraction medium like glass ($n = 1.5$) encounters a glass–air interface, and that the angle the ray makes with the normal to that surface is, say, $130^\circ (= \theta_1)$. (See section 9.2 for an explanation of how the directions of the normals to refractive index boundaries are defined.) Based on what has been said so far, we would expect the ray to be refracted as it crosses that surface, and Snell's law tells us that 1.5 $\sin(130^\circ)$ should equal $\sin \theta_2$, where θ_2 is the angle the ray will make with the same normal once it crosses the boundary. However, 1.5 $\sin(130^\circ) = 1.15$, and the largest possible value for $\sin \theta_2$ is 1.0. Clearly, Snell's law does not describe everything that can happen to a ray of light at refractive index boundaries! In fact, only if θ_1 is greater than 138.19°, the so-called ***critical angle*** *for this particular interface, will refraction occur.*

If θ_1 is smaller than the critical angle, the ray will be redirected back into the high-n medium by its encounter with the boundary, as though the boundary were a mirror. This phenomenon is called total internal reflection, and it was mentioned in section 3.8. Compared to reflection from the surface of a silvered mirror, total internal reflection is extremely efficient; essentially all the incident energy gets reflected.

When this phenomenon is analyzed more carefully from the point of view of wave optics, we discover that the electric field of the radiation incident on an index of refraction boundary at less than the critical angle extends across the interface into the low refractive index medium. However, it does not extend very far (e.g., ~100 nm), and its amplitude decays exponentially with distance. That portion of the incident field is called an "evanescent wave", and the evanescent wave phenomenon is related to the quantum mechanical phenomenon called "tunneling", which allows the penetration of wave functions into regions of space that, classically, would be forbidden to the particles those wave functions represent.

TIRF illumination is a convenient way to make the layer of some sample that is being observed by fluorescence microscopy much thinner than would otherwise be possible. This is advantageous when we wish to detect the fluorescence of single molecules in a sample in which their concentration is so high that signals generated by single fluorophores could not be resolved in the (x, y) plane if the illuminated layer were, say, a micron (or more) thick, simply because so many of them would be fluorescing.

It should be noted that the structures of the samples used for both STED and PALM imaging must not change over considerable lengths of time because both are very slow. No doubt, someone will invent a technique that makes it possible to do ultrahigh-resolution microscopy on time scales that are short enough so that the movements of materials that occur in live cells can be tracked in real time.

PROBLEMS

1. Focusing lenses all obey the lens law: $1/s_1 + 1/s_2 = 1/f$. If the distance between the lens and the object plane is changed from s_1 to $(s_1 + \delta)$, where δ is small compared to s_1, by how much will s_2 change? This question is most easily answered by taking the new value of s_2 to be $(s_2 + \Delta)$. What is the ratio of δ to Δ? This ratio is the longitudinal magnification of the lens.
2. One property of the light microscope that we have not talked about so far is its *field of view*, which is the diameter of the (circular) area of the sample that will recorded or seen when the stage is held in a fixed position. The field of view of a microscope is the *field number* of its ocular, which is the diameter of the circle in the image plane of the microscope that can be visualized using that ocular, measured in millimeters, divided by the magnification of the objective lens. Field numbers for oculars are typically around 20 mm.
   Consider a microscope with an ocular that has a field number of 20 mm that is operating with an oil immersion lens that has a focal length of 2 mm, a numerical aperture of 1.25, and a magnification of 100. Are the differences in angles subtended by the lens at the periphery of the field of view large enough relative to those at the center to make the resolution of images vary significantly across the field of view?
3. The shapes of peaklike functions are often characterized by their full widths at half maximum (FWHM) values. Suppose you are doing confocal microscopy with an instrument that is equipped with objective lenses that have numerical apertures of 0.707, and the distances are such that the magnifications of both are 100. To make calculations simple, assume the instrument is operating with monochromatic light that has a wavelength of 500 nm and with no oil immersion. Under ideal circumstances, what should the FWHM be of this microscope be along the optical axis? If an ordinary microscope were equipped with a similar objective lens, what would its FWHM be along the optical axis? In both cases, be sure the answer refers to the object plane, not the image plane.

FURTHER READING

Born, Max, and Wolf, Emil (1999). *Principles of Optics,* 7th ed. New York: Cambridge University Press.

Hell, S. W. (2010). Far field optical nanoscopy. In *Single Molecule Spectroscopy in Chemistry, Physics and Biology,* ed. A., Graslund, R., Rigler, and J., Widengren, 365–398. Berlin: Springer-Verlag.

Jonkman, J. E. N., Swoger, J., Kress, H., Rohrbach, S. A., and Stelzer, E. H. K. (2003). Resolution in optical microscopy. *Methods in Enzymology* 360, 416–446.

Slater, E. M., and Slayer, H. S. (1992). *Light and Electron Microscopy* Cambridge: Cambridge University Press. This text is a fine blend of both the practical and the theoretical aspects of optical and electron microscopy.

APPENDIX 10.1 THE DISTRIBUTION OF LIGHT ENERGY ON-AXIS, AND NEAR-FOCUS

The light emitted by an on-axis point source that is s_1 from an ideal focusing lens, where $s_1 > f$, will come to a focus on axis at a distance s_2 from the lens, which can be computed using the lens law. Under these circumstances, at the instant any wave front emerges completely from the lens, it will be a spherical surface of radius s_2 that is centered on its ultimate focal point. The question we address here is the distribution of light energy along the optical axis, near that focal point.

Consider a point on the optical axis that is displaced from the focal point by a distance Z, which is small compared to s_2. The distance from any point on the wave front to that location will be $[(s_2 \cos\theta + Z)^2 + (s_2 \sin\theta)^2]^{1/2}$, where θ is the angle between the optical axis and a line from the point of interest on the wave front and the focal point, and the difference between that distance and its distance to the focal point, which, by definition, is the same for all points on the front, δ, will be:

$$\delta = \left[(s_2 \cos\theta + Z)^2 + (s_2 \sin\theta)^2\right]^{1/2} - s_2.$$

When the expression inside the square brackets is evaluated, and terms that are second order in Z are eliminated, we find that:

$$\delta = s_2 \left(1 + (2Z/s_2)\cos\theta\right)^{1/2} - s_2,$$

and using the usual approximation for $(1 + x)^{1/2}$, it become apparent that if Z is small compared to s_2, then:

$$\delta \approx Z \cos\theta.$$

Thus, the phases of the rays arriving at $(s_2 + Z)$ depend only on θ. This implies that the amplitude and phase of the corresponding electromagnetic wave at $(s_2 + Z)$, A(Z), will be:

$$A(Z) = \int_0^{\theta_{max}} \exp[2\pi i (Z/\lambda)\cos\theta] 2\pi s_2^2 \sin\theta \, d\theta,$$

θ_{max}, is determined by the aperture of the lens. It is useful to normalize A(Z), by dividing it by the area of spherical surface that is being focused, which is: $2\pi s_2{}^2(1 - \cos\theta_{max})$.

The integral in the expression for A(Z) is easy to evaluate because $d(\cos\theta) = -\sin\theta \, d\theta$, when this is done, we discover that the normalized expression for it, $A'(Z)$, is:

$$A'(Z) = i\lambda \left[2\pi Z(1 - \cos\theta_{max})\right]^{-1} \left\{\exp\left[2\pi i (Z/\lambda)\cos\theta_{max}\right] - \exp\left[2\pi i (Z/\lambda)\right]\right\}.$$

The on-axis distribution of light energy, I(Z), is $A'(Z)[A'(Z)]^*$, and when that expression is evaluated, it is found that:

$$I(Z) = 2\lambda^2 \left[2\pi Z(1 - \cos\theta_{max})\right]^{-2} \{1 - \cos\left[2\pi (Z/\lambda)(1 - \cos\theta_{max})\right]\}.$$

Because $(1 - \cos\theta) = 2\sin^2(\theta/2)$, we can rewrite this equation as follows:

$$I(Z) = \left\{\sin\left[2\pi (Z/\lambda)\sin^2(\theta_{max}/2)\right] / \left[2\pi (Z/\lambda)\sin^2(\theta_{max}/2)\right]\right\}^2. \qquad \textbf{(10.1.1)}$$

In short, the on-axis distribution of light intensity near focus has the form $(\sin x/x)^2$. (Note: We have assumed here that the refractive index is 1 everywhere, except inside the lens.)

PART FIVE

Electron Microscopy

11

Lenses that focus Electrons

In 1897, J. J. Thomson discovered the electron using an instrument that was the progenitor of the modern cathode ray tube (CRT), which was the image-generating device that formed the heart of every television set manufactured prior to ~1995. Thomson's instrument consisted of an evacuated (glass) chamber that, at one end, had a device that produced a collimated beam of electrons of well-defined kinetic energy—an **electron gun**—and at the other end, a screen coated with a phosphor that glowed wherever the electron beam hit it. Thomson determined the ratio of the charge to the mass of the electron by measuring the effects that electric and/or magnetic fields had on the position of the bright spot on the screen of his apparatus. In the early decades of the 20th century, CRTs were developed that produced much smaller spots than Thomson's instrument could, and this improvement was achieved using magnetic and/or electrostatic devices that focused electron beams.

Any device that focuses radiation is a lens, and around 1930, it was realized that it might be possible to build a microscope based on CRT lenses that would image objects using electron radiation. In 1931, Ernst Ruska and Max Knoll built the first such instrument, and like every electron microscope (EM) constructed since, it was, basically, a CRT. Its optical track began with an electron gun. The beam produced by that gun passed through a series of magnetic lenses, and its flight path ended with an area detector for electrons, such as a phosphorescent screen or a piece of photographic film. The novel features of their CRT were the design and placement of its lenses, and the stage it included for holding specimens.

Both the magnification and resolution of the images produced by the first EM were modest by optical standards, but within a few years, instruments had been built that had far higher magnifications and far higher resolutions than any optical microscope can provide. By the late 1930s, Siemens was manufacturing EMs commercially, and after the Second World War, other firms entered the business. Among their customers were biologists whose interest had been piqued by electron micrographs of biological specimens that had been published before the war.

The impact of the EM on biology has been enormous. By the 1950s, images of cells and tissues were being routinely obtained that had much higher resolutions than those that light microscopes provide (~1 to 10 nm vs. ~200 nm). The information about the ultrastructure of tissues and cells gleaned from those images forms the foundation of modern cell biology. The field has now advanced to the point that under favorable circumstances, the three-dimensional structures of biological macromolecules can be determined at near atomic resolution (~0.1 nm) using EM data. There is every reason to believe that electron microscopy will continue to make major contributions to biological sciences far into the future.

Although EMs have been around for more than 70 years, they are still nowhere near as perfect, technically, nor as user-friendly as optical microscopes. As we already know, the multielement lenses in modern optical microscopes are effectively aberration-free, and their focal lengths are fixed at the time of manufacture. Once those lenses are installed in a microscope, which takes a few seconds, they are guaranteed to perform as advertised. By contrast, not only are the lenses in EMs comparatively poor optically, the instruments in which they are installed are complicated to operate. For example, the focal lengths of all the lenses in an EM are adjustable, and the settings the user chooses for them have a big impact on image quality, and focal lengths are not the only instrumental parameters that users must pay attention to if high-quality images are to be obtained. In fact, anyone aiming to produce EM images that will provide reliable structural information at resolutions in the low angstrom range must be prepared to push EM technology to the limit.

The objective here is not to make you an expert electron microscopist, but rather to provide you with a grounding in the fundamentals of that art that is good enough so that you understand how EM images are produced, and why experts worry about the issues they do.

11.1 BIOLOGICAL ELECTRON MICROSCOPY IS DONE USING TWO DIFFERENT KINDS OF EMS

Over the years, many different kinds of EMs have been developed, but only two are commonly used for biological research: the transmission electron microscope (TEM), and the scanning transmission electron microscope (STEM). The TEM and STEM are the electronic equivalents of the conventional optical microscope and the scanning optical microscope, respectively (see section 9.1). Both produce images that report on the effects that samples have on electron beams passing though them, but do so in ways that are different enough to be interesting. However, the TEM is by far the more important of the two, there being thousands of TEM papers in the biological literature for every STEM paper, and for this reason, only the TEM will be discussed here.

Figure 11.1 is a schematic diagram of a TEM. Its optical track begins with an electron gun, which is the EM equivalent of a light source, which is followed by

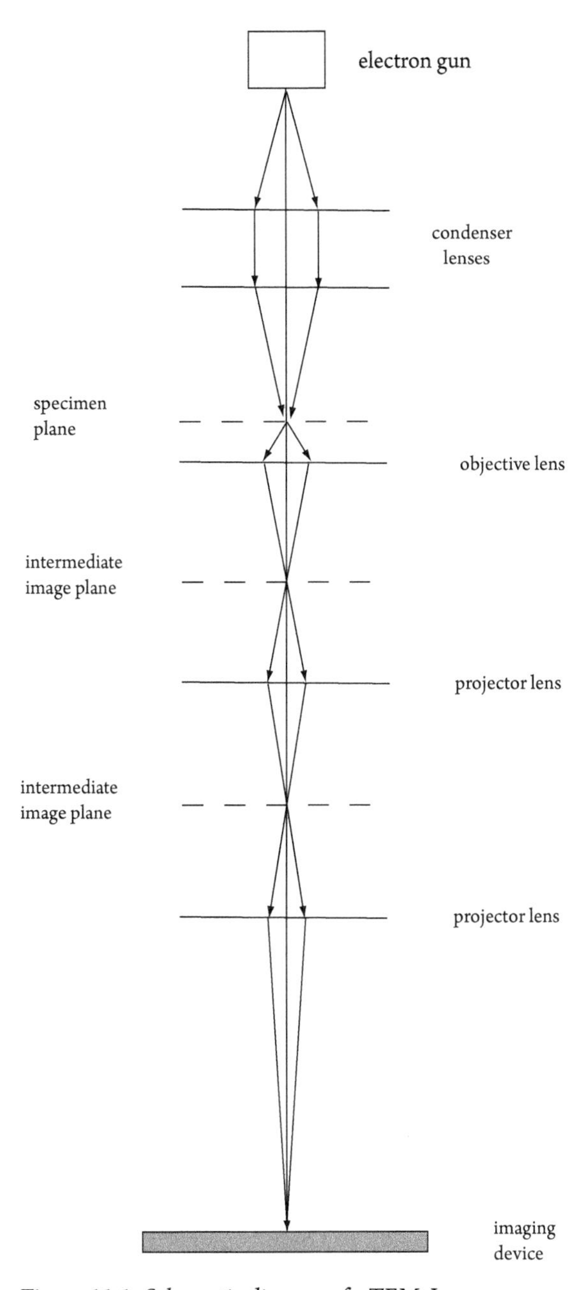

Figure 11.1 Schematic diagram of a TEM. Lenses are shown as solid lines, and optically significant locations are shown using dotted lines. The trajectories of electrons passing through the instrument are indicated using arrows.

one or two lenses that serve the same function as the condenser lenses in a light microscope. The radiation that emerges from the condenser system illuminates a specimen that is mounted just above the objective lens of the microscope, the performance of which largely determines the quality of the images the instrument creates, as usual. The objective lens is followed by two (or more) projector lenses, and then an image-recording and/or visualization device of some sort. The projector lenses in an EM play the same role as the ocular lenses in a light microscope; they magnify the image generated by the objective lens.

Conceptually, TEMs and conventional optical microscopes are exactly the same, but they differ operationally not only because electrons and photons differ in their interactions with matter, but also because electron lenses do not focus electrons the way glass lenses focus light. Thus, because all microscopes are built around lenses, a logical place to begin any discussion of EMs is with a description of how electron lenses work.

11.2 THE MAGNETIC FIELD OF A ONE-TURN COIL CAN FOCUS ELECTRONS

Beams of electrons can be focused both electrostatically and magnetically, and electrostatic lenses are easy to understand qualitatively. If the axis of a parallel beam of electrons is aligned so that it is collinear with the axis of a hole in a negatively charged plate, electrostatic interactions will push its peripheral electrons toward the center of the hole, where the electric field is 0, and thus make the beam converge, or focus. It is also easy to generate magnetic fields that will focus beams of electrons (or other charged particles). The field generated by a circular coil of wire that has a DC electric current flowing through it will do the job, but it is harder to understand why they work.

Figure 11.2 shows schematically what the magnetic field of a circular wire with a current running through it looks like. When a moving charged particle encounters a magnetic field, the force it experiences, **F**, is:

$$\mathbf{F} = \left(q\mathbf{v} \times \mathbf{B}\right),$$

where q is the charge on the particle, **v** is its velocity, and **B** is the magnetic field. Written using cylindrical polar coordinates (see figure 11.3), which are the best ones to use in this context, this equation becomes:

$$\mathbf{F} = q\left(v_\varphi B_z - v_z B_\varphi\right)\mathbf{1}_r + q\left(v_z B_r - v_r B_z\right)\mathbf{1}_\varphi + q\left(v_r B_\varphi - v_\varphi B_r\right)\mathbf{1}_z.$$

($\mathbf{1}_\varphi$ is the vector of unit length that is oriented tangentially to the circle of radius r, the origin of which is the point where **r** intersects that circle.) Because the field produced by the coil is cylindrically symmetric, B_φ, the component of **B** in the tangential direction, is 0 everywhere. Thus, in this instance:

$$\mathbf{F} = q\left(v_\varphi B_z\right)\mathbf{1}_r + q\left(v_z B_r - v_r B_z\right)\mathbf{1}_\varphi - q\left(v_\varphi B_r\right)\mathbf{1}_z.$$

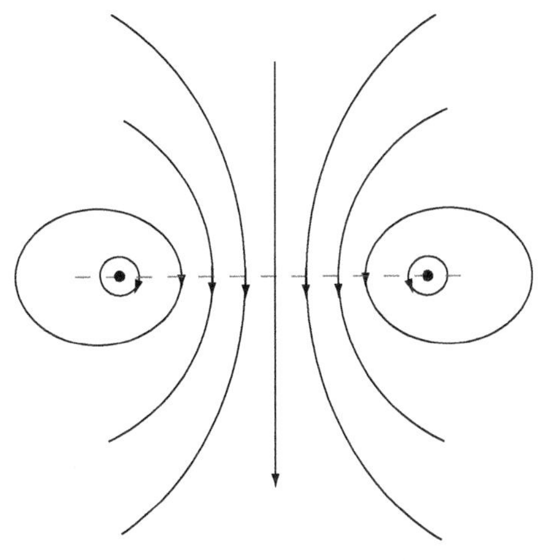

Figure 11.2 Schematic diagram of a cross section of the magnetic field produced by a circular wire that has an electric current flowing through it. The wire is indicated by the two large dots, and the current is flowing into the page on the left and out of the page on the right.

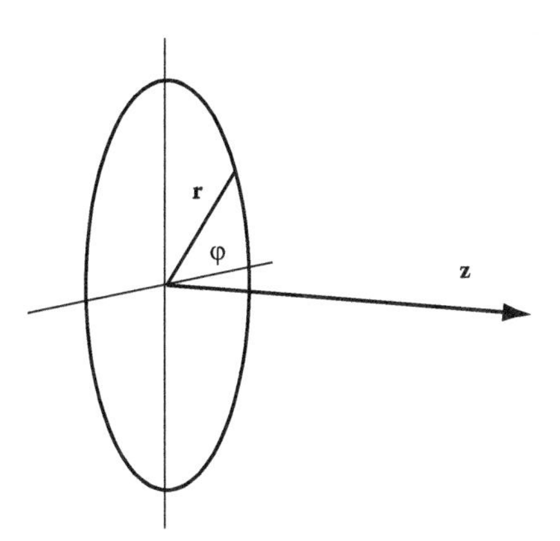

Figure 11.3 The coordinate system used below to describe magnetic field interactions.

Furthermore, because the tangential velocity of the charged particle, v_φ, is $r(d\varphi/dt)$, where r is the distance of the particle from the axis of symmetry of the coil, and φ is its azimuthal coordinate, we can also write:

$$\mathbf{F} = qr\,(d\varphi/dt)\,B_z\mathbf{1}_r + q\,(v_zB_r - v_rB_z)\,\mathbf{1}_\varphi - qr\,(d\varphi/dt)\,B_r\mathbf{1}_z.$$

Suppose a charged particle that originated at a point on the axis of symmetry of the coil enters it on a (straight line) trajectory that diverges from that axis. As it first encounters the magnetic field of the coil, the tangential component of its velocity, v_φ, will be 0, and hence the force it experiences will be:

$$\mathbf{F} = q\left(v_z B_r - v_r B_z\right)\mathbf{1}_\varphi.$$

Inspection of figure 11.2 reveals that the r- and z-components of the field of a coil, B_r and B_z, always differ in sign on the entry side of a coil, no matter what the direction of the current in the coil, and hence the two components that contribute to F_φ will initially have the same sign because v_z and v_r are both positive. Thus, the interaction of the particle with the field will accelerate it in a tangential direction, either clockwise or counterclockwise, depending on the direction in which the current is circulating in the coil and the sign of the particle's charge. Furthermore, its trajectory will be transformed from a straight line into spiral.

Once the trajectory of the particle begins to have a rotational component, the radial component of the force it experiences, $[qr(d\varphi/dt)B_z\mathbf{1}_r]$, will no longer be 0. You can easily verify that no matter what the charge on the particle or the sign of B_z, the radial component of the force it experiences will be negative, which means that the particle will be accelerated *toward* the axis. It should be noted that the further the particle is from the axis, the stronger that radial force will be, which is exactly the behavior required to focus a beam of charged particles.

Qualitatively, it is easy to understand how the trajectories of charged particles change as they pass through coils of this sort. In the first place, along the axis of the coil, B_r is everywhere 0. Thus, a particle traveling down that axis, which, by definition, must have a velocity in the r direction of 0 and an r-coordinate of 0, will experience no forces as it passes through the coil. Its trajectory will be a straight line, just like the trajectory of a ray of light traversing a glass lens along its optical axis. The trajectories of particles that encounter the coil off-axis are not straight lines, and it easiest to think about them one component at a time.

The tangential motion of any off-axis particle, which begins as the particle enters the field of the coil, interacts with the radial component of that field in a manner that initially makes it decelerate. However, as the particle crosses the plane of the coil, the sign of the axial component of that force changes because the sign of B_r changes, and it starts accelerating. By the time the particle leaves the neighborhood of the coil, its axial velocity will be what it was to begin with.

The motion of the particle in the radial direction is also easy to understand. Because the sign of the z-component of the magnetic field, the sign of r, and the sign of $d\varphi/dt$ do not change as a particle passes through the system, the r-component of its velocity will decrease continuously, and by the time it exits from the field, it may have become negative (i.e., focusing), provided the field strength of the magnet is appropriately matched with the (axial) velocity of the particle.

The tangential motions observed when particles traverse magnetic lenses are more complicated. As we pointed out earlier, the initial encounter of a charged particle with the magnetic field of a coil accelerates it tangentially. However, that acceleration becomes a deceleration once the particle crosses the plane of the coil because at that point, B_r changes sign, and if the field of the lens is strong enough to make the particle assume a convergent trajectory, v_r will have changed sign also (see figure 11.1). By the time the particle escapes from the magnetic field, its rotational velocity will be 0, but in the process, the value of φ that characterized its trajectory at the time it entered the field of the lens will have changed because φ will have been increasing (or decreasing) continuously during the particle's passage though the coil. For this reason, not only are the images formed by single magnetic lenses inverted relative to the objects they represent, just as is the case for the images formed by optical lenses, they are also rotated by some amount.

These qualitative arguments establish the plausibility of the idea that the magnetic fields generated by coils might focus beams of electrons. However, a quantitative understanding of the imaging properties of any particular magnetic (or electrostatic) lens can be obtained only by working out, in detail, its effects on the trajectories of electrons passing through it. This kind of analysis is the electronic equivalent of the ray tracing that has been an essential part of optical design since time out of mind. It can be carried out for magnetic lenses only if the magnetic fields they generate are accurately known at all values of r and z.

11.3 THE LENSES IN EMS ARE SOLENOIDS

Single-turn coils cannot generate magnetic fields strong enough to focus beams of electrons having kinetic energies in the range commonly used in TEMs; the heat generated by the current required to make them do so would melt them. For that reason, the lenses in EMs are (usually) water-cooled, multiturn solenoids of sub-

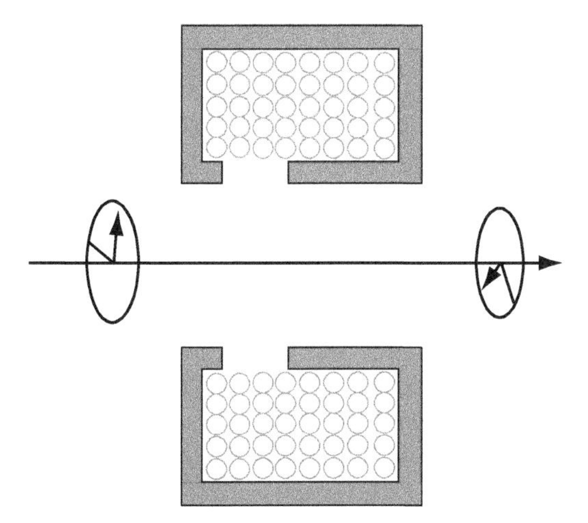

Figure 11.4 Schematic diagram of a magnetic lens in cross section. Its wire coils are encased in a ferromagnetic casing (gray) that has a gap on its interior side. The axial line points in the direction of the electron beam. The orientations of the arrows above and below the lens depict the image inversion and rotation that is characteristic of magnetic lenses.

stantial physical size. (In some EMs, superconducting lenses are used.) The coils in these lenses are invariably surrounded by a toroidal iron casing that has a narrow gap around its inner circumference. These casings shape and concentrate the magnetic field generated by the coils inside them so that charged particles that are traveling down the bore of the lens assembly are affected by the field only in the vicinity of the gap. The shapes of the magnetic fields in their gap regions are further optimized by iron **pole pieces** that are inserted in their bores. Figure 11.4 is a schematic diagram of such a lens from which the pole piece has been removed.

11.4 MAGNETIC LENSES HAVE FOCAL LENGTHS

A magnetic lens has a focal length, just like a glass lens. Its primary determinants are: (1) the magnetic field strength of the lens, (2) the way its magnetic field varies as a function of position along the optical axis, and (3) the kinetic energy of the electrons passing through it. Assuming that the lens is a thin lens in the usual optical sense, and that the variation in the strength of the axial component of the magnet field along its optical axis is a Lorentzian that has a half-width at half-height of a, the focal length of the lens, f, will obey the following equation:

$$1/f = (\pi/16)\left(q_e/mV\right) aB_o^2. \tag{11.1}$$

In this equation, q_e and m are the charge and the mass of the electron, respectively; V is the accelerating voltage in the gun; and B_o is the maximum magnetic field strength encountered along the axis of the lens (in Tesla). The mass assigned to electrons in equation (11.1) must be their relativistic masses, not their rest masses (see section 1.15) because the accelerating voltages in the TEMs that biologists normally use (~100 kV) impart velocities to them that are a significant fraction of the speed of light.

The objective lenses in TEMs have maximum field strengths of ~1 T, and the strength of the field varies along the optical axis in such a way that its half-width at half-height is ~2 mm. Operating with electrons that have energies of 100 keV, which is typical, a lens of this sort will have a focal length of ~1.5 mm. (Henceforth, we will refer to any magnetic lens that has a focal length of exactly 1.5 mm for 100 keV electrons as a *standard lens*.) Using the lens law, we find that if the image plane of such a lens is to be located 10 cm beyond its optical center ($= s_2$), which is typical of TEMs, and the magnification of the image formed there is to be ~65, which is again appropriate, the specimen must be mounted about 1.523 mm upstream of the optical center of the lens ($= s_1$). (Magnetic lenses are so bulky that specimens must often be mounted *inside* the bores of the objective lenses of TEMs to meet these specifications.)

The magnetic field of a solenoid is proportional to the current flowing through its coils; hence, the focal lengths of magnetic lenses are inversely proportional to current squared (equation (11.1)). Thus, the focal lengths of the lenses in EMs can be altered by adjusting the settings on their power supplies, and the operators of EMs adjust them all the time. Not only do these adjustments alter the magnification of images, they also (usually) make them rotate. (Some EMs are built so that the rotations caused by changes in the current in their objective lenses are compensated for by rotations in the opposite sense produced by lenses further down the optical track.) In this respect, the lenses in EMs are completely different from their optical counterparts; the only way that focal lengths can be changed in an optical microscope is by swapping out its objective lenses.

11.5 THE OPTICAL PROPERTIES OF EMS CAN BE WORKED OUT USING QUANTUM MECHANICS

As we already know, there are two ways to analyze what optical lenses do to light: (1) by tracing the trajectories of rays, or, equally appropriately, the trajectories of particles/quanta passing through them, or (2) by determining how they alter the shapes of the corresponding wave fronts. The ray tracing/particle mechanics approach is used in chapter 9 to derive the lens law, and we used it in section 11.2 to explain why solenoids focus electron beams. However, as noted in section 9.9, aberration is better approached starting from the wave front point of view.

What is the wave that corresponds to a moving electron? That question was answered in the 1920s by de Broglie, who hypothesized that moving particles behave like waves having wavelengths of (h/p), where h is Planck's constant, and p is the particle's momentum. The validity of his hypothesis was demonstrated experimentally a few years later. For the electrons in an EM, the expression for p that must be used to calculate wavelengths is the relativistic one, and hence:

$$\lambda = h\left[\left(2m_e q_e V_o\right)\left(1 + q_e V_o/2m_e c^2\right)\right]^{-1/2}. \qquad \mathbf{(11.2)}$$

Plugging the appropriate values into equation (11.2), we find that the de Broglie wavelength of a 100 keV electron is 0.037 Å.

The wave function associated with a moving object is a probability amplitude that varies with spatial location and time, and as every student of quantum mechanics knows, its square is the probability that the corresponding particle will be found at some point in space at some instant in time. Compared to many of the systems analyzed by quantum chemists, the EM is simple. Because the distances between successive electrons in the beam of an EM are often of the order of centimeters, interactions between electrons in the beam can be ignored. In addition, the electron guns in EMs produce radiation that is almost perfectly monochromatic (see section 11.11), and the trajectories of the electrons in the beam are all nearly parallel to the optical axis. Thus, to first order, a wave function appropriate for any single electron in the beam of an EM will be appropriate for them all, and the image produced by an EM is proportional to the sum of the squares of a large number of single-electron, wave functions that are nearly identical, evaluated at the image plane.

When the effects that magnetic (or electrostatic) lenses have on electron beams are worked out quantum mechanically, we find that there is a perfect correspondence between what happens to an electron's wave function as it passes through an EM and what happens to an electromagnetic wave front traveling through a light microscope. For example, the optical path length idea used in chapter 9 to analyze the aberrations in optical lenses can also be used to work out the aberration properties of electron lenses. For electrons, the optical path length of a trajectory is the integral over that trajectory of distance divided by (electron) wavelength, and the condition for focus remains the same. An EM will produce an image of a point object at some location downstream of its objective lens only if the optical path length from that object to its image is the same for an appreciable fraction of the trajectories electrons scattered by the object could take as they travel from the object to its image.

11.6 ABERRATION SERIOUSLY DEGRADES THE PERFORMANCE OF MAGNETIC LENSES

As we know, the images produced by optical microscopes are distorted by chromatic aberration and five different kinds of monochromatic aberration. The magnetic lenses used in EMs to form images display all the same faults, but from a practical point of view, that is about where the similarity between the lenses in EMs and optical microscopes ends. For about a century, the multielement objective lenses available for optical microscopes have been so close to ideal that aberration can be ignored by most users. By contrast, aberration has an impact on electron microscopy so large it can never be ignored.

The objective lenses found in most EMs today are the magnetic equivalents of single-element, optical lenses, which invariably have large aberration coefficients, and single-element magnetic lenses are much worse than single-element glass lens in this regard. Only recently have magnetic lens systems been built that are capable of generating images that are comparatively free of aberration artifacts (see section 11.15). Because only a handful of EMs are equipped with such lens systems today, now and for a long time to come, those interested in using EM images wisely will need to understand what aberration does to them. Surprisingly, it is not all bad news.

For reasons that will made clear shortly, the apertures of the magnetic lenses in EMs are deliberately made very small compared to their focal lengths, which is helpful to those interested in analyzing their performance for two reasons. First, because the apertures used are small, only the paraxial optics of magnetic lenses need be considered, and the second benefit follows from the first. If only the paraxial optics of a microscope are relevant, then the aberrations that will most seriously affect the quality of the images it produces will be chromatic aberration and spherical aberration. We will deal with spherical aberration first.

In section 9.9, we obtained an equation for difference in optical path length for off-axis and on-axis rays traveling between the same two points on the optical axis that is useful for estimating spherical aberration effects (equation (9.5)):

$$\Delta OP \approx (\alpha^2/2) f^2 [-1/f + 1/s_1 + 1/s_2] - (\alpha^4/4)\, C_s.$$

[NB: In EM world, the angle called θ by optical microscopists is usually designated α.] This equation is as useful for describing the properties of magnetic lenses as it is for dealing with glass lenses. Because the term that is second order in α represents ideal lens behavior, the component of ΔOP due to aberration, ΔOP_{ab}, must be:

$$\Delta OP_{ab} = -\left(\alpha^4/4\right) C_s + \ldots, \tag{11.3}$$

where C_s is the spherical aberration coefficient of the lens, as we already know. Appendix 9.2 provides an equation for C_s that is applicable to single-element glass lenses, but not to multielement glass lenses, let alone to magnetic lenses. In a pinch, rather than using some formula, we might have to measure C_s directly, but as is pointed out in chapter 9, however it is arrived at, C_s has the dimensions of length, and for converging lenses it is *always* positive.

11.7 SPHERICAL ABERRATION LIMITS THE RESOLUTION OF MAGNETIC LENSES

In section 9.9, it is pointed out that spherical aberration makes the focal length of lenses depend on α : $f(\alpha) = f - C_s(\alpha^2/2)$. This effect is referred to as *longitudinal spherical aberration* to distinguish it from the lateral effects of spherical aberration, which makes the images of points look like disks. Using the lens law, it is not hard to show that a ray leaving the point $(0, 0, -s_1)$ at the highest angle that the limiting aperture of the lens will accept, α_{max}, will cross the optical axis on the far side of the lens at a point that is $(s_2/f)^2 C_s(\alpha_{max}{}^2/2)$ short of the paraxial image plane, which is s_2 from the right-hand principal point of that lens. If α_{max} is small, the angle that ray makes with the optical axis at that point will be $\sim(\alpha_{max}/M)$, where M is the magnification of the image. Hence, it will intersect the image plane at a distance from the optical axis equal to $(s_2/f)^2 C_s(\alpha_{max}{}^3/2)(1/M)$. If M is large, as it always is for microscopes, $(s_2/f)^2 \approx M^2$; hence, the radius of the disk in the image plane that is illuminated by all the rays that entered the lens at angles $\leq \alpha_{max}$ will be $\sim MC_s(\alpha_{max}{}^3/2)$.

If the lens being used to form images was ideal, a disk in the image plane of some specified radius would correspond to a disk in the object plane that was smaller in radius by a factor of M. It follows that the radius of the disk in the object plane that corresponds to a disk in the image plane of radius $MC_s(\alpha_{max}{}^3/2)$ is $(\alpha_{max}{}^3 C_s/2)$. Clearly, neighboring points in the object plane will *not* be resolved in the images formed by a lens that has spherical aberration unless they are separated by a distance that is as great or greater than the larger of the two distances: $(\alpha_{max}{}^3 C_s/2)$, the radius of the spherical aberration disk, or $0.61\lambda/n\sin\theta$, the Rayleigh resolution.

The magnitude of the impact that spherical aberration can have on resolution is best understood by putting some numbers into the formulae just provided. Consider a lens made of glass that has a refractive index of 1.5, and spherical surfaces that have radii of 2 mm and −2 mm. (This lens will henceforth be our *standard optical lens*.) Ignoring its thickness, the focal length of this lens will be 2 mm, which is about the same as the focal length of our standard electron lens. If that lens is used as the objective lens of a microscope, its spherical aberration coefficient, C_s, will be about 3.3 mm (see appendix 9.1), and the implications of that fact for the resolution of the images this lens produces are startling. If the aperture of that lens, α_{max}, was limited to 0.174 rad ($\sim 10°$), which would be modest for the objective lens of an optical microscope, spherical aberration would reduce the resolution of the images it generated to roughly 9×10^{-6} m. By contrast, if an aberration-free lens of the same focal length were used the same way, only interference effects would limit the resolution of its images, and the resolution would be 2.8×10^{-6} m, which is more than three times better. (It is clearly a good thing that the multielement objective lenses in optical microscopes have spherical aberration coefficients much smaller than those of single-element glass lenses!)

For magnetic lenses, C_s is usually of the order of 1 mm, which sounds great when it is compared to the spherical aberration coefficient of the standard optical lens, but it is less wonderful than it might appear at first glance because the wavelength of the radiation used in TEMs is 0.037 Å, not 5,000 Å. Assuming for the sake of argument that α_{max} is again 0.174 radians, we find that whereas the Rayleigh resolution of an imaging system based on an aberration-free magnetic lens of that aperture would be 0.21 Å (2.1×10^{-11} m), spherical aberration would limit its resolution to about 2.6×10^{-6} m, a five order of magnitude disaster.

11.8 THE RESOLUTION OF THE IMAGES PRODUCED BY LENSES THAT HAVE LARGE SPHERICAL ABERRATION COEFFICIENTS CAN BE IMPROVED BY *REDUCING* THEIR LIMITING APERTURES

Improvements in lens design have rendered the spherical aberration problem moot for optical microscopy, as we know, but until recently, nothing so interesting was possible for electron microscopy. However, it is important to remember that the geometrical argument just made about the impact of spherical aberration on resolution is simple-minded. It is as true of electron optics as it is of light optics that resolution cannot be understood properly unless diffraction is taken into account, and the appropriate, diffraction-based analysis of resolution was done for EMs by O. Scherzer more than 60 years ago. The essence of his argument is easily grasped.

The "error" in optical path lengths produced by spherical aberration is $-(\alpha^4/4)C_s$, as we already know. The fact that spherical aberration is proportional to α^4 means that no matter what the value of C_s, the impact of spherical aberration on image formation will be negligible from $\alpha = 0$ out to some value that depends on the magnitude of C_s. Thus, out to the resolution implied by that scattering angle, the images formed by any lens will be usefully faithful, and standard, diffraction-based arguments about resolution will apply. When the problem is viewed from this perspective, it becomes clear that the key to understanding the true resolution of the images formed by lenses that have spherical aberration is estimating the value of α beyond which the effects of spherical aberration become intolerable, α_{limit}. If $\lambda/4$ is taken to be the maximum allowable error in the optical path, which seems plausible, then $\alpha_{\text{limit}} = (\lambda/C_s)^{1/4}$. Now for small α, Bragg's law is $1/d = \alpha/\lambda$, and thus the crystallographic resolution that corresponds to α_{limit} is $(\lambda^{3/4}C_s{}^{1/4})$. This expression is, basically, Scherzer's estimate for the resolution of EM images. (Depending on the transverse coherence of the electron beam in a microscope, and the details of the criterion used to decide when points are resolved, the estimates obtained for the resolution of EMs are Scherzer's expression multiplied by a constant having a value between ~0.6 and 1.0, which we will ignore.)

Using the formulae just provided, we find that α_{limit} for the standard magnetic lens is about 0.45°, that is, 7.9 mrad, and that the Scherzer resolution of an EM having the standard magnetic lens as its objective should be $\sim 4.7 \times 10^{-10}$ m (4.7 Å), a value far better than the estimate given in the previous section. The reason it is better is that α_{limit} is, paradoxically, much *smaller* than 10°, the value of α_{max} used previously to evaluate the impact of spherical aberration on resolution.

11.9 ELECTRON MICROSCOPES HAVE LARGE DEPTHS OF FIELD AND FOCUS

Now that we know something about the resolutions of TEM images, we can address their depths of focus and depths of field. Both are very large compared to the wavelength of the radiation used in EMs because the range of the scattering angles of the electrons that contribute to the images that these instruments produce are invariably very small; in other words, the numerical apertures of EMs are small. The Scherzer resolution of the images formed by the standard magnetic lens imply that all the useful information they contain is conveyed by electrons having scattering angles less than α_{limit}, which is only ~0.45°. Furthermore, the optical tracks of many TEMs include apertures, which are mounted beyond their objective lenses, to prevent electrons scattered at angles greater than ~1° to 2° from making any contribution whatever to the images they produce.

In section 9.16, it is shown that the depth of field of the image formed by a microscope is the resolution of that image divided by $\tan(\alpha_{\text{limit}})$. Thus, when a TEM, the objective lens of which is the standard lens, forms an image that in fact has the Scherzer resolution, the depth of field of that image will be ~600 Å. Clearly, the higher the resolution of a TEM, the lower its depth of field.

The depth of focus of a TEM image is its depth of field times the square of its magnification (see section 9.16). Because the magnification of the images formed by the objective lenses of TEMs are of the order of 100, their depths of focus are quite large at their focal planes, that is, ~0.5 mm. The projector lenses beyond that focal plane often increase the overall magnification by another factor of ~100 (at least), which means that depths of focus at the image plane are measured in meters. Not surprisingly, therefore, the quality of the images TEMs generate is totally insensitive to the accuracy with which the devices used to record them are positioned in the optical track, let alone to the thickness of the components of those devices that actually absorb the electrons that they detect, such as to the thickness of photographic emulsions.

11.10 FOCAL LENGTH VARIATION AND SPHERICAL ABERRATION ARE IMPORTANT DETERMINANTS OF THE CONTRAST TRANSFER FUNCTIONS OF EMS

As we just discussed, differences in optical path length between the on-axis and off-axis rays that contribute to an image correspond to phase differences, $\Delta\varphi = (2\pi/\lambda)\Delta OP$ (see equation (9.5)). Thus, we can write:

$$\Delta\varphi = (2\pi/\lambda)\left[(\alpha^2/2)f^2\left(-1/f + 1/s_1 + 1/s_2\right) - \left(\alpha^4/4\right)C_s\right].$$

In thinking about the application of this expression to electron microscopy, it is important remember that whereas the focal lengths of the objective lens of EMs can be easily be adjusted by their users, the locations of specimens (i.e., s_1), and image recording devices (i.e., s_2), cannot. It follows that the magnitude of the first term on the right-hand side of this equation will depend on how far the operating focal length of the objective lens is from the focal length that would be required to ensure that the (geometrical optic) image of a point in an object plane located at s_1 is a point in an image plane that is positioned at s_2. We will call that focal length f_o. It is easy to show that if Δf, the difference between f and f_o, is small, then:

$$\Delta\varphi = (2\pi/\lambda)\left[(\alpha^2/2)\,\Delta f - \left(\alpha^4/4\right)C_s\right]. \tag{11.4}$$

It follows that the contrast transfer function (CTF) of a lens that has spherical aberration, and/or other kinds of aberration, is the product of an aperture function, $A(\alpha)$, which is 1.0 if α is less than some value that will usually be as large as, or larger than, the Scherzer angle, α_{max}, and 0.0 if α is greater than that, and a function of the form $\exp[i\Delta\varphi(\alpha)]$, which accounts for phase distortions:

$$\mathrm{CTF}(\alpha) = A(\alpha)\exp\left[i\Delta\varphi(\alpha)\right]. \tag{11.5}$$

(CTFs may include additional terms, as we shall see later.)

Equation (11.5) implies that it is perfectly possible for the Fourier transforms of EM images to include components that have wavelengths *shorter* than the Scherzer resolution. Out to the resolution limit imposed by the size of the aperture of the objective lens of a microscope, all the CTF described by equation (11.5) does to the radiation passing through a microscope is alter the relative phases of the Fourier components of transmission functions. So if the Fourier transform of the transmission function of some object includes some short wavelength components that have significant amplitudes, so too will its image. However, because their contributions to that image may not have the right phase, the detail they add to the image may be misleading. It follows that the Scherzer resolution of an EM image is not a statement about the length scale of the finest details that may be discernable in some image, but rather it is a statement about the length scale of the finest details in that image that are *reliable*. It should be noted for future reference that the magnitude of the phase errors of the different Fourier components of some image *can* be controlled by the operator of an EM to some degree, because Δf is determined by the magnitude of the current in the objective lens of an EM, which is adjustable.

Finally, we need to be aware of the convention used to describe the Δf settings of EMs. If the current in the objective lens magnet of an EM is *less* than that required to make its focal length equal to f_o, its focal length will be larger than f_o (see equation (11.1)), and Δf will be positive. Under these conditions, if the position of the sample is held constant, which it is in TEMs because it is determined by their mechanical structures, the plane where the image of the specimen is in focus will be *further* down the optical track than it would be if the focal length of the lens were set at f_o. When an EM is operating this way, it is said to be **under focus**. When the lens current is "too high", the lens is **over focus**.

Comment: In the EM literature, equation (11.4) is sometimes written with the sign of its spherical aberration part the same as the sign of its defocus part. In that case, microscopes are under focus when Δf is negative, rather than being under focus when the sign of Δf is positive, which is the convention used here.

11.11 CHROMATIC ABERRATION IS A FOCAL LENGTH EFFECT

Both optical lenses and magnetic lenses display chromatic aberration because their focal lengths depend on wavelength. The energies of the electrons in the electron beam of an EM (i.e., their wavelengths) are determined primarily by the accelerating voltage that is applied across the cathode and anode of the electron gun at the top of the instrument (see chapter 12). At any instant in time, all the electrons in the beam produced by an electron gun will have the same energy, to a reasonably good first approximation (see following discussion). However, no power supply produces an output that is perfectly constant in time, and, thus, the energy of the electrons in the beam of an EM will vary slightly with time. That variation in energy implies a corresponding time dependence of the focal length of the objective lens (see equation (11.1)) and, hence, a variation in the position of the plane where images are perfectly focused. Because the time it takes to record EM images is invariably long compared to the time scale of these voltage fluctuations, the images produced by EMs are superpositions of images that differ slightly in Δf.

Starting with the lens law and equation (11.1), it can be demonstrated (see appendix 11.1) that if the energy of the electron beam in an EM changes by $\Delta E\ (= q_e \delta V)$ from its average value $E_o (= q_e V_o)$, and f_o is the focal length of its objective lens for electrons of energy E_o, the optical path length difference for the instrument, ΔOP, will change by $(\alpha^2/2) f_o (\delta V / V_o)$. The effect of this change on images is conveniently described by including a correction in the CTF of the microscope as follows:

$$\mathrm{CTF}\,(\alpha, t) = \mathrm{A}\,(\alpha) \exp\left[i \Delta \varphi\,(\alpha)\right] \exp\left(i \left(\pi f_o \alpha^2 / \lambda\right) \left[\delta \mathrm{V}\,(t) / V_o\right]\right). \quad \mathbf{(11.6)}$$

Note that the addition of a term for chromatic aberration to the expression for an EM's CTF makes its CTF a function of time.

11.12 CHROMATIC ABERRATION SUPPRESSES THE HIGH-RESOLUTION FEATURES OF IMAGES

As we already discussed, the images produced by EMs are time-averaged images, where the length of time over which averaging occurs, in other words, exposure times, are long compared to the time scale of the fluctuations in accelerating voltages responsible for chromatic aberration. Thus, what determines the effect of chromatic aberration on image quality is not the instantaneous value of the chromatic aberration contribution to the CFT of a microscope, but rather its time-averaged contribution. Thus, if voltage fluctuations in the electron gun were the only source of chromatic aberration (see the following discussion), then, leaving out all time-independent terms, their average effect on the CFT would be:

$$\langle \mathrm{CTF}\,(t)\rangle = (1/T)\int_0^T \exp\left(i\left(\pi f_o\alpha^2/\lambda\right)\left[\delta \mathrm{V}\,(t)\,/\,V_o\right]\right)dt,$$

where T is the exposure time. Happily, we already know how to compute averages of this sort (see appendix 5.1); hence, we can write down the answer directly:

$$\langle \mathrm{CTF}\,(t)\rangle = \mathrm{A}\,(\alpha)\exp\left[i\Delta\varphi\,(\alpha)\right]\exp\left[-(1/2)\left(\pi f_o\sigma\,/\,\lambda V_o\right)^2\alpha^4\right]. \quad \textbf{(11.7)}$$

In this equation, σ^2 is the variance of the fluctuations of the accelerating voltage.

As equation (11.7) shows, chromatic aberration does something to EM images that is completely different from what spherical aberration and defocus do to them. Spherical aberration and defocus can introduce false detail into images by altering the phases of the contributions made by the large $|\mathbf{S}|$ components of the transmission function of some object. Chromatic aberration does nothing to phases, but by systematically reducing the amplitudes of the contributions made to images by the short wavelength components of transmission functions, in other words, by their large $|\mathbf{S}|$ components, it reduces image resolution. To put it another way, chromatic aberration does not falsify the high-resolution detail in images, it eliminates that detail from images entirely by smoothing it away. This should not come as a surprise. Information about the fine structure of objects is certain to be lost when images that differ in focus are averaged, and the shorter the Bragg spacing of that detail, the more severe the effect will be.

11.13 CHROMATIC ABERRATION HAS MANY SOURCES IN EMS

Accelerating voltage instability is not the only source of chromatic aberration in EMs. For example, the focal lengths of magnetic lenses are proportional to the square of the currents circulating in their coils. Hence, instabilities in the power supply that energizes the objective lens of an EM will also make the position of its image plane vary with time. The effect of this kind of instability on images is indistinguishable from effect of accelerating voltage instabilities. In addition, the electrons produced by an electron gun do not all emerge from it with exactly the same kinetic energy because of differences in their thermal kinetic energy at the time they were emitted. Thus, the appropriate expression to use for chromatic aberration contribution to the CFT is:

$$\exp\left(i\left(\pi f_o \alpha^2/\lambda\right)\left[\delta \mathrm{E}(t)/E_o + \delta \mathrm{V}(t)/V_o + 2\delta \mathrm{I}(t)/I_o\right]\right),$$

where E is energy of the electrons in the system ($E_o = q_e V_o$) and δE is the thermal contribution to their energy. V refers to the accelerating voltage, as before, and I is the current in the objective lens.

If the fluctuations in E, V, and I are all independent, as is likely to be the case, then a more general form of equation (11.7) immediately follows:

$$\langle \mathrm{CTF}(t)\rangle = \mathrm{A}(\alpha)\exp\left[i\Delta\varphi(\alpha)\right]\exp\left(-(1/2)\left(\pi f_o \alpha^2/\lambda\right)^2 \times \left[(\sigma_{\mathrm{E}}/E_o)^2 + (\sigma_V/V_o)^2 + (2\sigma_I/I_o)^2\right]\right).$$

The effect that chromatic aberration has on EM images is best understood by feeding some numbers into the preceding equation. For the record, σ_{E} is of the order of 0.5 eV, and its magnitude depends on the type of source used to produce the electron beam in an EM (see chapter 12). For EMs that operate at 100 kV, σ_V is 1.0 to 0.1 V, as already noted, and for most EMs (σ_I/I) is $\sim 1 \times 10^{-6}$. If the standard lens is operated with $(\sigma_V/V_o) = 10^{-5}$, and the variances of all other sources of chromatic aberration are as stated, then the exponential term in the CTF will have a value of $1/e (= 0.367)$ at $\alpha = 0.0105$ rad, which corresponds to a Bragg spacing of 3.5 Å, which is only a little better than that implied by the Scherzer resolution of the standard lens. Furthermore, at scattering angles only slightly larger than that, that is, at a Bragg spacing of 2.95 Å, the value of the chromatic aberration term will be $(1/e)^2 (= 0.135)$. Clearly, there will be absolutely no information whatever about the structure of an object on 1-Å length scales in an image formed by an instrument that has this much chromatic aberration!

Note that if the voltage in the source were stabilized to one part in a million, and everything else left constant, the Bragg spacing at which the attenuation reached $(1/e)$ would improve to about 2.6 Å, which could help a lot when it came time to interpret electron density maps. Note also that if everything else could be held constant, it would also help to increase the accelerating voltage of the microscope. The reason is that as the wavelength of the electrons being used to form an image decreases, the maximum value of α required to achieve a given resolution falls. In this instance, a halving of the wavelength would move the $1/e$ resolution out to 2.50 Å.

These calculations provide an indication of the magnitude of the technical challenge faced by those interested in extracting atomic resolution structural information from the images produced by EMs. Furthermore, they demonstrate that chromatic aberration is a far more serious barrier to obtaining high-resolution EM images than defocus and spherical aberration. Chromatic aberration attenuates the amplitudes of the high spatial frequency components responsible for the fine detail in images, whereas defocus and spherical aberration merely alter their phases. You cannot recover the high spatial frequency detail that an image might otherwise have contained if it was lost at the time the image was produced due to chromatic aberration, but, as we will see later, you can hope to correct images for the phase errors introduced by defocus and chromatic aberration.

11.14 THE SCHERZER RESOLUTION OF AN IMAGE AND ITS INFORMATION LIMIT ARE NOT THE SAME

As the discussion just concluded suggests, resolution is a more complicated issue for electron microscopy than it is for light microscopy. When thinking about the resolution of EM images, it is important to realize that electron optics imposes two different kinds of resolution limits on the images (1) a Scherzer resolution, which we have already talked about; and (2) an **information limit**, which we have not. Furthermore, in addition to these two optically determined resolution limits, there is a third resolution limit that must be taken into account, the **specimen limit**, which is a property of the specimen, as its name implies.

As equation (11.7) shows, the magnitudes of the CTFs of EMs are determined by an exponentially decaying function that depends on the scattering angle raised to the fourth power. This means that beyond some value of α, which depends on the chromatic aberration of the microscope of concern, the amplitudes of the Fourier components of image transmission functions will be so severely attenuated by the CTF that that they no longer contribute effectively to image formation. That decay term falls so fast with increasing scattering angle that the value of α taken as the practical limit of the data is nearly independent of the level of attenuation that is deemed acceptable. The corresponding Bragg spacing is the information limit of an EM image, and it may be less than the Bragg spacing of its Scherzer resolution. The information limit of an EM corresponds to the length scale of the smallest features in objects that could be meaningfully imaged by that EM if its objective lens was free of spherical aberration and it was operated in perfect focus.

The third kind of resolution limit, the specimen limit, is easy to understand. For example, it is perfectly possible for both the Scherzer resolution and the information limit of the TEM image of some macromolecular crystal to be much *worse* than the resolution of the X-ray diffraction pattern that could be obtained from those same crystals, which is a measure of their internal order, (i.e., the resolution limit imposed by the ordering of the crystal). On the other hand, many of the procedures used to prepare biological specimens for TEM imaging disrupt their structures to such an extent that the structural limit of their TEM images is much *worse* than either the Scherzer resolution or the information limit of the microscope used to produce them. Clearly, the effective resolution of the image of a biological object is the worst of the three kinds of resolution relevant to it.

This line of reasoning suggests a method that might be used to estimate the true resolution of an EM image. One would start by computing the Fourier transform of that image, and then determining how the average value of $(F(\mathbf{S}))^2$ varies with

$|\mathbf{S}|$. Roughly speaking, we already know what a plot of $(F(\mathbf{S}))^2$ versus $|\mathbf{S}|$ will look like. It ought to decrease as $|\mathbf{S}|$ increases just as similar plots of crystallographic data do. The true resolution of the image would then be the reciprocal of the value of $|\mathbf{S}|$ at which the average value of $(F(\mathbf{S}))^2$ ceases to be distinguishable from noise. *Noise* in this case might be estimated using a portion of the image in question that corresponds to a region in the specimen that has no biological material in it.

11.15 BOTH THE CHROMATIC AND SPHERICAL ABERRATION OF MAGNETIC LENSES CAN BE CORRECTED INSTRUMENTALLY

For most biological applications, electron micrographs that have resolutions ~10 Å are good enough, and because magnetic objective lenses like our standard lens can routinely deliver resolutions that good, chromatic aberration not withstanding, conventional TEMs are satisfactory for such purposes. However, there has long been interest in using EMs to image biological structures at atomic resolutions, and for this purpose, images having resolutions of the order of 1 Å are essential.

There are two ways EM images having resolutions this good might be obtained. Section 11.14 suggested that it might be possible to extend the resolution of the images produced by conventional EMs into the near angstrom range by computational means, after the fact, an approach that will be addressed in chapter 12. The only alternative would be to construct TEMs that have Scherzer resolutions so much better than those of the conventional TEM that the images they produce need no correction. If the latter were possible, it would probably be the better option to pursue because, as a general rule, it is better to collect accurate data to begin with than to collect inaccurate data and correct them after the fact.

It has been known for decades that in theory, it should be possible to build EMs that have much lower aberration coefficients than those typical of the descendants of Ruska's first microscope, which is to say virtually all of the TEMs in existence today. However, the technical barriers that had to be overcome were/are formidable, and only in the last decade or so has success been achieved (e.g., Haider et al., 1998).

From the point of view of the casual observer, the principal difference between an ordinary TEM and the low-aberration TEMs now slowly appearing is that the latter have "extra" assemblies of magnetic and electrostatic elements inserted in their optical tracks both above and below their objective lenses. It would take us too far afield to explain how these assemblies work, so we will not do so here. Suffice it to say that TEMs that include such devices have spherical aberration coefficients in low micrometer range instead of the low millimeter range, and that the chromatic aberration effects seen in the images they produce are about 100-fold less than those typical of the standard TEM. Instruments of this sort can produce images that are faithful past Bragg spacings of 1 Å.

This remarkable technical advance has one obvious, practical drawback: its cost. Today, the inclusion of an aberration correction system of this sort in an EM increases its cost by about ~$\$10^7$. We can hope that the cost will come down as manufacturing experience increases, but it will be a while before instruments of this type become widely available.

PROBLEMS

1. Using equation (11.2), obtain an expression for $d\lambda/dV$. Use that expression to estimate the factor by which the wavelength of an electron changes if the accelerating voltage in the electron gun is increased from 200 kV to 250 kV.
2. What is the Scherzer resolution of the standard optical lens?
3. It is clear from section 11.12 that many of the instabilities that contribute to chromatic aberration could be reduced by appropriate alterations in the design and construction of the microscope.
   a. What would the temperature of the electron source in the electron gun of an EM have to be to reduce σ_E to 0.1 V? (It is normally about 0.5 V.)
   b. Supposing some way could be found to accomplish this, would any significant gains in performance result if σ_V/V_o was 10^{-5} in the EM in which that special gun was installed, and σ_I/I remained 10^{-6}? Which of these parameters should be tackled first?
4. Qualitatively, what does mechanical vibration do to the resolution of EM images? If you were to include a vibration term in the CTF of an EM, what might it look like?

FURTHER READING

Egerton, R. F. (2005). *Physical Principles of Electron Microscopy*. New York: Springer-Verlag.

Frank, J. (2006). *Three-Dimensional Electron Microscopy of Macromolecular Assemblies*. New York: Oxford University Press.

Haider, M., Rose, H., Uhlemann, S., Kabius, B., and Urban, K. (1998). Towards 0.1 nm resolution with the first spherically corrected transmission electron microscope. *J. Electron Microscopy* 47, 395–405.

Scherzer, O. (1948). The theoretical resolution limit of the electron microscope. *J. Appl. Phys.* 20: 20–29.

Slayter, E. M., and Slater, H. S. (1992). *Light and Electron Microscopy*. Cambridge: Cambridge University Press.

Spence, J. H. C. (2003). *High-Resolution Electron Microscopy*, 3rd ed., New York: Oxford University Press.

APPENDIX 11.1 ON THE CHROMATIC ABERRATION OF MAGNETIC LENSES

In any EM, the distance between the sample plane and the optical center of its objective lens, s_1, is a fixed at the time of manufacture. Thus, if the focal length of the lens varies with the energy/wavelength of the radiation passing through it, s_2, the distance from the center of the lens to the plane where that radiation forms a focused image plane will vary also. If the image plane distance for electrons of some energy, E_o, is s_2, then the image plane distance at a slightly different energy, s_2', will be:

$$s_2{}' = s_2 + \left(ds_2 / df_o\right) \left(df_o / dE\right) \Delta E.$$

When the lens law is differentiated, we find that $(ds_2/df_o) = (s_2/f_o)^2$, and thus:

$$s_2{}' = s_2 + \left(s_2 / f_o\right)^2 \left(df_o / dE\right) \Delta E.$$

Hence, the change in s_2, Δs_2, will be $(s_2/f_o)^2(df_o/dE)\Delta E$.

Following the reasoning outlined in section 11.7, at the paraxial image plane, the image formed by rays of radiation of energy $E_o + \Delta E$ that originated at a point at s_1 will be a disk, the radius of which will be $\Delta s_2(\alpha/M)$, where, as usual, M is the magnification. In the object plane, the radius of the corresponding disk, r_o, is $\Delta s_2(\alpha/M^2)$, or:

$$r_o = \left(s_2 / f_o\right)^2 \left(\alpha / M^2\right) \left(df_o / dE\right) \Delta E.$$

Now when focusing lenses are used as objective lenses for microscopes, $s_1 \approx f_o$, and M is s_2/s_1. Thus:

$$r_o \approx \alpha \left(df_o / dE\right) \Delta E.$$

Starting with equation (11.1), it is easy to show that $(df_o/dE) = (f_o/E)$, and hence:

$$r_o = \alpha f_o \left(\Delta E / E\right),$$

and:

$$\Delta OP = \left(\alpha^2 / 2\right) f_o \left(\Delta E / E\right).$$

12

Image Formation in the Electron Microscope

Now that we know how electron lenses work, we have the background needed to understand where the contrast in EM images comes from. Here, we discuss both the interactions between specimen and electron beams that generate image contrast, and the effects that the optics of the TEM have on contrast.

12.1 APERTURE AND PHASE EFFECTS ACCOUNT FOR MOST OF THE CONTRAST IN EM IMAGES

As we already know, any encounter of a quantum of radiation with an object can have three outcomes that are not trivial: (1) absorption, (2) elastic scatter, and (3) inelastic scatter. The first of the three, absorption, contributes little to the contrast in EM images of biological samples. The specimens biologists study in the TEM are often less than 10^{-7} m thick, and if unstained (see section 12.5), composed entirely of atoms of low atomic number. Virtually all the 100-keV electrons that impinge on such objects pass through them unabsorbed. It follows that scattering must account for the lion's share of the contrast in EM images.

Our discussion of light microscopy revealed that scattering can lead to two different kinds of contrast: aperture contrast (section 10.6), and phase contrast (section 10.8). Aperture contrast contributes to all the bright-field images of biological specimens produced by light and electron microscopes because: (1) biological specimens vary in light/electron scattering power from point to point, and (2) the apertures of all objective lenses are finite. Phase contrast, on the other hand, contributes little to the images of biological specimens formed by light microscopes unless those microscopes have been modified to enhance it. It is much more important in EM images because, as we will see, the spherical aberration characteristics of the objective lenses in TEMs make them phase microscopes, like it or not. We begin with a discussion of how aperture contrast works in EMs.

12.2 APERTURE CONTRAST IS BEST UNDERSTOOD USING CROSS SECTIONS

By optical standards, the apertures of the objective lens of TEMs are deliberately made very small because, paradoxically, this improves the resolution of the images they produce (section 11.8). However, even if electron lenses did not have large spherical aberration coefficients, most biologists would be content to use TEMs that have apertures that are tiny by optical standards. A 100-keV TEM equipped with an aperture so small that only radiation scattered at angles less than 2° can reach the image plane will still have a Rayleigh resolution of about 1 Å, which is more than good enough for most purposes.

Because the solid angles subtended by cones having half widths of a few degrees are very small, you might think that the aperture contrast would make an enormous contribution to TEM images, but in fact, even though large enough to be useful, it is not all that remarkable for reasons that have to do with the electron scattering cross sections of molecules.

Molecules, like atoms, have scattering cross sections. The intensity of the elastic scatter produced by some molecule at **S**, when it is in some particular orientation with respect to the incident beam of radiation, is proportional to the following (familiar) expression:

$$\mathrm{I}(\mathbf{S}) = \sum_{\text{all}} \sum_{\text{atoms}} \mathrm{f}_i(S)\, \mathrm{f}_j(S) \exp\left[-2\pi i \left(\mathbf{r}_i - \mathbf{r}_j\right) \cdot \mathbf{S}\right],$$

where here the f_i are the elastic electron scattering *lengths* of atoms (for 100-keV electrons), not X-ray scattering factors (see section 3.15). The corresponding elastic scattering cross section is thus:

$$\sigma_{\mathrm{mol}} = \lambda^2 \int \sum_{\text{all}} \sum_{\text{atoms}} \mathrm{f}_i(S)\, \mathrm{f}_j(S) \exp\left[-2\pi i \left(\mathbf{r}_i - \mathbf{r}_j\right) \cdot \mathbf{S}\right] dA_S. \qquad \mathbf{(12.1)}$$

The integration in equation (12.1) is over the entire Ewald sphere, and dA_S is the differential element of area on the surface of that sphere. The cross section for the (elastic) scattering of electrons at all angles larger than some minimum, α_{max}, can be computed by altering the range of integration in equation (12.1) so that it runs over all **S** such that $(2 \sin \alpha_{\mathrm{max}}/\lambda) \leq |\mathbf{S}| \leq (2/\lambda)$. If we call that partial cross section $\sigma_{\mathrm{mol}}(\alpha_{\mathrm{max}})$, then the probability that an electron encountering that molecule will be (elastically) scattered at an angle less than α_{max}, or not scattered at all, will be $[1 - \sigma_{\mathrm{mol}}(\alpha_{\mathrm{max}})]$.

Suppose some specimen is composed of a single kind of molecule, all in the same orientation, as they might be in a crystal, and many molecules thick. The probability that an electron passing such a specimen will not be scattered, or be scattered by an angle less than α_{max} will be $\Pi[1 - \sigma_{mol}(\alpha_{max})]$, where the product contains a term for each molecule encountered. If $\sigma_{mol}(\alpha_{max})$ is much, much less than 1.0, which it is certain to be (see section 3.15), then:

$$\prod_{1}^{N} [1 - \sigma_{mol}(\alpha_{max})] \approx \exp[-N\sigma_{mol}(\alpha_{max})],$$

where N is the number of molecules encountered by an electron as it passes through the sample. Because N is the number concentration of molecules in the sample, C, times the thickness of the sample, t, we can write:

$$\prod_{1}^{N} [1 - \sigma_{mol}(\alpha_{max})] \approx \exp[-C\sigma_{mol}(\alpha_{max})\, t].$$

Thus, the logarithm of the number of electrons available to form the image of some point in the specimen, divided by the number of electrons incident on that point will be (-1) times the product of a (number) concentration, a thickness, and a constant that depends on molecular properties. The similarity between this relationship and the Lambert-Beer law that describes light absorption is obvious.

Comment: It is clear from what has just been said that absorption contrast and aperture contrast have similar effects on the images formed by microscopes operating with monochromatic radiation. Experimentally, however, the two kinds of contrast are easily distinguished. Aperture contrast increases as apertures decrease, but absorption contrast does not.

$[C\sigma_{mol}(\alpha_{max})t]$ is a dimensionless number that will henceforth be referred to as μ, and, as we are about to see, the variations in its value in the plane of some specimen reports primarily on variations in specimen thickness and chemical composition. Because the number of electrons available to contribute to the image of the specimen will vary in the plane as μ varies, aperture limitation effects produce a point-to-point variation in image brightness (i.e., contrast).

12.3 THERE IS LITTLE STRUCTURAL INFORMATION IN HIGH ANGLE ELECTRON SCATTER

What could be learned from EM images if aperture contrast was the sole source of contrast? This question can be answered once the way $\sigma_{mol}(\alpha_{max})$ varies with α_{max} is understood.

Equation (12.1) describes the scatter produced by a molecule composed of atoms that have fixed positions, but, as we already know, the positions of atoms in real molecules change all the time due to thermal vibrations and rotations. Furthermore, the thermal motions of atoms within molecules are very fast compared to the speeds with which EM images can be recorded. Thus, the EM images of molecules are time-averaged images, and, thus, the cross sections we need to understand are time-averaged cross sections.

Assuming that the location of the center of mass of some molecule is fixed, and that it is not rotating as a whole, the average elastic, electron scattering pattern of some molecule, $\langle \mathrm{I}(\mathbf{S}, t)\rangle$, will be proportional to:

$$\langle \mathrm{I}(\mathbf{S}, t)\rangle = \sum_{\text{all}} \sum_{\text{atoms}} \mathrm{f}_i(S)\, \mathrm{f}_j(S) \exp\left[-2\pi i \left(\mathbf{r}_i - \mathbf{r}_j\right) \cdot \mathbf{S}\right] \langle \exp\left(-2\pi i \left[\boldsymbol{\delta}_i(t) - \boldsymbol{\delta}_j(t)\right] \cdot \mathbf{S}\right)\rangle,$$

where $\boldsymbol{\delta} i(t)$ is the ith atom's displacement from its average position at time t (see section 5.8). This equation can be rewritten:

$$\langle \mathrm{I}(\mathbf{S}, t)\rangle = \sum \mathrm{f}_i(S)^2 + \sum_i \sum_{j \neq i} \mathrm{f}_i(S)\, \mathrm{f}_j(S) \exp\left[-2\pi i \left(\mathbf{r}_i - \mathbf{r}_j\right) \cdot \mathbf{S}\right] \times \langle \exp\left(-2\pi i \left[\boldsymbol{\delta}_i(t) - \boldsymbol{\delta}_j(t)\right] \cdot \mathbf{S}\right)\rangle. \quad (12.2)$$

The first term on the right side of this equation depends *only* on the atomic composition of the molecule, whereas the second term depends on: (1) its three-dimensional structure, (2) its orientation with respect to the radiation beam, and (3) the pairwise fluctuations in the positions of its atoms.

The averaging called for in equation (12.2) is similar to the averaging that accounts for the temperature factors seen in diffraction patterns (see section 5.8); thus, we anticipate that in an instrument operating with 100-keV electrons, the magnitude of the structure-dependent term in equation (12.2) will fall rapidly as the scattering angle increases, which is so. In fact, as shown in appendix 12.1, at scattering angles beyond $\sim 5°$, $|\mathbf{S}| > 2.5\ \text{Å}^{-1}$, the structure-dependent term will be small, and:

$$\langle \mathrm{I}(\mathbf{S}, t)\rangle \approx \sum_{\text{all atoms}} \mathrm{f}_i(|\mathbf{S}|)^2.$$

It follows that:

$$\sigma_{\text{mol}}(\alpha_{\text{max}}) \approx \sum_{\text{all atoms}} \sigma_i(\alpha_{\text{max}}), \quad (12.3)$$

where $\sigma_i(\alpha_{\text{max}})$ is the appropriate partial cross section for the ith atom.

Comment: For the typical macromolecule, starting at around $|\mathbf{S}| = 0.5, \text{Å}^{-1}$, the average value of $\mathrm{I}(\mathbf{S})$ within any thin shell of constant $|\mathbf{S}|$ will obey Wilson statistics; in other words, they will be equal to $\Sigma f_i(S)^2$ (see section 5.12). The message here is that beyond $|\mathbf{S}| \sim 2.5, \text{Å}^{-1}$, $\Sigma f_i(S)^2$ is the not just the average value of $I(\mathbf{S})$ within any shell of constant $|\mathbf{S}|$, it is effectively the only value.

A similar analysis can be done for the inelastic electron scattering produced by macromolecules. In that case, the relevant atomic scattering lengths can be written $f_i(\lambda_0, \lambda, S)$, where λ_0 is the wavelength of the incident electrons, and λ is the wavelength of the inelastically scattered electrons. For each value λ in the inelastic scattering spectrum of the molecule, there will be a scattering profile that has the same qualitative properties as an elastic scattering profile, and the overall inelastic scattering profile of the macromolecule will be the (incoherent) sum of all such profiles. (NB: $\lambda_0 < \lambda$.) Thus, beyond some quite modest scattering angle (but quite high resolution measured in angstroms), both the elastic and the inelastic scattering profiles of a molecule are, effectively, sums of the corresponding scattering profiles of the atoms it contains.

12.4 APERTURE CONTRAST IMAGES HAVE BRIGHT BACKGROUNDS AND LOW RESOLUTIONS

The TEM image of a specimen is the incoherent superposition of the image formed by the electrons it scatters elastically and the image formed by the electrons it scatters inelastically. Because the number of 100-keV electrons scattered elastically by specimens composed of low atomic number atoms is about the same as the number scattered inelastically (see section 3.16), to a crude first approximation, the brightness of the two images should be about the same, but otherwise, they are quite different.

Compared to the elastic image of a specimen, its inelastic image will have much less aperture contrast because the inelastic scattering profiles of atoms are narrower than their elastic scattering profiles (see section 3.16). For example, roughly half of the electrons scattered elastically by a carbon atom in a 100-keV TEM will be intercepted by a 1° aperture, but only $\sim$10% of the electrons it scatters inelastically will be intercepted. Thus, aperture limitation generates contrast in elastic images more efficiently than it does in inelastic images.

In addition to being low contrast, inelastic images are badly blurred by defocusing. The energy losses electrons experience when they are scattered inelastically by biological materials, which are of the order of 10 to 50 eV, are $\sim$100 times bigger than the variations in energy that can be tolerated if the resolution of an EM image is not to be degraded unacceptably by chromatic aberration. Thus, the inelastic image of a specimen is a superposition of a series of images, each one of which is significantly out of focus with respect to the others, and even more seriously out of focus compared to the elastic image. Thus, the contribution inelastic scattering makes to the overall image of a specimen amounts to little more than a bright, featureless background that reduces its signal-to-noise ratio.

Comment: Included in the inelastic scatter will be contributions due to phonon scattering, which are not easily characterized.

This analysis of cross sections has a bearing on the resolution of the aperture contrast component of EM images. If all the electrons that transited any volume element of some specimen and passed through the limiting aperture of a TEM were brought to a focus at the corresponding region in the image plane, then the point-to-point variation in image brightness (i.e., image contrast) would be determined entirely by the point-to-point variations within the specimen of $\Sigma\,\sigma_i(\alpha_{max})$, which is a quantity that we now know is determined primarily by atomic composition and is insensitive to the way those atoms are arranged in three dimensions. It should also be noted that out to a resolution of around 3 Å, the images of specimens ~500 Å thick produced by 100-keV TEMs are projection images (see section 8.2). Furthermore, any cylindrical volume in a liquid or solid specimen that has a diameter of 3 Å and a depth of 500 Å will have a lot of atoms in it. Furthermore, the thickness and atomic composition of unstained EM specimens seldom vary all that radically from point to point. Thus, if the only contrast in an EM image of some protein was pure aperture contrast, we might be able to ascertain that the protein was, say, about 50 Å across and about 500 Å long, but we would not expect to be able to say much about the locations of its α-helices. Aperture contrast is more important at low resolution than it is at high resolution.

12.5 EM STAINS ENHANCE APERTURE CONTRAST

The contrast in EM images can be improved by staining. As we already know, the stains light microscopists use are dyes that absorb visible light, and almost all of them are **positive stains**, which is to say that they interact preferentially with DNA, but not with proteins or lipids, for example, and these days some are engineered to associate exclusively with a single species of protein or nucleic acid. By contrast, none of the stains used by electron microscopists are dyes. They are instead small

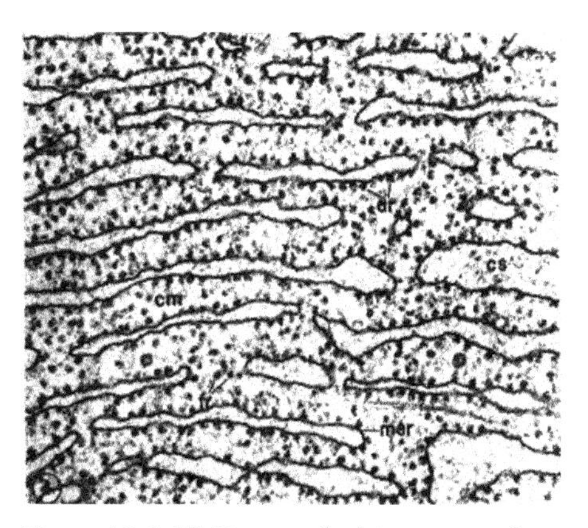

Figure 12.1 EM image of a thin section of pancreatic tissue showing the cytoplasm of a single cell. The stain used highlights acidic material—phospholipid membranes and ribosomes. (Reprinted from the Nobel Lecture of George Palade [1974] with permission from the copyright holder, the Nobel Foundation.)

molecules rich in heavy atoms, most of which show only modest selectivity in their interactions with biological molecules. The reason these substances are useful as EM stains is that the elastic scattering cross sections of high atomic number atoms are (much) larger than the elastic scattering cross sections of low atomic number atoms. Thus, the EM images of regions in a stained specimen that are rich in heavy atoms will appear dark due to enhanced aperture contrast. Figure 12.1 is an electron micrograph of a thin section of the endoplasmic reticulum of a pancreatic cell that has been stained with a substance that has a high affinity for anions. The most heavily stained components in this specimen were the phospholipid head groups of its lipid bilayers, and its ribosomes, which look like black dots.

Some EM stains are so nonspecific they can be used as **negative stains**. Negative stains are employed to assist in the visualization of small objects such as individual macromolecules or macromolecular complexes such as viruses. A negatively stained specimen typically consists of a modest number of particles embedded in an amorphous layer of a low molecular weight substance that is rich in high Z atoms, such as uranyl acetate or phosphotungstic acid. The images of negatively stained specimens are bright where a particle is present, but dark elsewhere, rather than being dark where particles are present and bright elsewhere, as is the case with positively stained images. The reason particles are bright in these images is that they are composed of weakly scattering atoms of low atomic number, and they displace the strongly scattering, heavy atom stain that would otherwise fill the volume they occupy. Thus, negatively stained images provide information about the shapes of the volumes occupied by macromolecular particles. Figure 12.2 is a negatively stained image of 70S ribosomes from *Escherichia coli*. As this figure shows, negative stains tend to "pool up" around macromolecules.

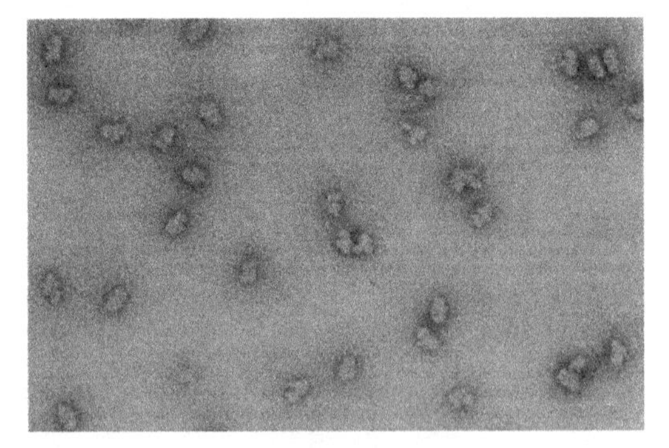

Figure 12.2 An image of 70S ribosomes negatively stained with uranyl acetate. The maximum linear dimensions of the particles shown are about 250 Å. (Reprinted from Lake, J. A. (1976), Ribosome structure determined by electron microscopy of *Esherichia coli* small subunits, large subunits and monomeric ribosomes, *Journal of Molecular Biology* 105, 131–159, with permission from Elsevier.)

12.6 INELASTIC SCATTERING DAMAGES SPECIMENS

From the standpoint of most biological users of TEMs, the chemical damage that inelastic scattering does to specimens is as important as the degradation in the signal-to-noise ratios it produces in their images. Most inelastic scattering events deliver an amount of energy to EM specimens that exceeds the energy of the average covalent bond (~5 eV) and, thus, are energetic enough to trigger chemical reactions. Exposure of biological materials to the electron beams in TEMs damages them chemically in a dose-dependent manner that resembles what X-rays do to macromolecular crystals. It follows that electron microscopists must balance the need to minimize radiation damage with the need to expose specimens to enough electrons so that the images of those specimens have decent signal-to-noise ratios. The higher the resolution of the image being sought, the more difficult this balancing act becomes, because, in general (but not necessarily always), the large $|\mathbf{S}|$ components of images tend to be weaker than their small $|\mathbf{S}|$ components and, hence, take more electrons to measure accurately. There is no free lunch.

Comment: Molecules also scatter X-rays inelastically, and the energy deposited in molecules by this mechanism can also cause radiation damage. Nevertheless, outright absorption accounts for the lion's share of the radiation damage that X-rays do to biological materials. The energy of a photon that has a wavelength of 1 Å is ~12 keV, and if it is absorbed by some sample, ***all*** *of its energy must be dissipated in that sample, not just the hundred electron volts (or less) that might have had to be dissipated if the photon had been scattered inelastically. It is the difference between a stick of dynamite and a firecracker. In fact, when the relevant cross sections are compared, we discover that the amount of structural information that can be obtained about some molecule, per unit of radiation damage, is far larger for 100 keV electrons than it is for 1 Å X-rays.*

12.7 THE PHASES OF ELECTRON WAVES CHANGE AS THEY PASS THROUGH OBJECTS

Before we can discuss the way phase contrast arises in TEMs, we need to be clear about what the term *phase* means in electron optics. The optical path length of the trajectory some electron takes as it passes from the electron gun at the top of an EM to some point in the image plane is $[1/\lambda(\mathbf{r})]d\mathbf{r}$ integrated over the entire trajectory. The wavelength in question is the de Broglie wavelength of the electron, and it will vary as an electron travels down the microscope because of the electron's interactions with both the specimen and the magnetic fields of the lenses. The quantities that are physically meaningful are the differences in optical path length between the trajectories that an electron might take between two points in space, a and b:

$$\Delta_{\text{opl}} = \int_a^b [1/\lambda(\mathbf{r}_1)]\, d\mathbf{r}_1 - \int_a^b [1/\lambda(\mathbf{r}_2)]\, d\mathbf{r}_2.$$

The corresponding phase difference is $2\pi\,\Delta_{opl}$ (in radians).

Phases change when electrons pass through matter because encounters with nuclei accelerate them, making wavelengths shorter, whereas encounters with electrons slow them down, making wavelengths longer. Because the sum of the charges carried by the electrons and nuclei in a bulk sample of matter is precisely 0, you might jump to the conclusion that, on average, the phase advances that an electron experiences as it transits some sample would be exactly balanced by the phase retardations, but this is not so. The fact that the sum of charges inside some substance is 0 does not mean that the average electric potential inside it is also 0. In fact, when the electrical potential distributions inside solids and liquids are worked out in detail, it is invariably found that the volume-average value of the electrical potentials in their interiors is of the order of +10 V. This means that, on average, the optical path length of the trajectory of an electron that has passed through matter will be *longer* than it would have been if there had been no matter present. Because this is also true of light passing through transparent solids, such as glass, the refractive index of bulk materials for electron radiation, like the refractive index of glass for light, must be greater than 1.0, but in this case it is only slightly so, for example, by $\sim$1 part in 10^4.

The magnitude of the phase change experienced by an electron passing through a specimen depends on where it crosses the specimen because the electrical potential of the specimen is determined by its molecular structure. Point-to-point variations in phase give rise to electron scattering—just the way spatial variations in refractive index cause light scattering. (Hidden behind these glib statements about the index of refraction of substances for electrons are some quantum mechanical subtleties that will not be explored here.)

The "thing" imaged by a microscope, whether it is an optical microscope or an EM, is not the specimen itself, but rather it is the transmission function of the specimen, $f(x, y)$. For an EM specimen that is thinner than the depth of field of the microscope being used to examine it:

$$f\left(x, y\right) = \exp\left[-\mu\left(x, y\right)\right]\exp\left[i\theta\left(x, y\right)\right],$$

where $\mu(x, y)$ is the aperture contrast contribution to $f(x, y)$, and $\theta(x, y)$ is the phase shift experienced by an electron passing through the specimen at (x, y) relative to some standard, which might be the direct beam. The contribution to $f(x, y)$ made by the objects of interest in the specimens that biologists image in TEMs at high resolution, which will be called "$\mu'(x, y)$" and "$\theta'(x, y)$", are usually quite small. Hence, the component of the transmission function of the specimen that relates to those objects, $f'(x, y)$, can be written as follows:

$$f'\left(x,y\right) \approx \left[1-\mu'\left(x, y\right)\right]\left[1+i\theta'\left(x, y\right)\right] \approx 1-\mu'\left(x, y\right)+i\theta'\left(x, y\right).$$

(This expression is the EM equivalent of the weak phase approximation used in section 10.8 to explain [optical] phase contrast microscopy.)

12.8 TEMS ARE NATURALLY PHASE CONTRAST MICROSCOPES

Electron microscopes are phase microscopes because of the strong dependence of their contrast transfer functions on scattering angle. As we already know, if defocus, spherical aberration, and chromatic aberration are the only faults of a microscope that need to be taken into account, its CTF will have the following form:

$$\mathrm{CTF}(\alpha) = \exp\left[i\Delta\varphi(\alpha)\right]\mathrm{Chr}(\alpha)\,\mathrm{A}(\alpha_{\max}).$$

The expression appropriate for $\Delta\varphi(\alpha)$ is equation (11.4), and $\mathrm{Chr}(\alpha)$, the chromatic aberration component of the CTF, will be an expression similar to equation (11.7). $\mathrm{A}(\alpha_{\max})$ is an aperture function of the usual sort.

It follows that if a specimen is coherently illuminated, in the back focal plane of the objective lens, the portion of the wave front that corresponds to $\mathrm{f}'(x, y)$ will be:

$$\mathrm{Q}(\xi,\eta) = \left\{\mathrm{FT}\left[\mathrm{f}'(x,y)\right]\right\}\exp\left[i\Delta\varphi(\alpha)\right]\mathrm{Chr}(\alpha)\,\mathrm{A}(\alpha_{\max}),$$

where ξ and η are coordinates in the back focal plane, and the image generated by the microscope, $\mathrm{I}(X, Y)$, will be (see section 10.2):

$$\mathrm{I}(X, Y) = \{\mathrm{FT}[\mathrm{Q}(\xi,\eta)]\}^2.$$

(To simplify notation, both the magnification of images, and their coordinate inversions are ignored here.)

It is tedious, but not hard to derive an expression for $\mathrm{I}(X, Y)$ that is appropriate for weak phase EM specimens, and when that is done (see appendix 12.2), we find that:

$$\begin{aligned}\mathrm{I}(X, Y) \approx 1 &- 2\mu'(x, y) * \mathrm{FT}\left(\cos\left[\Delta\varphi(\alpha)\right]\mathrm{Chr}(\alpha)\,\mathrm{A}(\alpha_{\max})\right) \\ &- 2\theta'(x, y) * \mathrm{FT}\left(\sin\left[\Delta\varphi(\alpha)\right]\mathrm{Chr}(\alpha)\,\mathrm{A}(\alpha_{\max})\right).\end{aligned} \quad \textbf{(12.4)}$$

Equation (12.4) has three implications worthy of note. First, because the spherical aberration coefficients of the objective lenses of most EMs are appreciable, $\theta'(x, y)$ is certain to contribute to the contrast seen in the images they produce. In other words, there will be phase contrast in them, no matter what the focal setting at which they are operated, because $\Delta\varphi$ will not be 0 for most α (see equation (11.5)). Second, if $\Delta\varphi$ were $\pi/2$ for all α, which it never is, the image produced by a TEM would be a pure, phase contrast image:

$$\mathrm{I}(X, Y) = 1 - 2\theta'(x, y) * \mathrm{FT}\left(\sin\left[\Delta\varphi(\alpha)\right]\mathrm{Chr}(\alpha)\,\mathrm{A}(\alpha_{\max})\right).$$

Third, even if the spherical aberration coefficient of the objective lens in an EM were 0, phase contrast effects could still be introduced into the images it forms by operating it out of focus.

Comment: The $A(\alpha_{max})$ *component of the CTF has no effect on* $\mu(x, y)$ *because* $\mu(x, y)$*, being an aperture contrast effect, already takes the diameter of the aperture into account.*

12.9 UNDERFOCUSED IMAGES ARE BETTER THAN IN-FOCUS IMAGES

As equation (12.4) indicates, the CTF for the aperture contrast contribution to EM images is $\cos[\Delta\varphi(\alpha)]\mathrm{Chr}(\alpha)\mathrm{A}(\alpha_{\max})$, whereas the CTF for its phase contrast portion is $\sin[\Delta\varphi(\alpha)]\mathrm{Chr}(\alpha)\mathrm{A}(\alpha_{\max})$. Furthermore, as is shown in section 11.10, $\Delta\varphi(\alpha)$ is the sum of two components, one proportional to Δf, and the other proportional to the spherical aberration coefficient of the objective lens, C_s. C_s is always positive, and like the chromatic aberration of a microscope, $\mathrm{Chr}(\alpha)$, cannot be controlled by users. However, users can control both the magnitude and the sign of Δf, and by so doing, vary the CTF of a TEM over a wide range.

Figures 12.3 and 12.4 compare the way the cosine and sine parts of the CTF of the standard lens vary with scattering angle for three different values of Δf, when the energy of the electrons in beam is 100 keV. No matter what the defocus, the value of the cosine part of the CTF is $+1.0$ when the scattering angle is small, but the value of its sine part is effectively 0. This means that $\mu'(x, y)$, aperture contrast, must account for the low spatial frequency components of any EM image. The cosine part

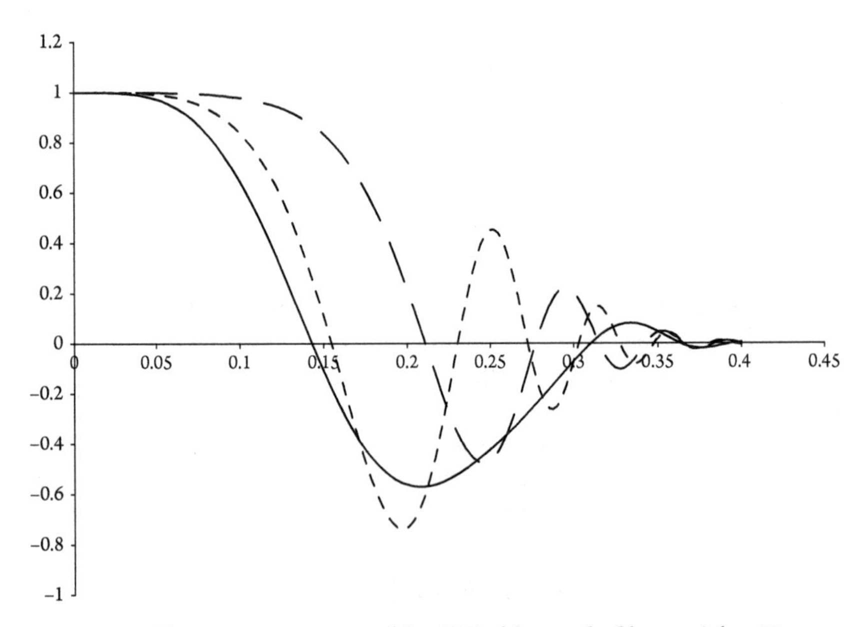

Figure 12.3 The cosine component of the CTF of the standard lens at $\Delta f = 80$ nm (solid line), perfect focus (broken line), and $\Delta f = -40$ nm (dashes). The standard error of the fluctuations that determine chromatic aberration is assumed to be 1.136 V (see section 11.12). The dimensions of the quantities horizontal axis are angstroms^{-1}.

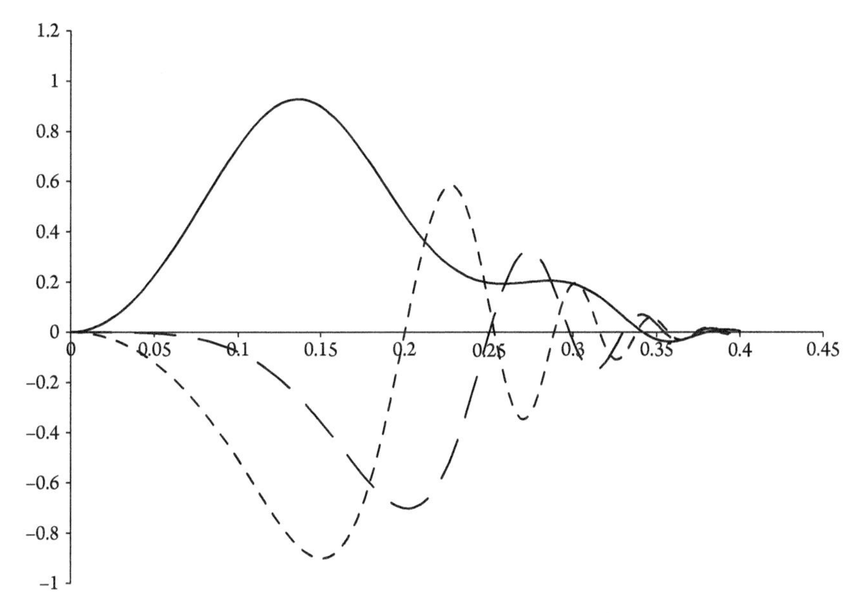

Figure 12.4 The sine component of the CTF of the standard lens at three levels of defocus. (See legend for figure 12.3.)

of the CTF dies off as the scattering angle increases and then begins to oscillate. Although the spatial frequency at which oscillation begins is determined by Δf, these oscillations usually have little effect on the EM images of biological objects because $|\mathrm{FT}[\mu'(x, y)]|$ is small beyond, say, $|\mathbf{S}| = 0.1\,\text{Å}^{-1}$ (see section 12.4).

The phase contrast component of the CTF, that is, its sine part, behaves quite differently. Regardless of the value of Δf, phase contrast starts contributing to image contrast in the standard microscope at Bragg spacings smaller than $\sim$20 Å, but not only does the magnitude of the phase contribution at any given Bragg spacing depend on Δf, but also its *sign*. If the microscope is under focus, the low angle part of the sine component of the CTF will be positive, and that means that regions of the specimen where $\theta'(x, y)$ is large will appear dark. However, if the microscope is operated in focus, or over focus, regions of the specimen where $\theta'(x, y)$ is large will appear bright. Now $\theta'(x, y)$ is certain to be large in those parts of a specimen where $\mu'(x, y)$ is large because both depend on the density of atoms and their scattering factors in similar ways. Thus, if the microscope is under focus, the phase contrast and aperture contrast features of images will reinforce each other, but if the microscope is operated in focus, or over focus, the two types of contrast will work against each other, reducing image contrast at low to intermediate resolution.

From the point of view of forming faithful, high-resolution images, the ideal CTF for an EM would have a cosine part that was 1.0 out to the spacings where the amplitude of $\mathrm{FT}[\mu(x, y)]$ fades away and a sine part that had a value of $+1.0$ at all spatial frequencies. That ideal is not hard to approximate as far as the cosine part of the CTF is concerned, but its sine part poses challenges. Of the three CTFs in figure 12.4, the one that is the best from this point of view is the CTF that corresponds to $\Delta f = +80\,\text{nm}$, but even it is still far from ideal. Nevertheless, as this analysis suggests, the resolution and fidelity of underfocus images should be better than the resolution and fidelity of overfocus images.

Scherzer derived a formula that can be used to estimate the amount of underfocus that will produce the best CTF an EM can provide, given the spherical aberration coefficient of its objective lens:

$$\Delta f_{\text{opt}} \approx 1.2\,(C_s\lambda)^{1/2}.$$

(For the standard lens, $\Delta f_{\text{opt}} = 73\,\text{nm}$.) When images are recorded with Δf_{opt} set at Scherzer's optimum value, the sine part of the CTF will first cross 0 at a spatial frequency that is the reciprocal of 0.6 times the Scherzer resolution, and for the standard lens, that spatial frequency would correspond to about $0.35\,\text{Å}^{-1}$, which is consistent with figure 12.4. Six-tenths of the Scherzer resolution is the number usually quoted as the resolution of TEMs, but because the value of the sine part of the CTF rapidly descends to 0 as that resolution is approached, it is an optimistic estimate.

In thinking about what CTFs do to EM images, it is important to remember that CTFs attenuate, or reduce to dead 0, the amplitudes of the contributions made to the image of an object by the Fourier components of its scattering length distribution that have spatial frequencies close to the 0 of CTFs. Furthermore, a change in sign in the CTF corresponds to a 180° change in the phase with which components of having some range of spatial frequencies contribute to an image. Thus, the features in images that are generated by Fourier components having spatial frequencies *greater* than the spatial frequency at which the relevant CTFs start oscillating are likely to be misleading because they will be making images brighter where they should be making them darker, and vice versa.

Comment: In the last few years, significant progress has been made in modifying TEMs so that they operate optically exactly the way phase contrast light microscopes do. It has long been understood that the advantages that might be gained by so doing could be considerable because the CTF for the phase component of the images produced by instruments that have been modified this way should be ~1.0 over a wide range of $|\mathbf{S}|$ *starting at* $|\mathbf{S}|$ *not much larger than 0, instead of being ~0 at small* $|\mathbf{S}|$*. The problem has always been constructing an appropriate phase plate.*

12.10 CTFS HAVE A BIG IMPACT ON HIGH-RESOLUTION EM IMAGES

The specimens that biological electron microscopists study at high resolution commonly consist of a layer of a dilute, aqueous solution of some macromolecule, or virus, for example, that is a few hundred angstroms thick that has been fast-frozen at liquid nitrogen temperatures. Aqueous layers that thin can be frozen so fast that the water in them does not have time to form ice crystals, and the frozen solid that results, which is called **vitreous ice**, is a glass having a liquid like structure at the molecular level. These frozen films are usually formed on, and supported mechanically by, thin films of amorphous carbon, which span the gaps in the

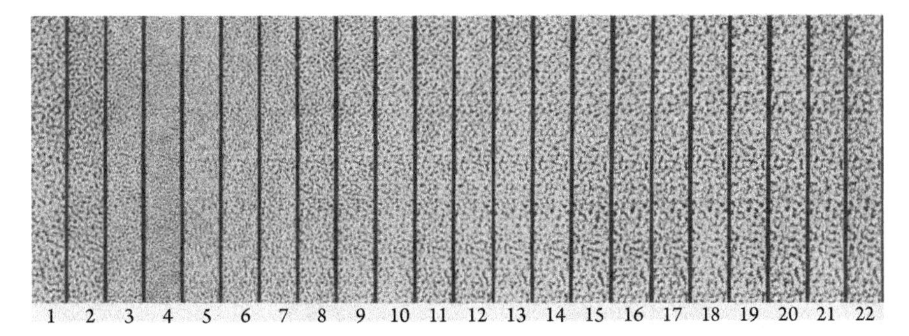

Figure 12.5 EM images of an amorphous film taken at different levels of defocus using a microscope operating with 100-keV electrons that has a spherical aberration coefficient of 4 mm. The levels of defocus of these images are approximately, left to right, −360 nm, −230 nm, −100 nm, −40 nm, +30 nm, +70 nm, +110 nm, +160 nm, + 180 nm, +200 nm, +220 nm, +250 nm, +270 nm, +280 nm, +310 nm, +330 nm, +350 nm, +370 nm, +390 nm, +420 nm, +450 nm, and +480 nm. (Reprinted with permission from Thon, F. [1966], Zur Defokussierungsabhängigkeit des Phasenkontrastes bei electronenmikroskopischen Abbildung, *Z. Naturforsch.* 21A, 476–478.)

disk-shaped copper screens (i.e., **grids**) that microscopists insert into their EMs. Once frozen, specimens of this type must be maintained at temperatures well below $-140\,^{\circ}$C to prevent the crystallization of the vitreous ice they contain, and this is why this type of electron microscopy is called **cryo-electron microscopy**.

There is nothing subtle about the effects that alterations in Δf have on EM images, and they are particularly obvious in the EM images of the parts specimens where only the support film is present. As figure 12.5 shows, the appearance of the EM image of such a region changes a lot as Δf is altered in small steps, starting with the image on the left, which is extremely over focus ($\Delta f \sim -350$ nm) and ending with the image on the right, which is far under focus ($\Delta f \sim +470$ nm). The image in this series that was taken closest to $\Delta f = 0$ is the fourth one from the left, and it has the smooth, fuzzy, low contrast appearance typical of EM images taken close to focus. The gain in contrast achieved by the sixth and seventh images in the series, which were taken at levels of underfocus that were close to optimum for the microscope used, is obvious. Both show much more detail than the in-focus image. The images at the two extremes look comparatively coarse.

12.11 THE CTFS RELEVANT TO AN EM IMAGE CAN BE DETERMINED AFTER THE FACT

The CTF relevant to an EM image can be determined after the fact using information obtained from its power spectrum. Analyses of this kind are particularly easy to perform on the regions of an image that correspond to areas in the specimen that lacked biological material (e.g., figure 12.5). To begin with, the length scale of the variations in structure that characterize the background regions of EM specimens is very short; it is of the order of the distance between adjacent atoms, that is, an angstrom or two. For that reason, all of the specklelike detail seen in the images of those regions is generated by phase contrast. Thus, leaving out the aperture term of CTFs for convenience, background images can be described as follows:

$$I_{bkg}(X, Y) \approx 1 - 2\theta_{bkg}(x, y) * FT\big(\sin[\Delta\varphi(\alpha)]\, Chr(\alpha)\big).$$

To first order in $\theta_{bkg}(x, y)$, the power spectrum of $I_{bkg}(X, Y), P(\xi, \eta)$ $\left(= FT[I_{bkg}(X, Y)]^2\right)$ is:

$$P_{bkg}(\xi, \eta) \approx \delta(\xi, \eta) - 4\Theta_{bkg}(\xi, \eta)^2 \sin^2[\Delta\varphi(\alpha)]\, Chr(\alpha)^2,$$

where $\Theta_{bkg}(\xi, \eta) = FT[\theta_{bkg}(x, y)]$. Thus, if the appropriate region of some EM image is digitized, and its Fourier transform computed, an experimental estimate of $\Theta_{bkg}(\xi, \eta)^2 \sin^2[\Delta\varphi(\alpha)]Chr(\alpha)^2$ can be obtained.

The component of this signal that we need to access is $\sin^2[\Delta\varphi(\alpha)]Chr(\alpha)^2$. Its $\Theta_{bkg}(\xi, \eta)^2$ component is of no interest, but its presence is less of a barrier to enlightenment than you might think. $\Theta_{bkg}(\xi, \eta)^2$ is the Fourier transform of the autocorrelation function of $\theta_{bkg}(x, y)$ (see section 2.10 and section 6.6). Because the structures of the background regions of EM specimens are basically isotropic, the autocorrelation function of $\theta_{bkg}(x, y)$ will depend primarily on radial distance, $r[= (x^2 + y^2)^{1/2}]$. In fact, it is proportional to the probability per unit area of specimen that a scattering center will be encountered at some distance, r, from any other scattering center. Like all autocorrelation functions, it will have its maximum at $r = 0$, and for all r greater than a few angstroms, its value will be a constant proportional to the average density of scattering centers in the specimen. $\Theta_{bkg}(\xi, \eta)^2$ will thus resemble the square of the transform of a Dirac delta function and, thus, will be roughly a constant for all (ξ, η), which we will call C. It follows that:

$$P_{bkg}(\xi, \eta) \approx \delta(\xi, \eta) - 4C \sin^2[\Delta\varphi(\alpha)]\, Chr(\alpha)^2.$$

By fitting an experimental $P_{bkg}(\eta, \xi)^2$ with a function of the form:

$$4C \sin^2\left((\pi/\lambda)\left[\Delta f\alpha^2 - (C_S/2)\,\alpha^4\right]\right) Chr(\alpha)^2,$$

we can obtain estimates of C, Δf, C_s, and σ_{elec}, the variance of the chromatic aberration function, and once those estimates have been obtained it is a simple matter to compute the CTF of an image.

Comment: If the currents in the projector lenses of an EM are set appropriately, the image it produces will correspond to the distribution of radiation in the back focal plane of a microscope's objective lens, rather than the distribution of radiation in the image plane of the objective lens. This is sometimes a useful thing to do, but, paradoxically, nothing can be learned about the CTF of a microscope from such an image because its CTF has no impact on the distribution of radiant energy in the back focal plane, $Q_{bkg}(\xi, \eta)^2$ (see equation (12.2.1) and question 2).

12.12 CTFS CAN BE EXAMINED USING OPTICAL DIFFRACTOMETERS

Long before computers became available, Sir Lawrence Bragg, who was one of the fathers of X-ray crystallography, pointed out that the power spectra of images can be observed directly using instruments called **optical diffractometers**. An optical diffractometer consists of a point source of monochromatic light that is located at the focal point of a focusing lens. The plane parallel radiation on the far side of that lens is used to illuminate a two-dimensional transparency of the image of interest, and it is best if the image is a negative, rather than a positive so that its background is dark rather than bright. If the light that emerges from the transparency is collected by a second focusing lens, the image it forms in the back focal plane of the second lens will be a representation of the power spectrum of the transparency, $(\mathrm{FT}[\rho(\boldsymbol{r})])^2$. Images of this sort are called **optical transforms**.

Figure 12.6 shows how the optical transforms of the images of EM background vary with focal level. The bright features in the centers of these images correspond to the aperture contrast part of background images, which we ignored earlier because it was irrelevant to our argument. It is perfectly obvious that the ring structures in these images are (somewhat) noisy renderings of $\sin^2(\Delta\varphi)\mathrm{Chr}(\alpha)^2$, as theory predicts, and as F. Thon first pointed out in the 1960s. For that reason, rings of this sort are called **Thon rings**, and the fact they look the way the weak phase theory of EM image formation predicts they should is strong evidence of the theory's validity.

12.13 CTF EFFECTS CAN BE REVERSED

It is now clear that the appearances of EM images are significantly altered by the CTFs of the microscopes that produced them, and that it is possible to determine what the CTF was of the EM that produced some image at the time the image was obtained by after-the-fact analysis of the image itself. These facts suggest that it might be possible to correct the CTF "errors" in the high-resolution components

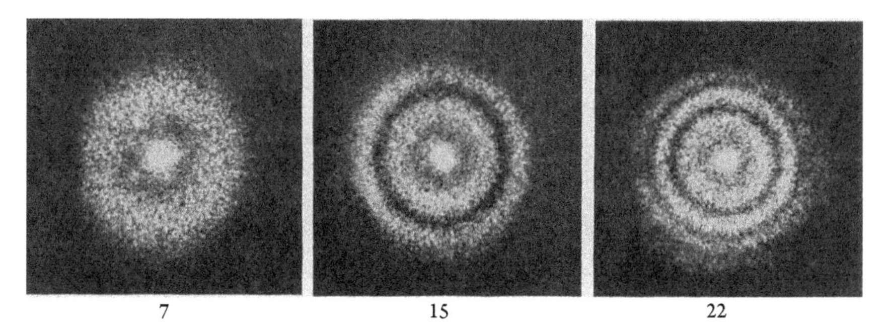

Figure 12.6 Optical transforms of background images taken at differing levels of defocus. The numbers under these transforms correspond to the numbers under the images in figure 12.5, and the levels of defocus they correspond to are: +110 nm, +310 nm, and +480 nm. (Reprinted with permission from Thon, F. [1966], Zur Defokussierungsabhängigkeit des Phasenkontrastes bei electronenmikroskopischen Abbildung, *Z. Naturforsch.* 21A, 476–478.)

of an EM image, by dividing its Fourier transform by the sine part of its CTF, and then back-transforming the result. You will instantly recognize that this process is a form of deconvolution (see section 10.11), and, thus, it is likely to be problematic because of its baleful effects on image noise.

There is no computational trick that can be used to extend the resolution of an EM image beyond its information limit, but the losses of information that occur at lower resolution in the neighborhood of the nodes in that image's CTF are addressable. If a specimen is imaged in a TEM at several different focal levels, the information about Fourier components available in one image can be used to fill the gaps in that information that occur in another. The reason this approach works is that the positions of the nodes in the CTF change as Δf is altered.

The power of this approach to image correction was spectacularly demonstrated over thirty years ago by P. T. N. Unwin and R. Henderson in their studies of the structure of purple membranes. Purple membranes are naturally occurring, two-dimensional crystals of bacteriorhodopsin, a protein carrying a purple chromophore (retinal) that is found in the cell membranes of some halophilic bacteria. The crystallinity of the purple membrane guaranteed that all the information about the (average) structure of bacteriorhodopsin contained in its EM images would be confined to the Bragg reflections observable in the Fourier transforms of those images.

The crystallinity of these samples had two important practical consequences. First, because of the amplification of molecular transforms that occurs in crystalline diffraction patterns, exceptionally low doses of electrons ($\sim$0.5 electrons per anstrom^2) could be used to image the purple membrane, which minimized the radiation damage done to specimens in the process. The left panel of figure 12.7, which is a low-dose image of the purple membrane, shows that the signal-to-noise ratios of EM images taken with electron doses that low are so poor that it is almost impossible to distinguish them from noise. The top right panel of figure 12.7 is an optical transform of image in the left panel, and the Bragg reflections it contains proves that there is useful structural information in the left panel. (The solid lines drawn on the top right panel display the reciprocal lattice of the crystal.) The lower right panel of figure 12.7 is the optical transform of an EM image of the same part of the same specimen taken at an electron dose so high that radiation damage has destroyed the order evident in the panel above it. That transform has no Bragg reflections in it, but its Thon rings are plainly evident.

The second practical consequence of the crystallinity of the samples Unwin and Henderson (1975) studied was that it made it easy to suppress the noise in the images they obtained. The transform of the image shown in left panel is full of features that have nothing to do with the underlying periodic structure. Much of it, presumably, is image noise. Thus, if we were to compute an image of the purple membrane using *only* the amplitudes and phases of the Bragg reflections in the transform obtained from an EM image of it, the computed image would lack almost all the noise contained in the original image. A similar outcome could have been obtained optically by punching small holes in a sheet of opaque plastic in exactly the location specified by the Bragg reflections shown in figure 12.7, top left. If that "mask" were then placed in the back focal plane of the optical diffractometer that was used to create figure 12.7, top left, only the Bragg components of the original image

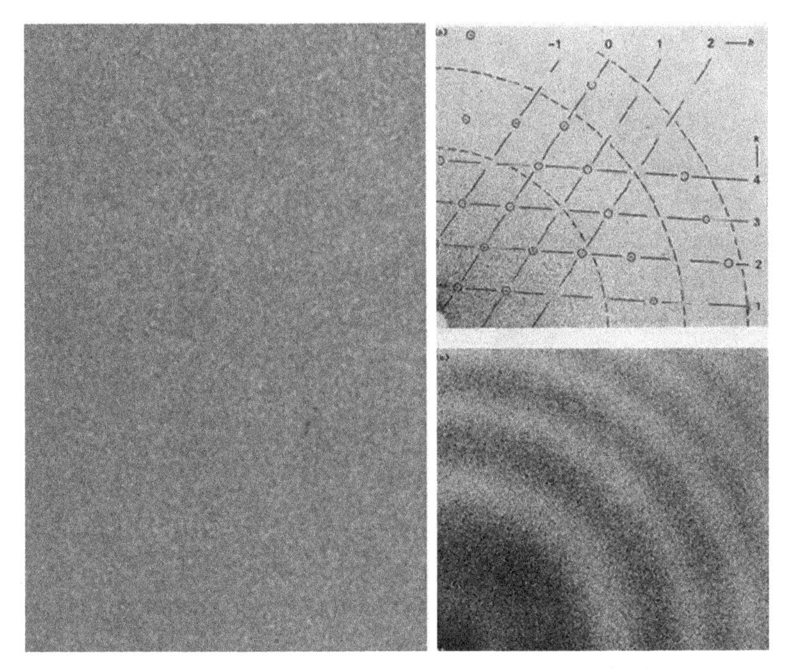

Figure 12.7 A low-dose image of a purple membrane sample (left) and its optical transform (upper right). The image on the lower right is an overexposed optical transform of a high-dose image of the same region shown on the left that was taken so that its Thon rings could be observed. The top and bottom right figures are not on the same scale. The dashed rings in the upper right image show the locations of the nodes of the Thon rings. (Reprinted from Unwin, P. N. T., and Henderson, R. [1975], Molecular structure determination by electron microscopy of unstained crystalline specimens, *Journal of Molecular Biology* 94, 425–432, with permission from Elsevier.)

would get past the back focal plane and form an image down stream. What would be seen there is a much less noisy image of the repeating part of the original micrograph. For obvious reasons, this kind of image enhancement is called **optical filtration**, whether it is done computationally or carried out using an optical diffractometer.

Figure 12.8 shows the outcome of a computational filtration experiment done using reciprocal space data obtained from several images of the same sample of purple membrane that were taken at different levels of defocus and corrected for CTF effects. Each of the ringlike structure in the electron density map shown represents three molecules of bacteriorhodopsin. A single molecule is three adjacent round blobs, which are α-helices seen end-on, and the arclike feature that adjoins them. The arc is the superimposed image of three helices that are more obliquely oriented to the plane of the membrane.

Comment: Data sets of all kinds are often manipulated/enhanced by systematic modification of their Fourier transforms. Apodization is the term used to describe all such procedures.

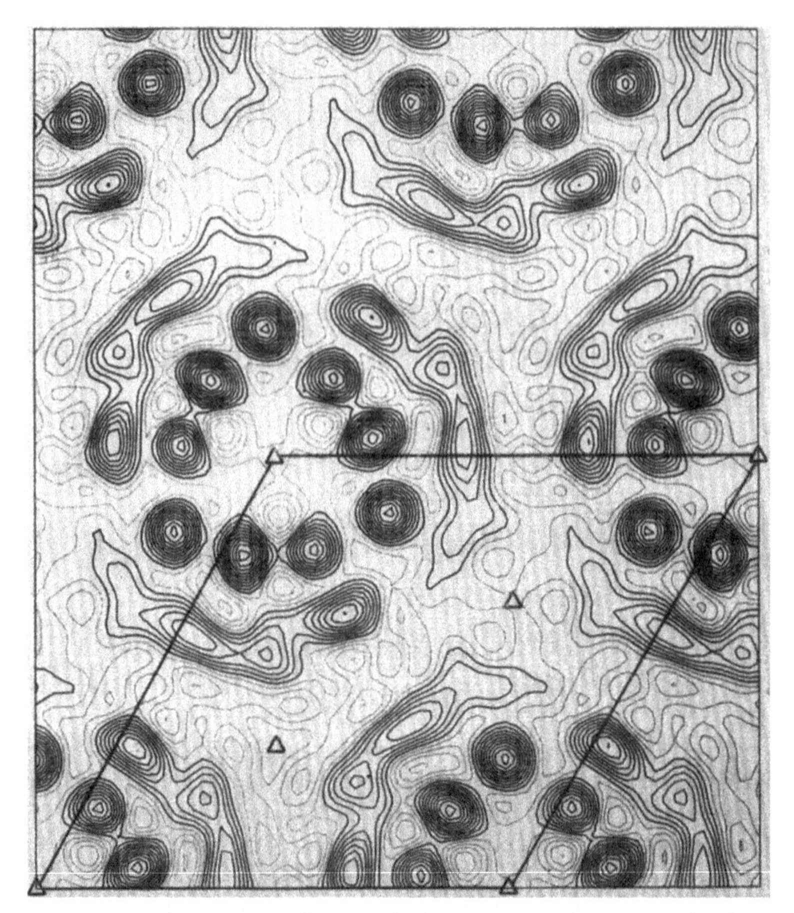

Figure 12.8 The structure of bacteriorhodopsin in a projection normal to the plane of the membrane at 7-Å resolution. The locations of the three-fold axes in the unit cell are displayed as are the unit cell axes (62 Å × 62 Å). (Reprinted from Unwin, P. N. T., and Henderson, R. [1975], Molecular structure determination by electron microscopy of unstained crystalline specimens, *Journal of Molecular Biology* 94, 425–432, with permission from Elsevier.)

12.14 THE TRANSVERSE COHERENCE OF THE ELECTRON BEAM AFFECTS IMAGE QUALITY

The theory of image formation presented earlier assumes that the radiation used to create images has a transverse coherence length that is larger than the linear dimensions of the objects in specimens whose structures are of concern. Only then will phase contrast effects contribute fully to the images obtained. However, this condition is not always met because the transverse coherence lengths of the electron beams in EMs can vary over a wide range and are determined in part by way they are operated.

If the transverse coherence length of the electron beam is small compared to the length scale of the smallest features in an object that needs to be imaged, then those features will be imaged in the EM the same way features are imaged in a light microscope operating with Koehler illumination. The image of a specimen will be an incoherent superposition of the images that would have been obtained if each point in that specimen had been imaged separately. Thus, it will be a sum of point spread functions, the relative brightnesses of which are determined entirely by differences in aperture contrast. Images of this kind are likely to have resolutions much poorer than those the same instrument would produce if it were operating with a coherent beam because so little information about the high-resolution detail in EM images is provided by aperture contrast.

Comment: Recall that in section 10.1, it is pointed out that, theoretically, the resolution of the images of incoherently illuminated specimens should be higher than the resolution of the images of coherently illuminated specimens, everything else being equal. For light microscopists, that theoretical advantage is often a real advantage because absorption commonly accounts for most of the contrast seen at all the length scales accessible to the light microscope. If this were true of aperture contrast in EM images, it would also make sense to operate EMs using incoherent radiation, but it is not.

By contrast, if the specimen is thin and the transverse coherence length of the electron beam is large, it will be imaged coherently. Moreover, when 100-keV electrons interact with such a specimen, most of the electrons in the beam will not be scattered, and the vast majority of those that are scattered will be scattered only once. Thus, the conditions of the first Born approximation will be fulfilled, and the images that emerge will be directly related to the Fourier transforms of the electrical potential distributions in entire three-dimensional objects, not just the point spread functions of their constituent points. Furthermore, resolution is likely to be high because phase contrast will contribute to images.

Comment: Even when the transverse coherence length of the beam in an EM is large, images that look like aperture contrast images will emerge if the specimen is so thick that the electrons passing through it are scattered many times.

12.15 PARTIAL COHERENCE SUPPRESSES IMAGE DETAIL

Unhappily, from the point of view of the theorist, electron microscopy is often done using beams that have transverse coherence lengths big enough to be significant, but not so big that incoherent imaging effects can be ignored (i.e., with partially coherent illumination). What does partial coherence do to EM images?

The transverse coherence length of the radiation in a microscope is determined by the spread in the angles that the rays reaching the specimen make with the optical axis. Thus, if the illumination of an EM has partial transverse coherence, the rays reaching the specimen will not be perfectly parallel and the resulting image will be a superposition of the images that could have been produced using perfectly parallel radiation, with the specimen in many, somewhat different orientations with respect to the axis of the beam.

It is easier to think about the consequences of partial transverse coherence in reciprocal space. When a microscope is operated with this kind of radiation, no point in its back focal plane corresponds to a unique location in the reciprocal space of the specimen. The radiation passing through any single point in the back focal plane of a microscope that is being operated this way will report an average value of the three-dimensional transform of the specimen, where the average is being taken over a disk-shaped region in the transform of the image that has a radius of roughly $|\mathbf{S}|\gamma$, γ being the half width of the range of the angles the rays in the beam make with the optical axis. Ray-to-ray differences in the physical location of the origin of reciprocal space will also contribute to the radius of that averaging disk, as well as giving it some thickness in the radial direction. Because the radius of the disk over which this averaging takes place increases with $|\mathbf{S}|$, the size of the volume in reciprocal space that is being averaged increases with resolution.

We know from our discussion of chromatic aberration that resolution always suffers when there is image averaging. Indeed, the effect that partial coherence has on the images an EM produces can be described by adding (yet) another multiplicative term to its CTF that resembles the chromatic aberration term. At low spatial frequencies, the magnitude of this function will close to 1.0, and the higher the resolution, the smaller it will become.

12.16 SOURCE BRILLIANCE SET A PRACTICAL UPPER LIMIT ON TRANSVERSE COHERENCE

The transverse coherence length of the beam in an EM is determined by: (1) the type of source in its electron gun, (2) the way that gun is operated, and (3) the settings of the condenser lenses that deliver the beam the gun produces to the sample. Because the user has some control over all of these aspects of EM operation, it is appropriate to say a few words about them here.

As section 1.10 shows, there is only one way to make the transverse coherence length of a radiation beam large, and that is to reduce the solid angle subtended by its source, as viewed from the specimen, because the smaller that solid angle is, the closer to parallel the trajectories of the quanta are that reach the specimen. That solid angle can be made small either by reducing the physical size of radiation source or by increasing the distance between it and the specimen. However, there are limits to what can be done because, everything else being equal, anything that increases the transverse coherence length of the beam will decrease its intensity at the sample, and the dimmer the beam is at the sample, the longer it takes to record an image. The terms of trade are poor; transverse coherence lengths increase linearly as solid angles shrink, but beam intensities drop quadratically.

The inverse square relationship that exists between transverse coherence length and illumination intensity means that, in practice, the maximum value of the transverse coherence length that can be achieved in an EM is determined by the **brilliance** of the source in its electron gun. The brilliance of a radiation source is the number of quanta it emits, per unit time, per steradian of solid angle, per unit area of source. If the brilliance of the source in an EM is high, its users will be able to record images in reasonable times when the instrument is operated so that the transverse coherence of its electron beam is large.

12.17 THE SOURCES USED IN ELECTRON GUNS DIFFER SIGNIFICANTLY IN BRILLIANCE

All electron guns have a **cathode**, which is the element of the gun that produces the electrons the gun emits, and an **anode**, which is basically a metal plate with a hole in the middle of it that is centered on the optical axis of the microscope. The anode of the gun, like most of the rest of the microscope, is grounded to reduce the probability that the user of the microscope will get electrocuted, and the electrical potential of the cathode of the gun (i.e., its electron sources), is maintained at a voltage that is tens to hundreds of thousands of volts *below* that of the rest of the instrument by its power supply. That voltage difference determines the kinetic energy of the electrons in the beam produced by the gun.

Two types of cathodes are used in EM guns today, **thermionic** cathodes and **field emission** cathodes. The cathodes in most older microscopes, like the cathode in Ruska's first EM, are thermionic, which means that electrons are "cooked" out of them. A cathode of this sort might consist of "vee"-shaped tungsten filaments that is heated to $\sim$2,000 K by passing an electric current through it. At those temperatures, thermionic cathodes will emit large numbers of electrons, provided there is an anode nearby that is maintained at a positive potential. Thermionic electron sources are cheap to make, and easy to operate, but lack brilliance.

The cathodes in the guns in most EMs used for high-resolution microscopy today produce electrons by a process called **field emission**, which is a manifestation of quantum mechanical tunneling. The cathode in a field emission gun is usually a sharply pointed needle made of a single crystal of tungsten. Electrons will be emitted from the point of a needle-shaped metal object if it is maintained at a comparatively modest negative potential with respect to its surroundings because voltage gradients are always much greater at the edges and corners of metallic objects than they are in flat regions. (Lightning rods exploit this fact.) Thus, the force driving electrons out of such a cathode is very large at its tip, and the probability of their escape by tunneling is correspondingly high. The emitting surface of a field emission source is usually much smaller than that of any thermionic source (linear dimensions: $\sim$10 nm versus $\sim$10 μm), and the emission per unit area is much higher. Consequently, field emission sources can be $\sim$1,000 times brighter than thermionic sources. An addition attraction of field emission sources is that the spread in energies of the electrons they produce is significantly smaller than the spread of energies of the electrons produced by thermionic sources because they operate at much lower temperatures. That reduction in energy spread, helps reduce chromatic aberration.

12.18 USERS CAN CONTROL THE TRANSVERSE COHERENCE LENGTHS OF EM BEAMS

In addition to a cathode and an anode, the electron guns in EMs include a third electrode that is used to control performance. The extra electrode in a thermionic gun does not work the same way as the corresponding electrode in a field emission gun does, but the end result of adjusting its potential is similar. By altering the potential on that electrode, users can control the size of the area of the cathode that contributes electrons to the beam. The smaller that area, the bigger the coherence width of the beam, but the fewer the number of electrons it contains, and, hence, the longer the exposures required to produce images.

Users also control the focal lengths of the condenser lenses in EMs that deliver the beam to the sample. Most EMs have two condenser lenses, and the settings of their currents determine the diameter of the illuminated region in specimens and affect the transverse coherence of the beam. The condenser lens closest to the gun (i.e., the first condenser lens), is invariably a short focal length lens ($\sim$2 mm), the optical center of which is tens of centimeters below the source. Thus, the plane where a focused image of the source appears will be only a few millimeters below its optical center, and its magnification may be $\sim$0.01; in other words, it will be substantially demagnified. If the current passing through the second condenser lens is held constant while the current in the first condenser lens is varied, the diameter of the illuminated area in the specimen will change.

The second condenser lens invariably has a long focal length and is usually equipped with a limiting aperture, the radius of which might be $\sim 50\ \mu\text{m}(= R)$. If the current in the second lens is set so that the distance between its optical center and the image formed by the first condenser lens exactly equals its own focal length, the trajectories of the electrons emerging from the second lens will be as close to parallel as they can be, the transverse coherence length of the beam will be large, and radius of the illuminated area will be R.

In practice, the transverse coherence length of the beam of the typical TEM equipped with a thermionic source can be as large as a few hundred angstroms. By contrast, an otherwise similar instrument equipped with a field emission source may produce a beam that has a transverse coherence length measured in tens of microns.

PROBLEMS

1. A sample is prepared that has a uniform thickness of 1,000 Å and consists of quasi-spherical protein complexes that have diameters of 500-Å embedded in vitreous ice. Suppose this sample is imaged using a TEM that has a limiting aperture that intercepts half the electrons elastically scattered by all the different kinds of atoms it contains.

   a. Assuming that all the contrast in this image is aperture contrast, by how much will the brightness of this image differ in locations where there is protein present from regions where there is only water present?
   b. Will protein look bright in this image, or will it look dark?

   The density of water is $1,000\ \text{kg/m}^3$ whereas that of bulk protein is about $1,350\ \text{kg/m}^3$. To make calculations simple, assume that the protein of interest is polyalanine (i.e., that its monomers have a molecular weight of 71 and an atomic composition of C_3H_5NO). Assume also: (1) that the elastic cross sections of carbon, nitrogen, and oxygen are all

8.5×10^{-23} m^2, which is not too far from the truth, and (2) that the elastic cross section of hydrogen is so small it can be ignored for these purposes, which is also reasonable.

2. Using equation (12.2.1), prove that there is no information available about Thon rings in images of the back focal plane formed directly in EMs.
3. Suppose the average electrostatic potential inside a slab-shaped object is exactly +10 V compared to ground, and suppose also that electrons having energies of 100 keV are passing through that slab parallel to its normal. How thick would such a slab have to be in order for the phase associated with the electrons emerging from it to differ by 180° from the phase of otherwise similar electrons that did not pass through it at all?
4. Suppose a phase plate were inserted into the back focal plane of a TEM that altered the phase of all the radiation passing through that plane at scattering angles greater than 0° by 90° (i.e., $\pi/2$). Show that the CTF applicable to the phase contrast component of the images produced by this modified TEM would be FT[cos $\Delta\varphi(\alpha)$] instead of FT[sin $\Delta\varphi(\alpha)$], which is the function appropriate for an ordinary TEM. At what level of focus should such a TEM be operated to maximize the resolution? (Hint: Examine figure 12.3.)

FURTHER READING

Burge, R. E., and Smith, G. H. (1962). A new calculation of electron scattering cross sections and a theoretical discussion of image contrast in the electron microscope. *Proc. Phys. Soc.* 79, 673–690.

Erickson, H. P., and Klug, A. (1971). Measurement and compensation of defocusing and aberrations by Fourier processing of electron micrographs. *Phi. Trans. Roy. Soc. B* 261, 105–118.

Reimer, L., and Kohl, H. (2008). *Transmission Electron Microscopy*, 5th ed. Berlin: Springer-Verlag.

Spence, J. H. C. (2003). *High Resolution Electron Microscopy*, 3rd ed. Oxford: Oxford University Press.

Thon, F. (1966). Zur defokussierungsbhängigkeit des phasekontrastes bei der elektronenmikroscopischen abbildung. *Zeit. für Naturforschung* 21, 476–478.

Unwin, P. T. N., and Henderson, R. (1975). Molecular structure determination by electron microscopy of unstained crystalline specimens. *J. Mol. Biol.* 94, 425–432.

APPENDIX 12.1 THE EFFECTS OF THERMAL MOTIONS ON MOLECULAR SCATTERING PROFILES

The average scatter produced by a molecule held in a fixed position and fixed orientation is (equation (12.2)):

$$\langle \mathrm{I}(\mathbf{S},t)\rangle = \sum_i \mathrm{f}_i(S)^2 + \sum_i \sum_{j\neq i} \mathrm{f}_i(S)\,\mathrm{f}_j(S)\exp\left[-2\pi i\left(\mathbf{r}_i - \mathbf{r}_j\right)\cdot\mathbf{S}\right]$$
$$\times \left\langle \exp\left(-2\pi i\left[\delta_i(t) - \delta_j(t)\right]\cdot\mathbf{S}\right)\right\rangle.$$

It is easy to show (see appendix 5.1) that:

$$\left\langle \exp\left(-2\pi i\left[\delta_i(t) - \delta_j(t)\right]\cdot\mathbf{S}\right)\right\rangle \approx \exp\left(-2\pi^2\left\langle\left(\left[\delta_i(t) - \delta_j(t)\right]\cdot\mathbf{S}\right)^2\right\rangle\right).$$

Now if two atoms in some molecule moved thermally as though they were rigidly connected, then $\|[\delta_i(t) - \delta_j(t)]\|$ would be 0 at all times, and the corresponding Gaussian would have a value of 1.0. However, there are no such pairs of atoms in any molecule because, if nothing else, the lengths of covalent bonds oscillate, and bonded atoms can "wag". If the motions of atoms i and j are completely uncorrelated, which will be the case, more or less, for most pairs of atoms in a macromolecule that are not covalently bonded, the Gaussian on the left side of the preceding equation will equal:

$$\exp\left[-\left(B_i + B_j\right)\left(S^2/4\right)\right],$$

where B_i and B_j are the temperature factors that reflect the motions of the two atoms, measured in the molecular coordinate frame.

Suppose some molecule is composed of 1,000 atoms. Its scattering profile will be the sum of a component having 1,000 terms that is not structure-dependent, and a component having 999,000 terms that is structure-dependent. If the B-factors of nonbonded atoms are around 10 Å^2, which implies relative motions of the order of 0.6 Å, which is small, but plausible, the contribution made by nonbonded pairs of atoms to the structure-dependent part of the average scattering profile of that molecule in a 100-keV TEM will be attenuated by a factor of $\sim 10^{-2}$ at a scattering angle of 2°, which corresponds to a resolution of about 1 Å, and by a factor of $\sim 5 \times 10^{-5}$ at a scattering angle of 3°, which corresponds to a resolution of about 0.7 Å. Thus, as the scattering angle increases, $\langle \mathrm{I}(\mathbf{S},t)\rangle$ will rapidly reduce to $[\sum \mathrm{f}_i(\mathbf{S})^2 +$ contributions from covalently bonded atoms$]$.

If each atom in the molecule of concern is bonded to two others, the covalently bonded atom component of the structure-dependent part of the average scatter will have 2,000 terms in it. If the amplitudes of the motions that result from bonds' length oscillations and wagging motions are of the order of 0.1 Å, which corresponds to B-factors $\sim$0.26 Å^2, the magnitude of the contribution bonded pairs make to the structure-dependent part of $\langle \mathrm{I}(\mathbf{S},t)\rangle$ will be attenuated by a factor of $1/e$ at scattering angles $\sim$6° (i.e., at a resolution around 0.36 Å) and will be even weaker at higher angles. It follows that the elastic scatter produced by a macromolecule exposed to 100-keV electrons will contain little or no information about its three-dimensional structure angles above $\sim$6° because for large $|\mathbf{S}|$,

$$\langle \mathrm{I}(\mathbf{S},t)\rangle \approx \sum_i \mathrm{f}_i(S)^2.$$

APPENDIX 12.2 THE IMAGES OF WEAK-PHASE OBJECTS

The transmission function to be imaged is:

$$f'(x, y) \approx 1 - \mu'(x, y) + i\theta'(x, y).$$

If FT$[\theta'(x, y)]$ is $\Theta(\xi, \eta)$, and FT$[\mu(x, y)]$ is $M(\xi, \eta)$, then in the back focal plane, ignoring the chromatic aberration and aperture components of the CTF, the transmission function will have become:

$$Q(\xi, \eta) = \delta(\xi, \eta) - \{M(\xi, \eta)\cos[\Delta\varphi(\alpha)] + \Theta(\xi, \eta)\sin[\Delta\varphi(\alpha)]\}$$
$$- i\{M(\xi, \eta)\sin[\Delta\varphi(\alpha)] - \Theta(\xi, \eta)\cos[\Delta\varphi(\alpha)]\}. \qquad \textbf{(12.2.1)}$$

Both terms in curly brackets will be multiplied by Chr(α) (see equation (11.7)) and by $A(\alpha_{max})$, the usual aperture function. Setting aside the magnification and inversion that occur when images form for the sake of notational simplicity, the distribution of intensity seen in the image plane will be:

$$I(X, Y) = \begin{bmatrix} \{1 - [\theta'(x, y) - i\mu'(x, y)] * \mathrm{FT}(\sin[\Delta\varphi(\alpha)]\,\mathrm{Chr}(\alpha)\,A(\alpha_{max})) \\ - [\mu'(x, y) - i\theta'(x, y)] * \mathrm{FT}(\sin(\Delta\varphi(\alpha)]\,\mathrm{Chr}(\alpha)\,A(\alpha_{max})) \end{bmatrix}^2$$

To evaluate $I(X, Y)$, the term in curly brackets on the right side of the preceding equation must be multiplied by its complex conjugate, and because both $\mu'(x, y)$ and $\theta'(x, y)$ are both small, only terms that are first order need to be retained. Furthermore, it is important to note that $\theta'(x, y)^* = \theta'(x, y)$ and $\mu'(x, y) = \mu'(x, y)^*$ because both functions are real. In addition, because $\cos[\Delta\varphi(\alpha)]\mathrm{Chr}(\alpha)A(\alpha_{max})$ and $\sin[\Delta\varphi(\alpha)]\mathrm{Chr}(\alpha)A(\alpha_{max})$ are both centrosymmetric in the back focal plane:

$$\{\mathrm{FT}(\cos[\Delta\varphi(\alpha)]\,\mathrm{Chr}(\alpha)\,A(\alpha_{max}))\}^* = \mathrm{FT}(\cos[\Delta\varphi(\alpha)]\,\mathrm{Chr}(\alpha)\,A(\alpha_{max})),$$

and:

$$\{\mathrm{FT}(\sin[\Delta\varphi(\alpha)]\,\mathrm{Chr}(\alpha)\,A(\alpha_{max}))\}^* = \mathrm{FT}(\sin[\Delta\varphi(\alpha)]\,\mathrm{Chr}(\alpha)\,A(\alpha_{max})).$$

Thus:

$$I(X, Y) \approx 1 - 2\mu'(x, y) * \mathrm{FT}(\cos[\Delta\varphi(\alpha)]\,\mathrm{Chr}(\alpha)\,A(\alpha_{max}))$$
$$- 2\theta'(x, y) * \mathrm{FT}(\sin[\Delta\varphi(\alpha)]\,\mathrm{Chr}(\alpha)\,A(\alpha_{max})). \qquad \textbf{(12.2.2)}$$

13

Electron Microscopy in Three Dimensions

Although many important biological questions have been answered using two-dimensional microscopic images of modest resolution, the ultimate goal of the structural biologist is to determine the three-dimensional structures of biological objects at atomic resolution. For decades, only X-ray crystallographers could do this, but they must now share this distinction with nuclear magnetic resonance spectroscopists and electron microscopists. The recovery of high-resolution, three-dimensional images from two-dimensional electron micrographs is called **three-dimensional reconstruction**, and under favorable circumstances, it can yield atomic resolution images.

If the resolution required is (very) modest, it may be possible to produce a satisfactory three-dimensional EM image of some biological structure by sectioning methods. It is much harder to obtain three-dimensional EM images of biological objects too small to section, but that too can be done, and only images of this type will ever have resolutions high enough so that they can be interpreted in atomic detail. This kind of three-dimensional image generation is what is usually meant by the phrase *three-dimensional reconstruction.*

The first three-dimensional reconstructions of macromolecular objects were reported in the mid-1960s, and the first protein structure was solved at atomic resolution using EM data about 25 years later. Nevertheless, the reconstruction field is still dominated by specialists, much the way macromolecular X-ray crystallography was until $\sim$ 1990, because theory and algorithm development are still important concerns. Another reason three-dimensional reconstruction has been slow in becoming a mainstream technique is that for a long time, most biochemists felt that the explanatory value of the three-dimensional images of molecules electron microscopists were producing did not justify the effort required to obtain them. However, there is a growing appreciation for the information that three-dimensional reconstructions can provide not only because the resolutions routinely attainable are improving, but also because of the steady growth in the number of atomic resolution crystal and nuclear magnetic resonance structures available for the components of macromolecular assemblies that can be used to interpret the lower resolution images of those assemblies EM can readily provide. The hybrid structural models produced by merging these two kinds of information are often very interesting.

Three-dimensional EM images of objects can be obtained without sectioning using three different strategies: (1) **tomography**, (2) **single particle reconstruction**, and (3) **electron crystallography**. *Tomography* is usually used to reconstruct the structures of objects such as entire bacterial cells, or organelles, no two of which have the same three-dimensional structure. *Single particle techniques* are commonly applied to smaller entities that vary little in structure from one copy to the next. The discussion here will be confined to these two techniques because the third, electron crystallography, does not differ qualitatively from X-ray crystallography, a technique we have already addressed at length.

13.1 THREE-DIMENSIONAL RECONSTRUCTIONS CAN BE DONE IN RECIPROCAL SPACE

We know what can be learned about the structure of a macromolecule from a single TEM image. The linear dimensions of most such objects are smaller than the depths of field of the images available of them, which are determined by both the specifications of the TEMs that produced them and the resolutions of those images. Thus, a micrograph of such a molecule is an image of its three-dimensional electric field projected onto a plane perpendicular to the incident electron beam, and the Fourier transform of that image is a central section of the three-dimensional Fourier transform of its three-dimensional electric field (see sections 8.2, and 9.16).

These facts suggest a four-step strategy that could be used to reconstruct three-dimensional images from two-dimensional micrographs. First, take EM micrographs of the object of interest in many different orientations relative to the electron beam. Second, compute the Fourier transforms of all those images. Third, place the Fourier coefficients so obtained in the three-dimensional reciprocal space of the object using the information available about the orientation of the object in each of its images. Fourth, invert the resulting three-dimensional Fourier transform to obtain a three-dimensional image of the object.

Comment: Step 2 will yield both amplitudes and phases; there is no phase problem in electron microscopy.
Comment: Three-dimensional reconstructions can be carried out without the use of Fourier transforms, but the Fourier approach is the only one that will be explored here because the problems that all reconstruction schemes must solve are more easily understood in reciprocal space than they are in real space.

The final step in the scheme will produce the desired result *provided*: (1) the preceding steps were properly executed, an obvious caveat that cannot be taken for granted; and (2) that phases are all computed relative to a common origin, which is usually easily done. Thus, the aspects of the scheme that appear problematic are: (1) determining which views of an object will be needed to obtain a three-dimensional image that has some specified resolution, (2) obtaining those images, and (3) determining the relative orientations of the corresponding central sections, if they are not known in advance, which they often are not. These practical problems are all aspects of a much larger issue that must now be explored.

13.2 INTERPRETABLE IMAGES CANNOT BE OBTAINED BY DIRECT INVERSION OF CENTRAL SECTION DATA SETS

As we already know, crystallographic data must be reduced before they are used to solve structures (see section 5.2). Except for the information they provide about background, the information in the pixels of each frame of data that do not correspond to Bragg reflections is of no interest, and the intensities of all reflections that have been measured more than once must be averaged. The end result is a nonredundant, indexed list of background-corrected intensities.

By contrast, it is far from obvious what electron microscopists can, or should, do to reduce their data. The objects they study are seldom crystalline; hence, the transforms of the images of those objects are not sampled on a lattice of points. It follows that every point in the transform of every image contains information about the structure of concern; thus, it is not clear that any of the data should be averaged, let alone discarded. Nevertheless, except under very unusual circumstances (see section 13.12), central section data sets *must* be reduced before they are inverted to produce images.

Central section data must be reduced because they do not sample the Fourier transforms of objects at points on an appropriate lattice in reciprocal space. Within each central section, the transform can be evaluated anywhere one wishes, but between sections, there will always be wedge-shaped regions in reciprocal space where there is no experimental information available at all. The real-space function that would result if all the Fourier coefficients obtained from a set of central sections were back-transformed would be a three-dimensional image of the object of interest convoluted with the Fourier transform of the array of points in reciprocal space where its transform had been evaluated. That array is certain *not* to be a lattice in reciprocal space; hence, its Fourier transform will not be a lattice in real space either. Consequently, the three-dimensional image that would emerge if the unreduced data were inverted directly would be the convolution of the transform of the sampling array with the density distribution function of the object under investigation. If interpretable at all, images of that sort are sure to be messy.

Figure 13.1 shows the result of a computational exercise undertaken to illustrate why central section data needs to be reduced. In this case, the object to be reconstructed was a two-dimensional, five-atom "molecule" of arbitrary structure. The images shown in the figure were computed by back Fourier transformation of data

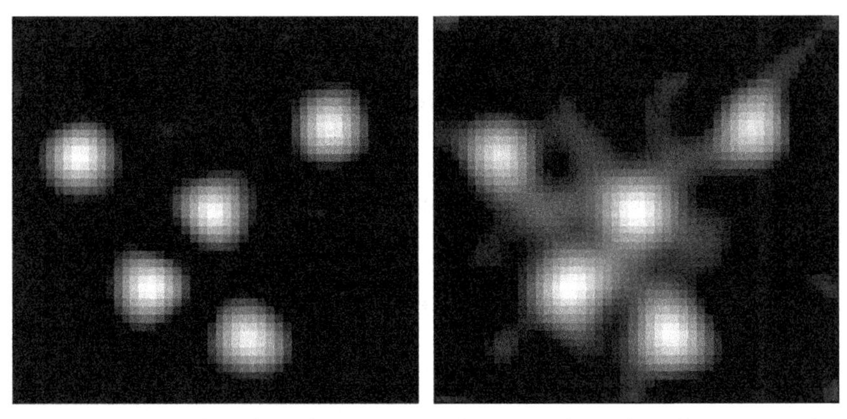

Figure 13.1 The effect of sampling in reciprocal space on the quality of reconstructed images. The left panel is the image of an arbitrary "5-atom molecule" obtained by back Fourier transformation of a set of Fourier coefficients that sample the (computed) molecular transform on a square lattice in reciprocal space. Each atom is Gaussian having a maximum amplitude of 1 and an amplitude of 0.37 at 0.56 Å from its center. The lattice spacing is $(1/6.0)$ Å^{-1}, and all the Fourier coefficients out to $|\mathbf{S}| = 1.0$ Å^{-1} are used. A 6 Å × 6 Å region of real space is shown (i.e., 1 unit cell). The grayscale for the image is linear, and, in arbitrary units, runs from +5 (black) to +70 (white). Negative density is not shown, but there are no negative features having amplitudes greater than ~ -5 in the image. The data transformed to obtain the image in the right panel sampled the molecular transform along lines radiating out from the origin of reciprocal space. The angular spacing between adjacent lines was 22.5°, and there were eight of them. The sampling interval along lines was $(1/6.0)$ Å^{-1}. All the data were included out to a resolution of $|\mathbf{S}| = 1.0$ Å^{-1}. The area shown in the right image is the same as that shown in the left image, and its grayscale is the same. This image has important negative features (~ -20), which are not shown.

sets that sampled the Fourier transform of the molecule's electron density distribution differently. The data from which the left-hand image was derived sampled the molecular transform on a square lattice in reciprocal space, the dimensions of which were set so that the unit cells of the corresponding real-space lattice would be large enough to accommodate the molecule comfortably. The data from which the right-hand image was derived sampled the molecular transform along lines radiating out from the origin of reciprocal space. (NB: In two dimensions, the equivalent of a central section is a line that passes through the origin of reciprocal space.) The sampling along radial lines was the same as the reciprocal space lattice spacing used to produce the data that gave rise to the left image. The angular interval between adjacent radial lines was the same, and as a group, they evenly sampled reciprocal space in an angular sense. Furthermore, the resolution limit of the two data sets were the same, and the number of radial lines was chosen so that the number of Fourier coefficients included in both Fourier syntheses would be the same.

It is obvious that the left-hand image faithfully depicts the molecule, and it is almost noise-free. It includes no positive features that are spurious, and the amplitudes of its negative features (not shown) are very low. By contrast, although it is certainly true that the positions of the five atoms are evident in the right-hand image, it leaves a lot to be desired. The shapes of the atoms are less circular, and they are connected by regions of positive density that are computational artifacts. In addition, in regions that are remote from the molecule, there are several large, irregularly shaped negative features (not shown) that have amplitudes a one-third those of the positive peaks that represent atoms. In short, the right-hand image is a mess.

EM data sets can be reduced in many different ways, the most obvious of which we will discuss here not because it is necessarily the method of choice, but because our exploration of it will uncover the important issues that are common to them all. In essence, no matter how they work in detail, all EM data reduction procedures use the experimentally determined Fourier coefficients of the transform of the images of some object, which are irregularly distributed in reciprocal space, to estimate the values of Fourier coefficients of that transform at points on a regular three-dimensional lattice in reciprocal space. The dimensions of the unit cells of that lattice, which, as we will see, the microscopist sets, must be chosen so that the unit cells of the corresponding real-space lattice are large enough to accommodate entire images of the object of interest. Inverse transformation of a set of Fourier coefficients that has been regularized this way will yield unoverlapped images of the object.

It follows that it is impossible to obtain a three-dimensional reconstruction from a set of two-dimensional images of some object unless the number and distribution of those images are such that at small $|\mathbf{S}|$, the number of lattice points per unit volume of reciprocal space is *smaller* than the number of measured Fourier coefficients those images provide. Data reduction eliminates the "extra" data available in that region using procedures akin to the averaging of reflection intensities crystallographers use to eliminate redundancy in their data sets. As $|\mathbf{S}|$ increases, the density of measured Fourier coefficients in reciprocal space will fall as $|\mathbf{S}|^{-2}$, but the density of lattice points will remain constant. Thus, at some value of $|\mathbf{S}|$, the density of observations will become less than the density of lattice points, and at larger $|\mathbf{S}|$, it will be impossible to obtain all the regularized Fourier coefficients required. The reciprocal of the value of $|\mathbf{S}|$ at which the measured data become insufficient is the resolution of the three-dimensional reconstruction that can be obtained from the data. Let us see how this works out in detail.

13.3 IF THE TRANSFORM OF AN OBJECT IS KNOWN ON A LATTICE OF APPROPRIATE DIMENSIONS, ITS TRANSFORM CAN BE EVALUATED ANYWHERE IN RECIPROCAL SPACE

As we already know, the density distribution of an object can be computed if the amplitudes and phases of the Fourier transform of that distribution are available on a rectangular lattice in reciprocal space that has the dimensions $(1/a_1,\ 1/a_2,\ 1/a_3)$, where a_1, a_2, and a_3 are all not less than the maximum linear dimensions of the object in the corresponding directions in real space. (NB: By the time any EM reconstruction experiment has progressed very far, the microscopist will have a good idea of what values of a_1, a_2, and a_3 ought to be because he or she will have many [real-space] images of the object of concern.) The relationship between density function and its transform can be written as follows:

$$\rho(\mathbf{r}) = \sum \mathrm{F}(\mathbf{h}) \exp(2\pi i\mathbf{h} \cdot \mathbf{r}),$$

where $\mathbf{r}$ is a location in real space specified using fractional unit cell coordinates, $\mathrm{F}(\mathbf{h})$ is the value of the Fourier transform of $\rho(\mathbf{r})$ at $\mathbf{h}$, and $\mathbf{h}$ is the location in reciprocal space specified using Miller indices: $h_1 = h$, $h_2 = k$, and $h_3 = l$.

The value of the transform of the object at any point in reciprocal space that is *not* a lattice point, $\mathrm{F}(\mathbf{h}')$, is:

$$\mathrm{F}(\mathbf{h}') = \int \rho(\mathbf{r}) \exp(-2\pi i\mathbf{h}' \cdot \mathbf{r})\, dV_r.$$

If $\rho(\mathbf{r})$ is replaced by its Fourier equivalent, we can show:

$$\mathrm{F}(\mathbf{h}') = \sum \mathrm{F}(\mathbf{h}) \int \exp\left[-2\pi i(\mathbf{h}' - \mathbf{h}) \cdot \mathbf{r}\right] dV_r,$$

where the sum is over all $\mathbf{h}$, and the components of $\mathbf{h}'$ might not be integers.

The value of the integral in this equation depends *only* on the dimensions of the unit cell, the orientations of its axes, and the locations in reciprocal space specified by $\mathbf{h}$ and $\mathbf{h}'$. Because none of these parameters has anything to do with the structure of the object(s) inside the unit cell, the integral can be evaluated for all $\mathbf{h}$ and all $\mathbf{h}'$ once the unit cell has been defined and the locations in reciprocal space determined where data are available. If $\mathrm{B}(\mathbf{h}', \mathbf{h})$ is the number produced when that integral is evaluated for a particular pair of values of $\mathbf{h}'$ and $\mathbf{h}$, then the equation for $\mathrm{F}(\mathbf{h}')$ can be written in an even more compact form:

$$\mathrm{F}(\mathbf{h}') = \Sigma \mathrm{B}(\mathbf{h}', \mathbf{h})\mathrm{F}(\mathbf{h}). \qquad (13.1)$$

In this case, $\mathrm{F}(\mathbf{h}')$ is any one of the (many) values of the Fourier transform of some object that can be obtained from its EM images. It is convenient to represent the set of all available $\mathrm{F}(\mathbf{h}')$ as a column vector, $\mathbf{F}'$, that might have tens of thousands of components. It will include components out to values of $\mathbf{h}'$ that correspond to $|\mathbf{S}|_{\max}$, the inverse of the Bragg spacing beyond which the $\mathrm{F}(\mathbf{h}')$ are indistinguishable from noise. A similar vector can be constructed for the $\mathrm{F}(\mathbf{h})$, which will be designated $\mathbf{F}$, and the number of its components, n, will be:

$$n = (4\pi/3)|\mathbf{S}|_{\max}{}^3/(2/a_1a_2a_3).$$

(The factor 2 in the denominator reflects Friedel's law.)

Comment: If the components of $\mathbf{F}'$ *are all effectively 0 beyond* $|\mathbf{S}|_{max}$, *all the components of* $\mathbf{F}$ *will perforce be 0 beyond* $|\mathbf{S}|_{max}$ *also. Thus, operationally, the sum in equation (13.1) is a finite sum.*

The set of $B(\mathbf{h}', \mathbf{h})$ relevant to any given problem can be represented as a matrix, **B**, each entry of which is a $B(\mathbf{h}', \mathbf{h})$ coefficient that has only to do with geometry of data collection process, not the actual data itself. The number of columns in this matrix will equal the number of components of **F**, *n*, and the number rows will be the number components of **F**$'$, *m*. Using this notation, equation (13.1) can be summarized as follows:

$$\mathbf{F}' = \mathbf{BF}.$$

Like equation (13.1), this equation indicates that each $\mathbf{F}'$ is a linear function of all the **F**s, and this implies that EM data reduction is a problem in linear algebra.

13.4 B IS A PRODUCT OF SINC FUNCTIONS

We already know how to evaluate the components of **B** (see section 2.7 and 2.8). If the dimensions of the real-space lattice are a_1, a_2, and a_3, then:

$$B(\mathbf{h}', \mathbf{h}) = \int \exp\left[-2\pi i\left(\mathbf{h}' - \mathbf{h}\right) \cdot \mathbf{r}\right) dV_r = \prod_1^3 \sin\left[\pi a_i\left(S_i' - S_i\right)\right] / \pi a_i\left(S_i' - S_i\right)$$

$$= \prod_1^3 \sin\left[\pi\left(h_i' - h_1\right)\right] / \pi\left(h_i' - h_i\right).$$

The products in these equations are products of three terms, one for each axis of the unit cell.

> *Comment*: $B(\mathbf{h}', \mathbf{h})$ *has several interesting mathematical properties. First, even if you knew the values of all of the components of* **F** *for some object out to* $|\mathbf{S}|_{max}$ *except for just one—the one at* **j**—*you would still not know enough about that object's transform to determine* $F(\mathbf{j})$ *because* $B(\mathbf{j}, \mathbf{h}) = 0$ *whenever all the components of both* **j** *and* **h** *have integer values, and that would be true for every* $B(\mathbf{j}, \mathbf{h})$. *Thus, the components of* **F** *are* **linearly independent**. *Furthermore, equation (13.1) indicates that the value of the Fourier transform at any point in reciprocal space inside of* $|\mathbf{S}|_{max}$ *can be expressed as a weighted sum of the values of the components of* **F**, *which means that the components of* **F** *constitute a* **(complete) basis set** *for the Fourier transform of the object; in other words,* **F spans** *that transform. Finally, if we represent the value of the Fourier transform of the object at two arbitrarily chosen points in reciprocal space as vectors,* **a** *and* **b**, *where* $\mathbf{a} = \Sigma a_h F(\mathbf{h})$ *and* $\mathbf{b} = \Sigma b_h F(\mathbf{h})$, *then their (inner) product,* $\mathbf{a} \cdot \mathbf{b}$, *will be* $\Sigma a_h b_h$. *Clearly, the inner product of* $F(\mathbf{h}_1)$ *and* $F(\mathbf{h}_2)$ *is 0 unless* $\mathbf{h}_1 = \mathbf{h}_2$. *Thus, not only is* **F** *a basis set for the Fourier transform in question, it is an* **orthogonal** *basis set.*

The functional form of the components of the **B** matrix has important consequences. When $\mathbf{h}'$ does *not* fall on a lattice point, a large number of the $B(\mathbf{h}', \mathbf{h})$ coefficients are likely not to be 0, often every single one of them. Thus, $F'(\mathbf{h}')$ cannot be estimated simply by averaging the values of the components of $F(\mathbf{h})$ that are close to $F(\mathbf{h}')$ in reciprocal space.

Any data reduction problem of this sort can be solved using many different reciprocal space lattices. Any lattice in reciprocal space that has unit cells smaller than those implied by the smallest unit cell in real space that will accommodate the entire object will do. In addition, there is no need to describe the density distributions of objects using Cartesian coordinates, and when you do not, the mathematical expressions used to compute transforms will change. What will not change, however, is the value of $F(\mathbf{S})$ at any given point in the reciprocal space of any particular object. Thus, it is simply a matter of convenience which kind of coordinate system you use, and the computations required to reduce central section data sets can often be radically simplified if a coordinate system is used that either reflects the symmetry of the structure under investigation or responds appropriately to the way the data were collected.

13.5 F CAN BE ESTIMATED BY MATRIX INVERSION

The equation $\mathbf{F}' = \mathbf{BF}$ would accurately describe the relationship between $\mathbf{F}'$ and $\mathbf{F}$, if the images used to compute $\mathbf{F}'$ were free of noise, but they never are. When real images are Fourier transformed, each estimate of F' will include a contribution of unknown magnitude due to noise, $\delta(\mathbf{h}')$. If those noise contributions are written as a column vector, $\boldsymbol{\delta}$, then an equation can be written that does describe the relationship between the experimental data, $\mathbf{F}'$, and the noise-free, reduced data, $\mathbf{F}$, we would like to know about:

$$\mathbf{F}' = \mathbf{BF} + \boldsymbol{\delta}. \tag{13.2}$$

(NB: In this equation, $\mathbf{F}'$ and $\mathbf{B}$ are known, but $\mathbf{F}$ and $\boldsymbol{\delta}$ are not.)

Because of noise, all we can hope to recover from real data are *estimates* for the components of $\mathbf{F}$ that will include some error. Estimation problems of this sort are commonly solved using procedures that find $\mathbf{F}_{est}$, the $\mathbf{F}$ that minimizes the quantity $\sum[F(\mathbf{h}')_{calc} - F(\mathbf{h}')_{obs}]^2$, which is the sum over all the measured data of the square of the difference between the $F(\mathbf{h}')$ we calculate by feeding $\mathbf{F}_{est}$ into equation (13.1) and their observed values. Estimates of that sort are called *least squares* estimates, and so we will designate this $\mathbf{F}_{est}$ "$\mathbf{F}_{lsq}$".

As every textbook on linear algebra shows (see appendix 13.1):

$$\mathbf{F}_{lsq} = [(\mathbf{B}^T\mathbf{B})^{-1}\mathbf{B}^T]\mathbf{F}'_{obs}. \tag{13.3}$$

The $\mathbf{F}'_{obs}$ are known from experiment, and though the algebra may be painful to execute, once $\mathbf{B}$ has been determined (see section 13.4), the matrix $[(\mathbf{B}^T\mathbf{B})^{-1}\mathbf{B}^T]$ can be evaluated also. Thus, formally at least, $\mathbf{F}_{lsq}$ can be obtained from $\mathbf{F}'_{obs}$, which is a complicated way of saying that there is a rational way to reduce experimental EM data sets.

Comment: In practice, the computations implied by these equations are seldom carried out as written because the matrix inversion called for is often unmanageable computationally.

As we already know, if **F** were known, it would possible to evaluate the Fourier transform of the density distribution of the object of interest anywhere in reciprocal space (see section 13.4); hence, the same must be true of $\mathbf{F}_{lsq}$ with respect to our best estimate of that transform. Furthermore, on the face of it, equation (13.3) implies that every component of $\mathbf{F}_{lsq}$ can also be written as a weighted sum of the components of $\mathbf{F}'_{obs}$. This suggests that $\mathbf{F}'_{obs}$ ought to be just as useful as $\mathbf{F}_{lsq}$ in this regard, but it almost never is. In fact, given the irregular way central section data are distributed in reciprocal space, it would be astonishing if usefully accurate estimates of the values of all the components of $\mathbf{F}_{lsq}$ out to the resolution limit of some data set could be produced by feeding $\mathbf{F}'_{obs}$ into equation (13.3). It is much more likely that the information $\mathbf{F}'_{obs}$ provides will suffice for the estimation of some of the components of $\mathbf{F}_{lsq}$, but not others.

13.6 ERROR PROPAGATION DETERMINES RESOLUTION

The components of $\mathbf{F}_{lsq}$ that are well determined by $\mathbf{F}'_{obs}$ can be identified by examining the impact that experimental errors in $\mathbf{F}'_{obs}$ have on $\mathbf{F}_{lsq}$. Because there are errors in $\mathbf{F}'_{obs}$, it is certain that $\mathbf{F}_{lsq} \neq \mathbf{F}$, $\mathbf{F}$ being the noise-free transform of the structure of the object of interest. For anyone intending to use $\mathbf{F}_{lsq}$ to compute a three-dimensional image—what else would you do with it?—nothing could be more important than knowing which components of $\mathbf{F}_{lsq}$ should be paid attention to because they are well-determined by the data, and which should be ignored because they are not.

The first step in any error analysis is estimating the magnitude of the errors in the input data. This can be done by using $\mathbf{F}_{lsq}$ to compute the values of the Fourier transform that would be observed at $\mathbf{h}'$, $\mathbf{F}'_{calc}$, if $\mathbf{F}_{lsq}$ corresponded exactly to $\mathbf{F}$, and then comparing those values with what was actually observed. This comparison will yield a set of residuals, $\mathbf{r}$, that obey the following relationship:

$$\mathbf{F}' - \mathbf{F}'_{calc} = \mathbf{F}' - \mathbf{B}\mathbf{F}_{lsq} = \mathbf{r}.$$

If $\mathbf{F}_{lsq}$ is a reasonable approximation of $\mathbf{F}$, then $\mathbf{r}$ will approximate $\boldsymbol{\delta}$.

For the sake of simplicity, assume that each component of $\boldsymbol{\delta}$ is independent of all the others, and that they are all samples drawn from the same normal, error distribution. If that is so, the variance of the error distribution of the data, σ^2, that is, the average value of δ^2 will be (approximately):

$$\sigma^2 = \langle \delta^2 \rangle \approx (m - n)^{-1} \sum_i r_i r_i^*,$$

where r_i is the residual associated with the ith observation, and the sum is over all observations. (Derivations of this expression may be found in any textbook on statistics.) Note also that n, the number of regularized Fourier coefficients sought, must be less than m, the number of measured Fourier coefficients, if there is to be any chance at all of arriving at useful estimates for all the components $\mathbf{F}_{lsq}$, and, obviously, the smaller n is relative to m, the better.

The second step in this analysis is determining how errors in $\mathbf{F}'_{obs}$ propagate into $\mathbf{F}_{lsq}$. As appendix 13.2 shows, information about how errors propagate in systems of equations like this one is to be found in their **normal matrices**; in this case, $\mathbf{B}^T\mathbf{B}$. Specifically, the variance of the estimate obtained for each component of $\mathbf{F}_{lsq}$, $F_{lsq}(\mathbf{h})$, $\sigma^2(\mathbf{h},\mathbf{h})$, will be:

$$\sigma^2(\mathbf{h},\mathbf{h}) = \sigma^2\left(\mathbf{B}^T\mathbf{B}\right)^{-1}(\mathbf{h},\mathbf{h}). \qquad (13.4)$$

The hope is that $(\mathbf{B}^T\mathbf{B})^{-1}(\mathbf{h},\mathbf{h})$ will be less than 1 for many of the components of $F_{lsq}(\mathbf{h})$, in other words, that the data over determine their values. Every such $F_{lsq}(\mathbf{h})$ will thus be, effectively, an average of several components of $\mathbf{F}'$. However, there will usually be other components for which $(\mathbf{B}^T\mathbf{B})^{-1}(\mathbf{h},\mathbf{h})$ is greater than 1, and sometimes *much* greater than 1. This will be true of any component of $\mathbf{F}_{lsq}$, $F_{lsq}(\mathbf{j})$, having the property that *all* $F_{obs}(\mathbf{h}')$ can be well approximated by the expression $\sum B(\mathbf{h}', \mathbf{h})F_{lsq}(\mathbf{h})$ when the $F_{lsq}(\mathbf{j})$ contribution to those sums is omitted. In practice, "well-approximated" might mean that the full sum estimate for $F_{obs}(\mathbf{h}')$ differs from that obtained when the $F_{lsq}(\mathbf{j})$ term is omitted from that sum by less than $\boldsymbol{\sigma}$. The reason this happens, when it happens, is that because of accidents of geometry, none of the components of $\mathbf{F}_{obs}$ contains much (any) information about the part of the transform represented by $F_{lsq}(\mathbf{j})$.

Clearly, all components of $\mathbf{F}_{lsq}$ that are poorly determined should be excluded when it comes time to compute density distribution maps because the errors associated with them are likely to be so large that their inclusion will add more noise than useful information to those maps. In fact, the conservative thing to do would be to use only those components of $\mathbf{F}_{lsq}$ for which $\boldsymbol{\sigma}^2(\mathbf{h},\mathbf{h}) < \boldsymbol{\sigma}^2$ and to limit the resolution of maps by excluding all components of $F_{lsq}(\mathbf{h})$ that fall outside the sphere in reciprocal space inside of which *all* $F_{lsq}(\mathbf{h})$ are well determined by the data. Once this is done, the redundancy in the experimental data will have been dealt with, its effective resolution determined, and the data reduction process will be complete.

13.7 TILT DATA SETS ARE BEST DESCRIBED USING CYLINDRICAL POLAR COORDINATES

Tomography is the simpler of the two approaches used to reconstruct the images of single particles. The reason it is simpler is that the relative orientations of the central sections that correspond to the images used in a tomographic reconstruction are known before the reconstruction process begins. They do not have to be determined by image analysis after the fact, as they often must be for single particle reconstructions.

EM tomography depends on the availability of specimen stages that make it possible to rotate specimens around an axis perpendicular to the electron beam. The orientation of an object mounted on such a stage can be altered relative to the beam (within limits) by angles that are determined in advance by the microscopist, and the object can be imaged at each setting.

A set of micrographs of an object accumulated this way corresponds to a set of central sections in the corresponding three-dimensional Fourier transform that intersect along a single line in reciprocal space, called the **common line**. The common line passes through the origin of reciprocal space and is parallel to the tilt axis.

Tomographic data collection is not as easy this description makes it sound. First, when the stage in an EM is tilted, the optical track of the microscope will cease to have cylindrical symmetry, and its optical performance may suffer. Second, unless the object being imaged is located on the tilt axis of the stage, its center of gravity will move up and down in the microscope as the tilt angle of the stage is altered. Thus, the CTFs appropriate for the different images in a tilt series are unlikely to be the same and may even vary significantly across a *single* image. Third, in the direction of the beam, the thickness of the specimen is bound to change with the tilt angle, which will alter image contrast, if nothing else. Fourth, for mechanical reasons, there will always be a maximum tilt angle beyond which imaging is impossible. Finally, tomographic reconstructions require that the same object be imaged many times, and although it is by no means necessary that every image taken of it have a high ratio of signal to noise, the total electron dose it receives in the process of obtaining the images required may be enough to make radiation damage a concern.

Because the central sections produced when a set of tilt images is Fourier transformed differ in their angular orientation around a single axis, it makes sense to compute the Fourier transform of each image on a two-dimensional lattice, one axis of which corresponds to the common line. That direction in reciprocal space will be called **Z**, and the corresponding axis in real space, **z**. The sampling interval along **Z** must be at least d_z^{-1}, where d_z is the maximum linear dimension of the object in the **z** direction. In the direction in each plane perpendicular to **Z**, which we will call R, the spacing should be not less than $r_{\max}^{-1}$, where $r_{\max}$ is the maximum width of the object perpendicular to the tilt axis observed in *any* of the images available. Sampling intervals should be the same in the transforms of all images.

If the axes of the reciprocal space lattices of all the images in a tilt series are defined this way, then the location of any point in any of one of its central sections in the three-dimensional reciprocal space of the object can be specified conveniently using cylindrical polar coordinates. Its coordinates will be R_i, Θ_i, and Z_i, where Θ_i is the tilt angle of the image from which that particular value of the transform came.

13.8 TILT DATA CAN BE REDUCED ONE PLANE AT A TIME

If the density distribution function of *any* object is described using cylindrical polar coordinates, in any plane perpendicular to **z**, $\rho(r, \theta, z)$ at constant r will be a periodic function of θ because $\rho(r, \theta, z)$ must equal $\rho(r, \theta + 2\pi, z)$. Thus, it will *always* be possible to represent the density distribution of the object in each plane of constant z as a Fourier series in θ that has the following form:

$$\rho\,(r,\,\theta,\,z) = \sum_{-\infty}^{+\infty} g_n\,(r,\,z)\,\exp\,(in\theta).$$

For exactly the same reason, it will also always be possible to write $F(R,\,\Theta,\,Z)$ the same way:

$$F\,(R,\,\Theta,\,Z) = \sum_{-\infty}^{+\infty} G_n\,(R,\,Z)\,\exp[in\,(\Theta - \pi\,/2].\qquad \textbf{(13.5)}$$

[The $(n\pi/2)$ factor in equation (13.5) is included to simplify subsequent arithmetic; it has no deep physical significance.] As appendix 13.3 demonstrates, the g_n and the G_n values are related to each other by an operation called a **Fourier-Bessel transformation**:

$$g_n\,(r,\,z) = \int_{-S_{max}}^{+S_{max}} \left[\int_0^{S_{max}} G_n\,(R,\,Z)\,J_n\,(2\pi rR)\,2\pi R dR\right] \exp\,(2\pi izZ)\,dZ,$$

where in practice, S_{max} is the resolution limit of the images available of the object in question.

When tilt data are processed in the manner just described, Fourier coefficients will be available on planes of constant Z that are separated by an interval in reciprocal space that is not greater than d_z^{-1}. Because that spacing meets the Nyquist criterion, the integral over Z in the equation for $g_n(r,\,z)$ can be replaced by a sum with no sacrifice in the quality of the rendering of $\rho(\mathbf{r})$ that ultimately emerges:

$$g_n\,(r,\,z) = \sum \left[\int_0^{S_{max}} G_n\,(R,\,Z)\,J_n\,(2\pi rR)\,2\pi R dR\right] \exp\left(2\pi izZ_j\right).\qquad \textbf{(13.6)}$$

The sum in equation (13.6) is over all planes of constant Z that contain useful data: all $|Z| \leq |\mathbf{S}|_{max}$. It follows that $\rho(r,\,\theta,\,z)$ can be recovered once the information provided by the transforms of the images available has been used to estimate the relevant G_n values, which is not hard to do because:

$$G_n\left(R,\,Z_j\right) = (1/2\pi)\int_0^{2\pi} F\,(R,\,\Theta,\,Z)\,\exp\left[-in\,(\Theta - \pi/2)\right] d\Theta.$$

One problem remains. $F(R,\,\Theta,\,Z_i)$ is known experimentally only along a finite number of lines of constant Θ, where the central sections available intersect the Z_i plane. By design, the sampling in R along all those lines is guaranteed to be adequate in the Nyquist sense, but the same might not be true for the sampling in Θ.

It is clear from sections 13.3 to 13.6 how problems of this sort should be dealt with, and the product obtained when they are applied here will be a set of least squares best estimates of the $G_n(R, Z_i)$, as well as their likely errors. When error is taken into account, within each plane of constant Z, it will be true that:

$$F_{obs}(R_i, \Theta_i) = \sum B(\Theta_i, n)\, G_n(R_i) + \delta(R_i, \Theta_i),$$

where $B(\Theta_i, n) = \exp(-in(\Theta_i - \pi/2)$. The $F_{obs}(R_i, \Theta_i)$, the $G(R_i, n)$, and the $\delta(R_i, \Theta_i)$ expressions can all be written as a column vectors: $\mathbf{F}_{obs}$, $\mathbf{G}$, and $\boldsymbol{\delta}$, and the set of $B(\Theta_j, n)$ constitutes a matrix, $\mathbf{B}$. Thus:

$$\mathbf{F}_{obs} = \mathbf{BG} + \boldsymbol{\delta},$$

and, once again, the task is to estimate $\mathbf{G}$ given $\mathbf{F}_{obs}$ and $\mathbf{B}$. As we already know, the least squares best estimate of $\mathbf{G}$ will satisfy the following equation:

$$\mathbf{G}_{lsq} = \left(\mathbf{B}^T\mathbf{B}\right)^{-1}\mathbf{B}^T\mathbf{F}_{obs},$$

and:

$$\sigma^2\left(G_{lsq}(R_i, n)\right) = \left(B^T B\right)^{-1}(n, n)\,\sigma^2.$$

In each plane of constant Z in reciprocal space, the procedures these equations require must be carried out for all values of R_i starting at $R_i = 0$ and continuing out to the largest R_i that is less than $(|\mathbf{S}|^2_{max} - Z_i^2)^{1/2}$.

There are substantial practical advantages to be gained by using cylindrical polar coordinates instead of Cartesian coordinates to analyze tomographic data. The Cartesian data reduction process discussed in sections 13.3 to 13.6 could easily require the inversion of a matrix that has dimensions in excess of $1,000 \times 1,000$. That huge, unwieldy inversion might well be replaced by ~ 50 much smaller (50×50) matrix inversions if the problem were reformulated using cylindrical polar coordinates. However, the (large) gain in computational speed and convenience that results would come at a price. Cartesian data reduction makes better use of the available data than any of the approaches to data reduction that are more tractable computationally.

13.9 ONLY A FINITE NUMBER OF G_N VALUES THAT MUST BE TAKEN INTO ACCOUNT IN A TOMOGRAPHIC RECONSTRUCTION

Bessel order and resolution interact with each other in a useful way because of the way Bessel functions behave when their arguments are small. Bessel functions resemble sine and cosine functions to some degree because their amplitudes tend to oscillate between positive and negative values in a quasi-regular way as their arguments increase. Like $\cos(x)$, the Bessel function of order 0 (i.e., $J_o(x)$), has a value of 1.0 at $x = 0$. However, all other Bessel functions have a value of 0 at $x = 0$, like $\sin(x)$. Moreover, the larger the (absolute) order of a Bessel function (i.e., the larger $|n|$), the larger its argument at the point where its (absolute) value reaches its first maximum, and oscillatory behavior begins. That first maximum occurs at $x \approx (|n| + 2)$, and for x much smaller than that, $|J_n(x)| \sim 0.0$.

From appendix 13.3, we know that:

$$F(R, \Theta, Z) = 2\pi \sum_n \exp\left[in(\Theta - \pi/2)\right] \int \exp(-2\pi izZ)\, dz \times \int g_n(r, z) J_n(2\pi rR)\, r\, dr.$$

Furthermore, we know that $F(R, \Theta, Z)$ is effectively 0 for $R > |\mathbf{S}|_{max}$ because $|\mathbf{S}|_{max}$ is the resolution at which the data cease being significant. Thus, if the maximum radius of the object of concern is r_{max}, the largest value of the argument of J_n that will ever need to be taken into account in the expression above is $(2\pi r_{max}|\mathbf{S}|_{max})$. It follows that for all $|n|$ much greater than $\sim [(2\pi r_{max}|\mathbf{S}|_{max}) - 2]$, the integral over r in the preceding equation will be ~ 0 because its $J_n(2\pi rR)$ component is ~ 0 for all values of $(2\pi rR) \leq (2\pi r_{max}|\mathbf{S}|_{max})$. Thus, instead of running from $n = -\infty$ to $n = \infty$, the sum over Bessel orders needed to represent $F(R, \Theta, Z)$ accurately out to the resolution specified need run only over the range $\pm[(2\pi r_{max}|\mathbf{S}|_{max}) - 2]$.

What are the practical implications of this fact? Suppose we wished to carry out a tomographic reconstruction of the image of an object that has a maximum radius of 100 Å in the direction perpendicular to the tilt axis, and that the micrographs available have a resolution of 0.1 Å^{-1}. The maximum value of the argument of the Bessel functions that would have to be considered would be ~62.8, which implies that all Bessel orders between, roughly, −61 and +61 would have to be taken into account. Thus, in the $Z = 0$ plane, at least, the sum over n would have to include (not less than) 123 terms ($= 61 \times 2 + 1$, for the zero-order term). The number of terms required for other planes would be less because the larger the value of $|Z|$ of some plane, the smaller the number of Bessel orders that have to be dealt with when reducing the data in that plane because of the smaller value of R at which $(Z^2 + R^2)^{1/2} = |\mathbf{S}|_{max}$.

Suppose it were possible to collect tilt data at equal intervals from $\Theta = 0$ to $\Theta = \pi$. How many images would be required to determine the structure just described to the resolution that the images available permit? The range of angular frequencies for which G_n values were required would be 0 to $\pm 2\pi/61$. Equation (2.2) indicates that each annulus in the $Z = 0$ plane of reciprocal space would have to be sampled not less than 123 times to achieve this goal, which is consistent with the Nyquist theorem. However, every central section in a tilt series intersects every annulus in every plane of constant Z twice. Hence, the number of images required would be 62. More generally, for an object of arbitrary structure, N_{views}, the minimum number of views required to reconstruct its structure to a resolution of $|\mathbf{S}|_{max}$ is:

$$N_{views} \approx 2\pi r_{max}|S|_{max}.$$

If the views available of some object are not evenly spaced, the number of views needed to attain a given resolution will exceed this minimum, and if the range of angles spanned by the views available is insufficient, it may not be possible to obtain a reconstruction of that resolution no matter how many views are available. (Similar expressions for the relationship between particle size, resolution, and number of images required emerge no matter how the data reduction problem is formulated.)

13.10 SYMMETRY REDUCES THE NUMBER OF IMAGES REQUIRED TO RECONSTRUCT THE STRUCTURE OF AN OBJECT TO ANY GIVEN RESOLUTION

Data collection is easier for objects that have internal symmetry than it is for objects that do not. For example, if an object has a four-fold axis of rotation, then a single image of that object taken perpendicular to its four-fold axis will be identical to three other images that could have been obtained of that same object. If such an object were to have a radius of 100 Å, and its four-fold axis was aligned with the rotation axis of the tilt stage of the microscope used to image it, then the number of images needed to solve its structure to a resolution of $0.1\ \text{Å}^{-1}$ would drop from $\sim$61 to $\sim$16. Thus, the higher the symmetry of an object, the fewer images it takes to reconstruct its structure to a given resolution. It is natural to wonder, therefore, if an object could have a symmetry so high that its structure could be reconstructed from the data obtained from a single micrograph. The answer to this question is a qualified "yes". The structures of objects having helical symmetry can often be reconstructed to limited resolution using the information provided by a single image.

13.11 SYMMETRY DETERMINES THE ORDERS OF THE BESSEL FUNCTIONS THAT CONTRIBUTE TO EACH LAYER PLANE IN A HELICAL DIFFRACTION PATTERN

Before you can understand why a single image might suffice to reconstruct the image of a helical object, you must understand how the diffraction patterns of helices differ from those of nonhelical objects. A helix is a spiral-like structure that has a long, straight central axis and can be thought of as a one-dimensional crystal because its structure repeats in the axial direction. The simplest conceivable helix would be a thin helical wire wound in a regular spiral that has a rise per turn along the helix axis—a pitch—equal to p, and a (constant) radius of r. In cylindrical polar coordinates, the trajectory of that helical wire would obey the following equations:

$$x = r\cos\left(2\pi z/p\right); y = r\sin\left(2\pi z/p\right); z = z.$$

If the length of the helical wire along z is much larger than its diameter, its transform will differ from 0 only on, or very close to planes perpendicular to the axis in reciprocal space parallel to $\mathbf{z}$, which we will call $\mathbf{Z}$, when $|\mathbf{Z}|$ is an integer multiple of $(1/p)$. On those planes, which are called **layer planes**, the Fourier transform of the helix will be (see appendix 13.3):

$$F\left(R,\Theta,n/p\right)=\int \exp\left[-2\pi i\left(rR\cos\left[\left(2\pi z/p\right)-\Theta\right]+zn/p\right)\right]dz.$$

The result obtained when this integral is evaluated is remarkably simple:

$$F\left(R,\Theta,Z\right)=\exp\left[in\left(\Theta-\pi/2\right)\right]J_n\left(2\pi rR\right)\delta\left(Z-n/p\right). \qquad \textbf{(13.7)}$$

(NB: n can be both positive and negative.)

Thus, the distribution of diffracted intensity will be exactly the same in all the central sections of the diffraction pattern of a helical wire that include the Z axis, which is referred to in "helix-speak" as the **meridian**. On the line perpendicular to Z that includes $(0, 0, 0)$ (i.e., along the **equator** of the pattern), there will be a peak at $(0, 0, 0)$ that is flanked by secondary maxima, and on all other **layer lines**—lines for which $Z = n/p$—there will be two strong peaks symmetrically placed on either side of the meridian, and beyond them subsidiary maxima extending to higher resolution. The larger the value of n, the greater the distance between the meridian and either of the two dominant peaks on a layer line because the argument of J_n at which its first maximum occurs increases with n, as we already know. In short, the diffraction pattern of a helical wire looks like the letter X (figure 13.2).

The helices biochemists care about are made of atoms, and the simplest atomic helix we could imagine would consist of point atoms evenly distributed along the trajectory traced out by a helical wire. That distribution of atoms can be described as the product of a helical wire function and a set of equally spaced parallel planes, the normals of which are parallel to $\mathbf{z}$ (and $\mathbf{Z}$). The spacing between planes, d, determines the difference in the z coordinate of adjacent atoms in the helix. It follows that the Fourier transform of an atomic helix will be the convolution of the transform described by equation (13.7) with the transform of the set of planes. The transform of such a set of planes is a row of delta functions that fall on $\mathbf{Z}$, each of which is separated from its nearest neighbors by a distance of $(1/d)$. When the convolution called for is carried out, each meridional delta function will become the origin of a (continuous) helical diffraction pattern (figure 13.3).

It is clear that the transform of an infinitely long atomic helix can be nonzero only on planes of constant Z where $Z = n/p + m/d$. In this equation, n is the order of a Bessel function that contributes intensity to the layer plane, the $\mathbf{Z}$ coordinate of which is Z, and m is any integer whatever. (m/d is the location in reciprocal space of the origin of the X-pattern to which the layer plane of order n "belongs" that contributes intensity to the layer plane at Z.)

Now suppose that the atomized helix repeats exactly after u turns of the helix, and v atoms, where u and v are both integers. (A helix of this sort is said to be an integer helix, which by no means all biological helices actually are.) Then $up = vd = P$, where P is the overall repeat distance of the helix. Thus, we can write:

$$Z=n/p+mv/pu,$$

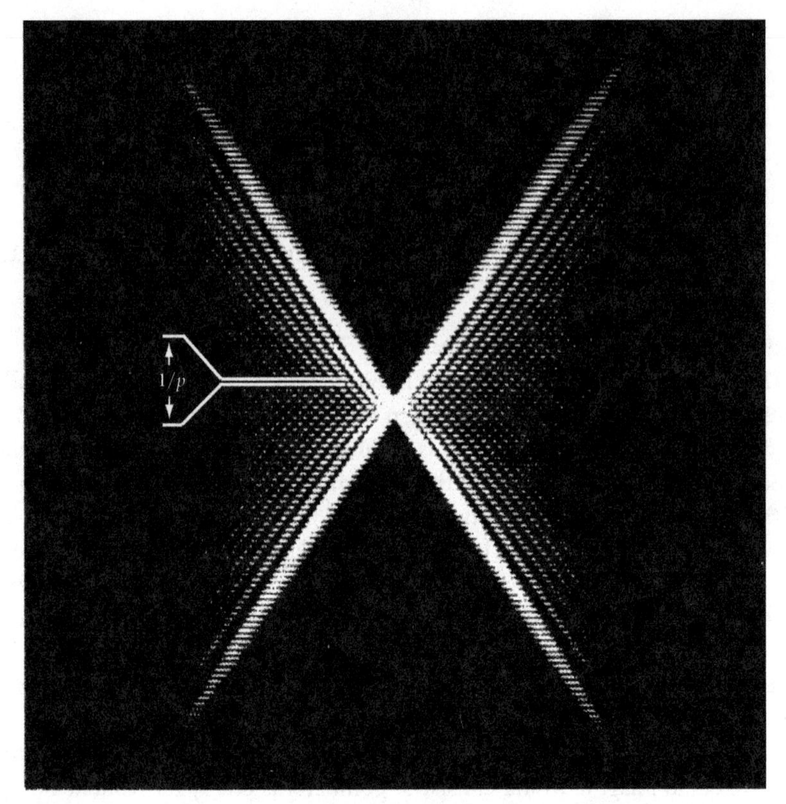

Figure 13.2 A central section of the diffraction pattern of a continuous helical wire. (Reproduced from Harburn, G., Taylor, C. A., and Welberry, T. R. (1975), *Atlas of Optical Transforms*, Ithaca, New York: Cornell University Press, with permission of the copyright holders, G. Harburn and T. R. Welberry.)

which implies that:

$$Zpu = nu + mv.$$

However, $pu = P$, the repeat distance of the atomic helix, and the Z-coordinate of any plane in the diffraction pattern where intensity is seen, must be an integer multiple of $1/P$, which we will call the **layer line number**, l. It follows that the Bessel orders that contribute intensity to the lth layer plane must satisfy the following equation:

$$n = (l - mv)/u. \qquad \textbf{(13.8)}$$

Equation (13.8) is an integer equation, which means that its only solutions are those for which n is an integer, and m can be any integer whatever. Equations like equation (13.8) are called **selection rules**.

Figure 13.3 Diffraction pattern of a helix composed of point atoms. The pitch of this helix is the same as that in figure 13.2, and the number of atoms per turn is about 50. (Reproduced from Harburn, G., Taylor, C. A. and Welberry, T. R. (1975), *Atlas of Optical Transforms*, Ithaca, New York: Cornell University Press, with permission of the copyright holders, G. Harburn and T. R. Welberry.)

Comment: What happens if p/d is not a ratio of integers, that is, if the helix is a noninteger? The diffraction pattern of such a helix will resemble that of an integer helix for which $v/u \approx p/d$. Because the diffraction pattern of that related helix is that of an integer helix, there will be several different Bessel orders contributing to each of its layer planes. In the diffraction pattern of the corresponding noninteger helix, for every layer plane in the integer helix, there will be several layer planes having Z-coordinates close to that of the corresponding layer plane of the integer helix, each separated from its neighbors by a spacing much less than $1/P$. Each such "subplane" will have a single Bessel order contributing to it, and each such set of subplanes will include all the Bessel orders that contribute to the diffraction pattern of the integer helix at the corresponding Z-spacing.

It follows that the Fourier transforms of helices differ from those of objects that have no symmetry in two ways. First, transforms of helices are nonzero only on layer planes because they are one-dimensional crystals, which asymmetric objects are not. Second, when the transform of an asymmetric object is expressed in cylindrical polar coordinates, we find that, in theory, *all* Bessel orders can contribute to every plane of constant Z (see equation (13.5)). However, for helices, only *some* Bessel orders will contribute to any layer plane, namely the ones compatible with equation (13.8).

The best way to understand the implications of equation (13.8) for helical reconstructions is to feed some numbers into it. Suppose some helix repeats after two turns, and that there are seven atoms per repeat. Its selection rule will be:

$$n = (l - 7m)/2,$$

and the only Bessel orders contributing intensity to the equatorial plane of its transform will be: 0, ±7, ±14, ... On the $l = 1$ plane, the orders contributing will be: . . . − 10, −3, +4, +11, . . .

Comment: If the repeating elements in a helix of are entire molecules rather than point atoms, then equation (13.7) must be replaced by:

$$\mathrm{F}\,(R, \Theta, n/d) = \exp\left[in\,(\Theta - \pi/2)\right] \sum f_j J_n\left(2\pi r_j R\right) \exp\left(in\theta_j\right)$$
$$\exp\left[-2\pi i z_j\,(n/d)\right],$$

where r_i, θ_i, and z_i are the coordinates of the ith atom relative to the atom in the molecule that is taken as the reference atom. The reference atom is the atom that occupies the same location as the point atoms in the corresponding, idealized atomic helix.

13.12 AT MODEST RESOLUTIONS, A SINGLE BESSEL ORDER MAY ACCOUNT FOR THE INTENSITY OBSERVED ON EACH PLANE IN A HELICAL DIFFRACTION PATTERN

As we already know, all Bessel function except J_o are 0 when their arguments are 0, and the larger their absolute order, $|n|$, the larger their arguments must be before their values deviate significantly from 0. The significance of these facts becomes clear when they are applied to the helix selection rule of that we just determined. The orders contributing to its equatorial plane are 0, ±7, ±14, . . ., and that means that for all R such that $(2\pi r_{\mathrm{max}} R)$ is less than the value at which J_7 starts to have appreciable value, there will be only one Bessel order contributing, J_0. If the selection rule of some helix is such that there is an appreciable region in the inner part of all the layer planes in its transform where only a single Bessel order contributes, it will be possible to reconstruct its structure in three dimensions out to the resolution where Bessel order superposition begins on the basis of a single image taken perpendicular to its helix axis.

Even when the lowest order contributions to the layer planes of the transform of some helix overlap, single image reconstruction may still be feasible. For example, the two lowest orders contributing to the $l = 1$ layer plane of our test helix are

-3 and $+4$. The maxima of these two orders will not be sufficiently separated in reciprocal space for their contributions to be to be distinguished on that basis alone. However, the transform of a single image of that helix taken perpendicular to its axis provides values for both $F(R, \Theta, Z_1)$ and $F(R, (\Theta + \pi), Z_1)$. Because -3 is an odd number and 4 is an even number, the G_4 contribution to F will be the same on both sides of the meridian, but the G_3 contribution will change sign, and that should be enough to make it possible to separate the contributions they make to that layer plane. More complicated overlap issues can be resolved, or the resolutions of helical reconstructions can be extended using data obtained from multiple images of helices that differ in tilt around the helix axis. Thus, helical reconstruction really should be thought of as a special case of tomographic reconstruction.

Comment: The structures of helical biological objects are seldom reconstructed from EM images today using the logic just described because most of them do not display perfect helical symmetry. Instead, they vary irregularly in structure along their lengths to at least some degree and are flexible enough so that the images of straight, multirepeat segments that classical helical reconstruction methods require are hard to obtain. Instead, their structures are reconstructed using single particle methods of the sort we are about to describe. The reason helical reconstruction is discussed in detail here is that the first three-dimensional reconstructions ever done were helical reconstructions, and the thinking that motivated that enterprise still strongly influences the field.

13.13 THE IMAGES USED FOR SINGLE PARTICLE RECONSTRUCTIONS OFTEN HAVE VERY LOW SIGNAL-TO-NOISE RATIOS

As we already know, tomographic reconstructions are usually done using sets of images taken of single objects in many different orientations. By contrast, single particle reconstructions are done by merging the information obtained from the images of many *different* objects, and that merger will lead to useful results *only* if the three-dimensional structures of all the objects imaged are the same. (How similar two structures must be in order to be considered "the same" for these purposes is an important question that is still under investigation.) If this condition is met, it will be possible to explain *all* the images obtained of some particle as projections of the same three-dimensional density distribution function. Thus, the objective in a single particle reconstruction experiment is to find the three-dimensional density distribution that best explains all the images available.

Single particle reconstructions are commonly done using images that have very low signal-to-noise ratios. In the 1960s and 1970s, when the field was in its infancy, the images analyzed were usually those of negatively stained specimens, which had high signal-to-noise ratios because of the contrast generated by staining and because the electron doses used to produce them were large. Because the resolutions of these images were often worse than 20 Å, all that researchers could hope to extract from them was information about particle shape. In the 1970s, it was discovered that far higher resolution structures can be obtained from low-dose EM images of particles embedded in thin layers of vitreous ice, in other words, by doing cryo-electron microscopy

The contrast in most cryo-EM images is intrinsically low because per unit volume, the average cross sections of water and protein for 100-keV electrons are not all that different (see chapter 12, question 1). Hence, per unit electron dose, the signal-to-noise ratios of these images are much lower than those of the images that would have been obtained from otherwise similar specimens that were negatively stained. The signal-to-noise limitations of these images are often further aggravated by the tendency of microscopists to limit radiation damage by minimizing the electron doses used to produce them. The justification for working this way is that low-dose images of cryo-specimens, noisy as they are, can convey information not only about molecular shape, but also about internal structure. The three-dimensional density distributions produced by the analysis of such images can have atomic resolution.

The noise in low-dose, low-contrast images comes from many sources. For example, there is statistical noise due to the small numbers of electrons used to produce them, as well as statistical noise added by the devices used to record them. In addition, each image includes contributions due to the structure of the glass in which objects are embedded, and the supporting film, if any. The only way the impact these irrelevant features have on the three-dimensional structures derived from such images can be controlled is by averaging.

13.14 IMAGE ORIENTATIONS CAN BE DETERMINED BY THE RANDOM–CONICAL TILT METHOD

Both conceptually and practically, the most difficult reconstructions to do are single particle reconstructions because at the outset of most such reconstructions, little or nothing is known about the orientations of the particles that have been imaged relative to the beam. Thus, determination of image orientations is *the* major issue in single particle reconstruction, and it is the aspect of the process that we will concentrate on because once image orientations have been determined, the rest is easy.

The expectation is that the particles in any cryo-specimen are randomly oriented. If that is so, the larger the number of single particle images processed, the more dense the population of experimental Fourier coefficients available will be in the reciprocal space of the structure of the object. However, it is not uncommon for macromolecules to orient in EM specimens. For example, long, cylindrical molecules are much more likely to be imaged side-on than end-on, and preferred orientations may also be observed for isometric particles because of their noncovalent interactions with supporting films, or even air–liquid interfaces. If a specimen shows this kind of bias, there will be regions in reciprocal space where no information is available no matter how many images get processed. The random–conical tilt technique for image reconstruction, which we are about to describe, was developed to solve this problem, but it is also used when orientational bias is not an issue.

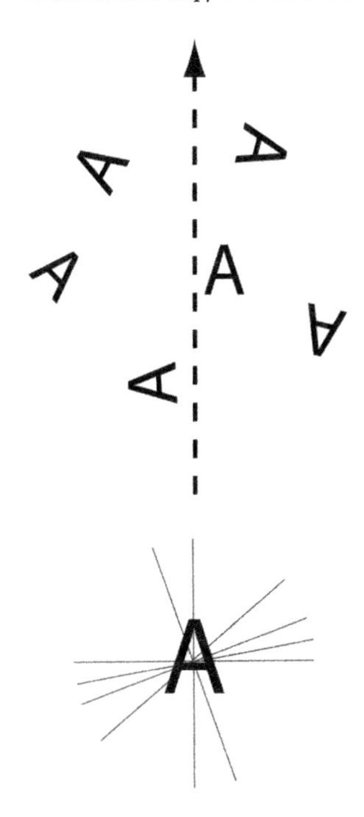

Figure 13.4 Schematic diagram that explains the rationale for random–conical tilt reconstructions. The top portion of the figure displays the images of six identical molecules, all of which bind to the grid surface in an orientation such that, in projection, they all look like the letter A. If the grid is tilted around an axis parallel to the dotted arrow, the direction of the tilt axis with respect to the molecules that gave rise to the six images in the top part of the figure will differ. Their orientations relative to the images of those molecules are shown, superimposed in the lower part of the figure.

Suppose that in all the EM samples prepared of some macromolecule, the molecule always orients the same way relative to the plane of the grid. If this is so, when micrographs are taken of specimens of this kind with the grid perpendicular to the beam, as is usually the case, then the images of individual particles in those micrographs will convey exactly the same information about the structure of that macromolecule because they will differ only trivially, that is, in their rotational orientations around an axes parallel to the optical axis and in the locations of their centers of gravity. If, on the other hand, a micrograph is taken of such a specimen with the stage tilted, the images of individual molecules will no longer look the same because the orientation of the tilt axis relative to their structures will differ. Thus, by tilting specimens, it is possible to obtain different projection images of the molecule, even though it always adheres to the surface of the grid in the same orientation.

These facts suggest a strategy for reconstructing the structures of objects that orient. Take two micrographs of the same field of particles on the same grid, one with the plane of the grid normal to the beam, and the other with the grid tilted by a large, but known amount, say 45°. By comparing the images of the particles in the first micrograph, we can determine their relative orientations about an axis normal to the plane of the grid. We can use that rotational information and the tilt angle to work out the relative orientations of the central sections that correspond to the molecular images obtained when the specimen was tilted (see figure 13.4). This approach to populating reciprocal space is referred to as the **random–-conical tilt method**.

It is not easy to determine tilt axis orientations when the images to be analyzed look like noise smudges, as low-dose micrographs of single macromolecules commonly do, but it can be done. The statistical techniques that have been developed to solve this problem are quite interesting, but it would take us too far afield to discuss them here.

The rendering of the three-dimensional reciprocal space of an object that emerges from the analysis of random–conical tilt data sets often leaves a lot to be desired. In directions parallel to the plane of the grid, the density of central section data will be high, but normal to the grid, the density will be poorer, especially along the vector parallel to the beam direction that passes through the origin of reciprocal space. In fact, on either side of the central section that is parallel to the plane of the (untilted) grid, there will be two cone-shaped regions where no data is available at all. Their angular widths will be the difference between 90° and the tilt angle, times 2. The defects this systematic absence of information creates in constructions is referred to by the *cognescenti* as the **missing cone problem**.

If the molecules of interest are randomly oriented in the specimens that are prepared of them, there is a way random–conical tilt reconstructions can be done with them that will produce images of them that do not suffer from missing cone artifacts. Two images are taken of the specimen, one at a tilt angle of −45° and the other at a tilt angle of +45°. One of the two micrographs is (arbitrarily) chosen as the reference image. Using well-established statistical methods, the images of single molecules contained in that micrograph are grouped according to their similarity of appearance, and the relative rotational orientations about the direction of the electron beam worked out for every image within each group. Once this has been done, the images of the members of each group of particles contained in the second micrograph can be used as input for a random–conical tilt reconstruction of the usual sort. However, because the difference in tilt angle between the two micrographs is 90°, instead of being restricted to some value (often) well below 90° by the mechanics of the stage, there will be no missing cones in three-dimensional Fourier data sets that emerge. In addition, the three-dimensional reconstructions obtained from each group of images in the reference micrograph should be the same, save for differences in rotational orientation.

Usefully accurate three-dimensional reconstructions of macromolecular structures have been produced using the random–conical tilt method, even when the missing cone problem either was not or could not be avoided. NB: If a macromolecule has two (or more) preferred orientations in EM specimens, it may be possible to obtain an overall reconstruction of its structure that is largely free of missing cone artifacts by merging the reconstructions obtained for the molecule in each of its different orientations, even though both reconstructions are suffer from missing cone artifacts.

13.15 COMMON LINES CAN BE USED TO DETERMINE IMAGE ORIENTATIONS

A surprisingly large amount can be learned about the relative orientations of the particles imaged in micrographs of fields of randomly oriented particles by comparing images. Because, out to some limiting resolution, the transform of any image of a single particle is a central section of the particle's three-dimensional Fourier transform, the central sections that correspond to the images of two identical particles must intersect along a single, common line in reciprocal space that includes the origin. Along that common line, the coefficients of the Fourier transforms of two images should be identical, save for noise. The common line of two central sections corresponds to the axis around which either one of the two particles would have had to be rotated in order to make its image identical to the image of the other particle.

What cannot be extracted from such pairwise image comparisons are estimates of the magnitude of the rotation that would have been required to make two images look alike.

If a third view of an identical particle is available, not only can all the common lines be determined that relate pairs of images, but also the rotation angles that would have been required to make the images of the three particles look alike. The central section that corresponds to the third image will intersect the transforms of the other two images along common lines, both of which are different from the common line that relates the transforms of the first two images. Once all three common lines have been identified, the problem is solved because there are only two ways three planes that intersect at the origin of reciprocal space can be arranged so that they will intersect along those lines. The two arrangements are mirror images of each other.

In theory, the relative orientations of the central sections that correspond to hundreds or even thousands of different images of some molecule that a micrograph of a single EM specimen might contain could be worked out using the common line method. Once this was done, it would be possible to reduce all the data those central sections provide and reconstruct the structure of the particle in the usual way. The absolute hand of the reconstruction that emerged would be arbitrary, but left-right ambiguities can be resolved by doing tilt experiments. This approach to single particle reconstruction is called **angular reconstruction**.

Angular reconstruction is an elegant solution to the single particle reconstruction problem, but it is hard to implement. Not surprisingly, the principal issue is signal to noise. How reliably can the common line that relates the transforms of two noisy images be identified? Furthermore, if common lines are determined for a set of images in a serial manner starting with three noisy images, errors made in the beginning of the process may concatenate disastrously. However, common line algorithms are being developed today that may make it possible to determine the relative orientations of entire sets of images in a single, self-consistent, statistically sound computation. If this proves practical, angular reconstruction could become the method of choice for single particle reconstruction.

13.16 ORIENTATION REFINEMENT IS AN IMPORTANT PART OF MOST SINGLE PARTICLE RECONSTRUCTIONS

Most high-resolution single particle reconstructions are done today by iterative procedures that begin at low resolution and work their way gradually to higher resolution. Once such a process gets started, each additional cycle uses the reconstruction generated by the prior cycle as a reference object to obtain more refined estimates of the relative orientations of images. For that reason, reconstructions of this sort are described as **reference-based**.

The sine qua non of any reference-based reconstruction is an initial, low-resolution structure for the object under investigation. For signal-to-noise reasons, it is often useful to base that first-approximation structure on images that have been averaged in two dimensions in real space, rather than on unaveraged images of single particles. Powerful algorithms exist for classifying (two-dimensional) images on the basis of similarity in appearance, and once a set of images has been divided into classes this way, it is easy to generate class-average images that have much higher levels of signal to noise than any single image. Random–conical tilt methods, or common line approaches, can be used to reconstruct structures starting from sets of averaged images of this sort, and both processes are likely to perform well because signal-to-noise ratios of the images used are high. That said, the drawback to this modus operandi is obvious. Some of the noise eliminated by averaging will represent real differences between images caused by small differences in particle orientations within classes. Thus, the averaging process is bound to limit the resolution of reconstructions, and if there are no preferred orientations in the specimens, the images of which are being analyzed, the limitation can be severe.

Once an initial reconstruction has been obtained, refinement can begin. For example, if an initial reconstruction was based on 50 averaged images, in the next iteration, we might use that reconstruction to compute the (approximate) appearance of the particle in 1,000 different orientations that evenly sample orientation space. The orientation associated with each single particle image could then be estimated by determining which picture in that gallery of 1,000 images it most closely resembles. Once the orientations associated with each image were (re)determined this way, a more accurate, higher resolution reconstruction of the particle of interest could be computed, and the cycle could be repeated until the quality of reconstructions stopped improving.

Iterative procedures for image reconstruction suffer from a liability that is conceptually similar to a problem crystallographers face when they are refining structures, namely the tendency of errors in initial models to persist (see section 7.10). The low-resolution reconstructions used as the starting points of iterative processes are bound to include spurious features due to the noise in individual images or the misalignment of images. Because the galleries of possible views created from these initial structures will include those features, they may appear in the models that emerge at the end of the iteration processes.

13.17 IT IS EASY TO VALIDATE RECONSTRUCTIONS AND TO ESTIMATE THEIR RESOLUTIONS

Two questions are certain to arise about any reconstruction. Is it right? What is its resolution? The "is it right" question is harder to answer for EM reconstructions than it is for crystallographic structures because the resolutions of EM reconstructions are seldom so high that the chemical structure of the molecules being imaged can be recognized unambiguously. Nevertheless, both questions can be addressed.

In the first place, no reconstruction can be valid unless it is consistent with the images on which it is based, and it is easy to find out whether that is true or not because all the images on which a reconstruction is based must be explicable as projections of that reconstruction, due allowance being made for noise. However, it is important to realize that all that is verified by demonstrating that this is true is that the reconstruction algorithm was executed properly.

The most compelling of the validation methods commonly used demands that the operator divide the single particle images available randomly into two sets of equal size, and then use both sets of images to do independent three-dimensional reconstructions. Not only should the two reconstructions obtained this way be identical, within error, but it should also be possible to determine the resolutions of both of them by comparing them appropriately.

It is obvious that the agreement between the two independent reconstructions of the same particle ought to depend on resolution because their long-length scale features are more likely to have been well determined by the data than their short-length scale features. Thus, if the two structures that are to be compared are superimposed (i.e., translated and rotated in real space so as to maximize the correlation between their density distribution functions) and their Fourier transforms computed, the agreement between their transforms ought to be high at small values of $|\mathbf{S}|$ and decrease as $|\mathbf{S}|$ increases.

The agreement between pairs of structures is commonly assessed by computing a **Fourier shell correlation (FSC) curve.** The transforms of the two structures, F_1 and F_2, are divided into shells of equal radius $|\mathbf{S}|$, and of equal thickness, ΔS, and within each shell, one computes:

$$\mathrm{FSC}\,(S) = \mathrm{Re}\left[\sum \mathrm{F}_1\,(\mathbf{S}_i)\,\mathrm{F}_2\,(\mathbf{S}_i)\right] / \left[\sum \mathrm{F}_1\,(\mathbf{S}_i)^2\,\mathrm{F}_2\,(\mathbf{S}_i)^2\right]^{1/2},$$

where the sums run over all values of $\mathbf{S}$ that lie inside that shell of radius S and thickness ΔS. If $F_1 = F_2$ for all $\mathbf{S}$ inside some shell, then FSC($|\mathbf{S}|$) will be 1.0. Typically, at low $|\mathbf{S}|$, FSC is close to 1.0, and it invariably declines as resolution increases. Many microscopists take the reciprocal of the value of $|\mathbf{S}|$ at which the FSC curve of a reconstruction passes through 0.5 as its resolution. Figure 13.5 is a schematic diagram of an FSC curve that is prettier than many. By crystallographic standards, the 0.5 criterion for resolution is rather stringent; on paper, at least, the crystallographic resolution of a reconstruction corresponds to the value of $|\mathbf{S}|$ where the FSC falls to 0.2.

It may surprise you that anyone would feel it necessary to estimate the resolution of an EM reconstruction by the laborious method just described. Why is it not enough to estimate resolution using the normal matrix approach discussed earlier? The reason is that the resolution estimated for a reconstruction from its normal matrix is the *best* resolution the data could possibly provide, rather than the resolution it really does provide. One reason real resolutions are worse than theoretical resolutions is that the relative orientations of the central sections that correspond to noisy images can never be determined with perfect accuracy, and inaccuracies in image orientations reduce the reliability of short spatial wavelength features in reconstructions, that is, their resolutions. The great virtue of the FSC approach to estimating resolution is that it takes account of *all* the sins committed during the reconstruction process; nothing is left out.

Doubtless, you have noticed that nothing has been said in this chapter about CTFs so far. The reason is that their inclusion in the preceding discussions above would have made an already challenging topic even more so. The fact is, however,

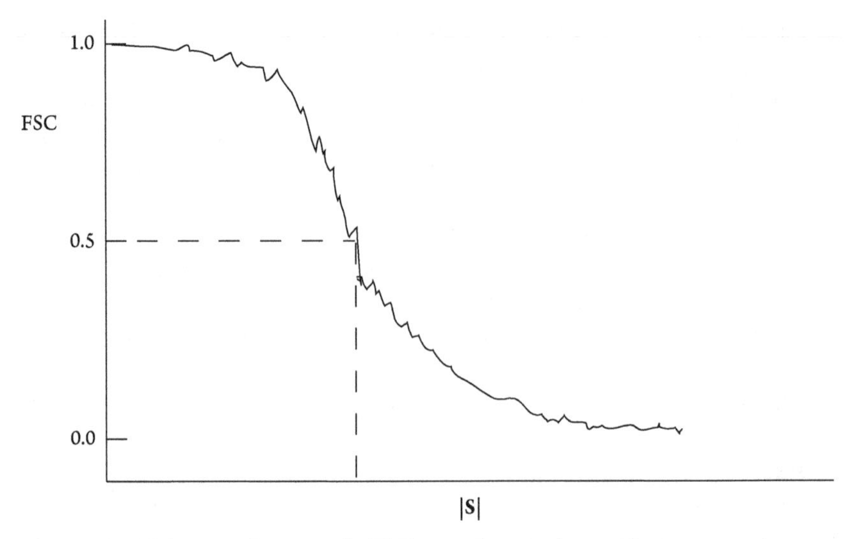

Figure 13.5 Schematic diagram of a FSC curve for a single particle reconstruction. The value of |**S**| at which the FSC falls to 0.5 (dotted lines) is usually taken as the resolution of a reconstruction.

that if we hope to reconstruct the image of a biological object at a resolution better than ~10 Å, we must pay attention to CTFs. Experience shows that the image artifacts caused by CTFs are best addressed by taking micrographs of the particles of interest at several different levels of focus. Once a sufficient number of images have been obtained at each level of focus, there are a number of ways we can proceed. For example, the images obtained at each focal level can be used to generate independent reconstructions, which should agree well at low resolution, but diverge at higher resolutions, due to CTF differences. A CTF-corrected version of the three-dimensional transform of the particle of interest can then be generated by averaging the transforms of those structures appropriately in the regions of reciprocal space where the absolute values of the CFTs of their parent images are large, and "bridging" the shells in the transforms of any reconstruction where the absolute value of the CTF of its parent images is small (or 0) using data from the corresponding shells of the transform of reconstructions derived from images having CTFs that have large absolute values at those Bragg spacings. (The relevant CTFs can be obtained using the Thon ring approach described in section 12.11.)

We close by presenting some pictures that illustrate what is being achieved today using particle reconstruction procedures. The first pictures (figure 13.6) come from a 6.7-Å resolution image of the ribosome from *Escherichia coli* (MW $\sim 2.5 \times 10^6$) with three tRNA molecules and EF-Tu (elongation factor temperature unstable) bound (left panel). It was generated by a reference-based procedure. The middle panel shows the density its authors have assigned to EF-Tu with the X-ray crystal structure of that protein superimposed on it. There can be no doubt that they have interpreted that part of their density map correctly. The second picture comes from a single particle reconstruction of the inner core of a rotavirus, the so-called

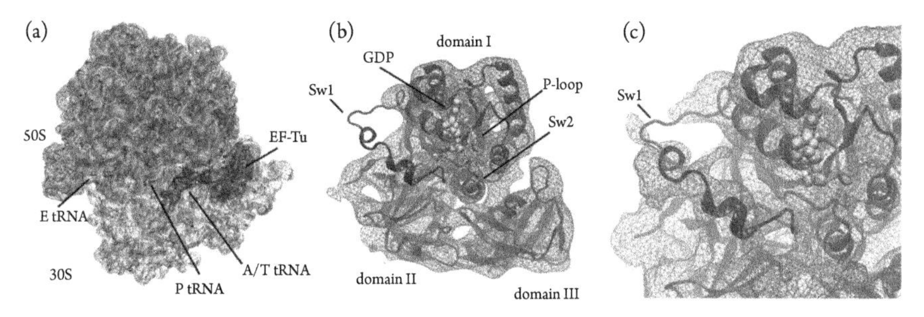

Figure 13.6 A 6.7-Å resolution image of the 70S ribosome from *Escherichia coli* with three tRNAs and EF-Tu/GDP (guanosine 5′-diphosphate) bound. Panel A is the global view. Panel B shows the density of EF-Tu with the crystal structure of that protein superimposed (ribbon). Panel C shows the detail of the same region of the density map as in panel B contoured at a lower level so that the Sw1 loop can be visualized. (From Villa, E., Sengupta, J., Trabuco, L. G., LeBarron, J., Baxter, W. T., Shaikh, T. R., Grassucci, R. A., et al. (2009), Ribosome induced changes in elongation factor Tu conformation control GTP hydrolysis, *Proc. Nat. Acad. Sci. U.S.A.* 106, 1063–1068. Reproduced with permission of the copyright holder, J. Frank.)

double-layer particle, having a resolution of 4 Å. (This reconstruction benefited significantly from the imaging averaging made possible by the icosahedral symmetry of these particles.) Small portions of the electron density map obtained of the virus's protein shell are shown in figure 13.7. Not only can the polypeptide backbone be followed unambiguously in these maps, even in beta sheet regions, where the interpretation of electron density maps is often difficult (panel c), but also bulky side chains are clearly visible. Panels a and b compare a portion of the electron density map obtained of this same particle by X-ray crystallography at a resolution of 3.8 Å (a), with the corresponding part of the EM map (b). As one of the pioneers in the EM image reconstruction field once (notoriously) remarked, "Who needs crystals anyway?"

PROBLEMS

1. What is the scattering angle that corresponds to a Bragg spacing of 3 Å in an EM that is operating at 100 kV? What is the distance between the surface of the sphere of reflection and the corresponding central section at that scattering angle? How thick would an object have to be for that separation to result in significant difference between the 3-Å resolution data obtained from images formed that way and the data that would have been measured at that resolution if EM images really did correspond to sections?
2. The ribosome is a (roughly) spherical object that has a radius of about 125 Å. Suppose you wanted to produce a three-dimensional reconstruction of the ribosome that has a resolution of 3 Å, which would be enough to enable you to fit sequences into the resulting density map. What is the minimum number of images that would be required to achieve that goal,

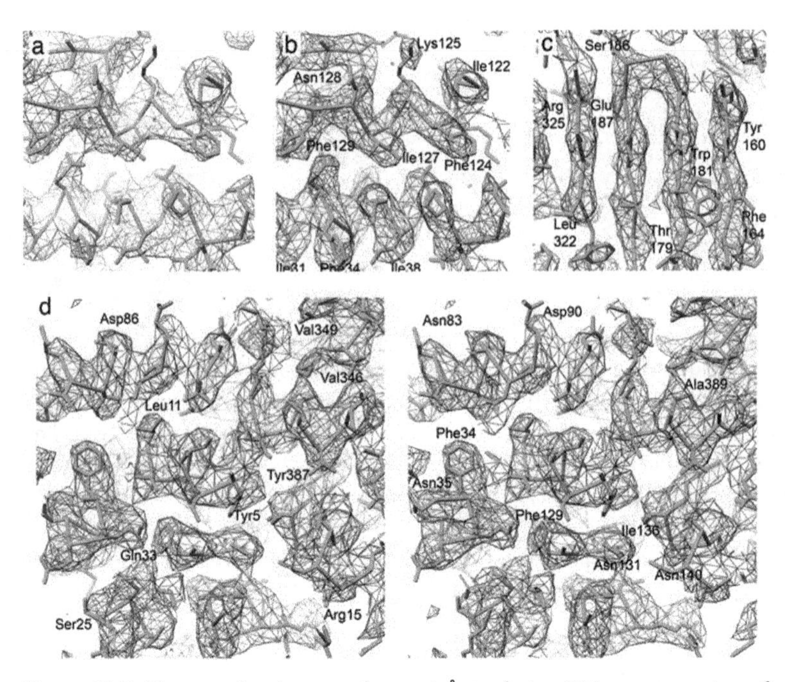

Figure 13.7 Electron density maps from a 4-Å resolution EM reconstruction of a double-layer particle from a rotavirus. The electron density shown corresponds to the protein that forms the icosahedral shell of this particle. Panel A displays a portion of a 3.8-Å resolution electron density map of the protein in question that was obtained crystallographically. Panel B shows what the same part of that protein looks like in the EM map. Panel C shows the EM electron density obtained for a region of the protein that contains beta sheet. Panel D is a stereo pair that illustrates the overall quality of the electron density map. In all panels, the model the resulted from the interpretation of this map is superimposed on the density. (From Zhang, X., Settembre, E., Xu, C., Dormitzer, P. R., Bellamy, R., Harrison, S. C., and Grigorieff, N. (2008), Near atomic resolution using electron cryomicroscopy and single particle reconstruction, *Proc. Nat. Acad. Sci. U.S.A.* 105, 1867–1872. Copyright [2008] National Academy of Sciences, USA.)

assuming every image used had an in-plane resolution at least that good, and that no CTF corrections were needed? What would the angle be between the normals of two adjacent images in that set of images? (Two images are adjacent in this context if they differ in orientation by the minimum amount required to populate reciprocal space properly.)

3. The most famous biological helix of them all is double-stranded DNA. It consists of two polynucleotide strands that wind helically around a common axis. Like so many other biological helices, it is not perfectly helical. The geometry of the helix depends measurably on the sequences of its two strands, and it is flexible enough so that its axis is never perfectly straight. Its structure was solved using X-ray diffraction patterns of bundles of helical molecules that resemble figure 13.2.

The helix parameters of B-form DNA, which is the form of the molecule that is most commonly encountered in living organisms, are the following. Its pitch is 33.8 Å, and there are 10 base pairs per repeat of the double helix. At what Bragg spacing should the first meridional reflection be seen in a DNA diffraction pattern? How many layer planes should there be between the equator and that first meridional reflection? What is the difference in Bragg spacing between adjacent layer planes? If DNA were a single-stranded helix that had the same pitch and number of bases per turn, what would the orders be of the Bessel functions having the lowest three orders that contribute to the equator and layer planes 1 and 2 of the transform of that helix?

FURTHER READING

Crowther, R. A., DeRosier, D. J., and Klug, A. (1970). The reconstruction of a three-dimensional structure from projections and its application to electron microscopy. *Proc. Roy. Soc. Lond.* A317, 319–340. (This classic paper provides an authoritative exposition of the theoretical issues that surround image reconstruction by EM.)

APPENDIX 13.1 SOLVING OF LEAST SQUARES PROBLEMS

Problems arise in many branches of science that are best solved by fitting measured data with functions in a manner that minimize the sum over all observations of the squares of the differences between the measured data and what the functions predict the data should be. The procedures used to solve such problems are straightforward.

Let R be the sum of the squares of the differences between the calculated data and and observation, which in this case implies that:

$$R = \sum \left[\mathrm{F}\left(\mathbf{h}'\right)_{calc} - \mathrm{F}_{obs}\left(\mathbf{h}'\right)\right]^2,$$

where the sum runs over all the locations in reciprocal space where F was observed. Using equation (13.1), we can write:

$$R = \sum_{\mathbf{h}'} \left(\sum_{\mathbf{h}} \mathrm{B}\left(\mathbf{h}', \mathbf{h}\right) \mathrm{F}\left(\mathbf{h}\right) - \mathrm{F}_{obs}\left(\mathbf{h}'\right) \right)^2.$$

If $\mathrm{F}(\mathbf{h})$ is to be the least squares best estimate for the true value of that component of $\mathbf{F}$, $\mathrm{F}_{\mathrm{lsq}}(\mathbf{h})$, it must be true that $[\partial R/\partial \mathrm{F}_{\mathrm{lsq}}(\mathbf{h})] = 0$, which is the condition that must be met by all maxima and minima. Thus, each component of $\mathbf{F}_{\mathrm{lsq}}$, $\mathrm{F}_{\mathrm{lsq}}(\mathbf{j})$, must obey the following equations:

$$\partial R/\partial \mathrm{F}_{lsq}\left(\mathbf{j}\right) = 0 = 2 \sum_{\mathbf{h}'} \left(\sum_{\mathbf{h}} \mathrm{B}\left(\mathbf{h}', \mathbf{h}\right) \mathrm{F}_{lsq}\left(\mathbf{h}\right) - \mathrm{F}_{obs}\left(\mathbf{h}'\right) \right) \mathrm{B}\left(\mathbf{h}', \mathbf{j}\right),$$

which will be true only if:

$$\sum_{\mathbf{h}'} \mathrm{B}(\mathbf{h}', \mathbf{j}) \sum_{\mathbf{h}} \mathrm{B}(\mathbf{h}', \mathbf{h})\mathrm{F}_{lsq}(\mathbf{h}) = \sum_{\mathbf{h}'} \mathrm{B}(\mathbf{h}', \mathbf{j})\mathrm{F}_{obs}(\mathbf{h}').$$

Now:

$$\sum_{\mathbf{h}'} \mathrm{B}(\mathbf{h}', \mathbf{j})\mathrm{F}_{obs}(\mathbf{h}') = \sum_{\mathbf{h}'} \mathrm{B}^{T}(\mathbf{j}, \mathbf{h}')\mathrm{F}_{obs}(\mathbf{h}'),$$

where $\mathbf{B}^{\mathrm{T}}$ is the transpose of $\mathbf{B}$, which means that the (i, j)th element of $\mathbf{B}^{\mathrm{T}}$, $\mathrm{B}^{\mathrm{T}}(i, j)$, is the (j, i)th element of $\mathbf{B}$, $\mathrm{B}(j, i)$.

Comment: If B is complex, the transpose required will be the Hermitian transpose, or $B^{T}(i, j) = B^{*}(j, i)$.

Similarly:

$$\sum_{\mathbf{h}'} \mathrm{B}(\mathbf{h}', \mathbf{j}) \sum_{\mathbf{h}} \mathrm{B}(\mathbf{h}', \mathbf{h}) = \sum_{\mathbf{h}'} \sum_{\mathbf{h}} \mathrm{B}^{\mathrm{T}}(\mathbf{h}', \mathbf{j})\, \mathrm{B}(\mathbf{h}', \mathbf{h}).$$

Thus, the entire set of equations that must be satisfied by the components of $\mathbf{F}_{\mathrm{lsq}}$ can be summarized as follows:

$$\mathbf{B}^{\mathrm{T}}\mathbf{B}\mathbf{F}_{lsq} = \mathbf{B}^{\mathrm{T}}\mathbf{F}'_{obs}. \tag{13.1.1}$$

It follows that:

$$\mathbf{F}_{lsq} = \left(\mathbf{B}^{\mathrm{T}}\mathbf{B}\right)^{-1}\mathbf{B}^{\mathrm{T}}\mathbf{F}'_{obs} \tag{13.1.2}$$

where $(\mathbf{B}^{\mathrm{T}}\mathbf{B})^{-1}$ is the inverse of $(\mathbf{B}^{\mathrm{T}}\mathbf{B})$. This is so because by definition $(\mathbf{B}^{\mathrm{T}}\mathbf{B})^{-1}(\mathbf{B}^{\mathrm{T}}\mathbf{B}) = \mathbf{1}$.

APPENDIX 13.2 ERROR PROPAGATION IN LINEAR SYSTEMS

The relationship between $\mathbf{F}_{\mathrm{lsq}}$, the set of least squares best estimates of $\mathrm{F}(\mathbf{h})$, and $\mathbf{F}'$, the observed data, is provided by equation (13.3): $\mathbf{F}_{\mathrm{lsq}} = [(\mathbf{B}^{\mathrm{T}}\mathbf{B})^{-1}\mathbf{B}^{\mathrm{T}}]\mathbf{F}'_{\mathrm{obs}}$. It follows that an error in the observation made at $\mathbf{h}'$, $\delta(\mathbf{h}')$ will result in an error in the estimate of $\mathrm{F}_{\mathrm{lsq}}(\mathbf{h})$, $\varepsilon(\mathbf{h}, \mathbf{h}')$, of:

$$\varepsilon\left(\mathbf{h},\,\mathbf{h}'\right)=\left(\mathbf{B}^{\mathrm{T}}\mathbf{B}\right)^{-1}\mathbf{B}^{\mathrm{T}}\left(\mathbf{h},\,\mathbf{h}'\right)\delta\left(\mathbf{h}'\right).$$

Summing this expression over all $\mathbf{h}'$, we get as estimate of the total error in the hth component of $\mathbf{F}_{\mathrm{lsq}}$,

$$\varepsilon\left(\mathbf{h}\right)=\left(\mathbf{B}^{\mathrm{T}}\mathbf{B}\right)^{-1}\mathbf{B}^{\mathrm{T}}\left(\mathbf{h}\right)\boldsymbol{\delta}.$$

(Note that $\mathbf{B}^{\mathrm{T}}(\mathbf{h})$ is the hth row of $\mathbf{B}^{\mathrm{T}}$.) If the column vector of the errors in $\mathbf{F}_{\mathrm{lsq}}$ is designated, ε, the matrix that contains the products of all the errors in $\mathbf{F}_{\mathrm{lsq}}$ will be ε^2:

$$\varepsilon^2=\sum\varepsilon\varepsilon^T.$$

The elements of ε^2 will include terms of the form $\varepsilon(\mathbf{h}_i)^2$, which clearly relate to the variances of the components of $\mathbf{F}_{\mathrm{lsq}}$, as well as terms of the form $\varepsilon(\mathbf{h}_i)\varepsilon(\mathbf{h}_j)$, which have to do with covariances, in this case, the tendency of the values of $\mathrm{F}_{\mathrm{lsq}}(\mathbf{h}_i)$ and $\mathrm{F}_{\mathrm{lsq}}(\mathbf{h}_j)$ to vary in a correlated way. It follows that:

$$\begin{aligned}\varepsilon^2\left(\mathbf{h}_i,\mathbf{h}_j\right)&=\left(\mathbf{B}^{\mathrm{T}}\mathbf{B}\right)^{-1}\mathbf{B}^{\mathrm{T}}\left(\mathbf{h}_i\right)\delta\left(\left[\left(\mathbf{B}^{\mathrm{T}}\mathbf{B}\right)^{-1}\mathbf{B}^{\mathrm{T}}\left(\mathbf{h}_j\right)\right]\delta\right)^{\mathrm{T}}\\&=\left(\mathbf{B}^{\mathrm{T}}\mathbf{B}\right)^{-1}\mathbf{B}^{\mathrm{T}}\left(\mathbf{h}_i\right)\left(\delta\,\delta^{\mathrm{T}}\right)\mathbf{B}\left(\mathbf{h}_j\right)\left[\left(\mathbf{B}^{\mathrm{T}}\mathbf{B}\right)^{-1}\right]^{\mathrm{T}}.\end{aligned}$$

However, by hypothesis, the components of $\boldsymbol{\delta}$ are independent of each other, and sample a single normal distribution, which has a variance of σ^2. Thus, the matrix of the most probable values of $\boldsymbol{\delta\delta}^T$, $\langle\boldsymbol{\delta\delta}^{\mathrm{T}}\rangle$, has terms only on its diagonal, and they are all equal:

$$\langle\boldsymbol{\delta}\,\boldsymbol{\delta}^T\rangle=\sigma^2\mathbf{I},$$

where $\mathbf{I}$ is the $(m\times m)$ identity matrix; in other words the $(m\times m)$ matrix that has 1.0 values on its diagonal, and 0s everywhere else. Thus:

$$\begin{aligned}&\langle\varepsilon^2\left(\mathbf{h}_i,\,\mathbf{h}_j\right)\rangle=\sigma^2\left(\mathbf{h}_i,\,\mathbf{h}_j\right)=\sigma^2\left(\mathbf{B}^{\mathrm{T}}\mathbf{B}\right)^{-1}\mathbf{B}^{\mathrm{T}}\left(\mathbf{h}_i\right)\mathbf{B}\left(\mathbf{h}_j\right)\left[\left(\mathbf{B}^{\mathrm{T}}\mathbf{B}\right)^{-1}\right]^{\mathrm{T}},\\&\sigma^2\left(\mathbf{h}_i,\,\mathbf{h}_j\right)=\sigma^2\left(\mathbf{B}^{\mathrm{T}}\mathbf{B}\right)^{-1}\left[\mathbf{B}^{\mathrm{T}}\left(\mathbf{h}_i\right)\mathbf{B}\left(\mathbf{h}_j\right)\right]\left(\mathbf{B}^{\mathrm{T}}\mathbf{B}\right)^{-1}.\end{aligned}$$

[Because $\mathbf{B}^{\mathrm{T}}\mathbf{B}$ is a symmetric matrix, $(\mathbf{B}^{\mathrm{T}}\mathbf{B})^{-1}$ is symmetric also, and the transpose of a symmetric matrix equals that matrix.] Now, $(\mathrm{B}^{\mathrm{T}}(\mathbf{h}_i)\mathrm{B}(\mathbf{h}_j))$ is simply the $(\mathbf{h}_i,\mathbf{h}_j)$th term of the matrix $\mathbf{B}^{\mathrm{T}}\mathbf{B}$. It follows that:

$$\boldsymbol{\sigma}^2=\sigma^2\left(\mathbf{B}^{\mathrm{T}}\mathbf{B}\right)^{-1}. \qquad \mathbf{(13.2.1)}$$

$(\mathbf{B}^{\mathrm{T}}\mathbf{B})^{-1}$ is called the covariance matrix for this system because: $(1/\sigma^2)\sigma^2 = (\mathbf{B}^{\mathrm{T}}\mathbf{B})^{-1}$. In addition, the variance of $\mathrm{F}_{\mathrm{lsq}}(\mathbf{h})$, which is its statistical characteristic of most immediate concern, will be $\sigma^2(\mathrm{B}^{\mathrm{T}}\mathrm{B})^{-1}(\mathbf{h},\mathbf{h})$.

APPENDIX 13.3 THE FOURIER TRANSFORM IN CYLINDRICAL COORDINATES

Cylindrical polar coordinates are the best ones to use when dealing with systems that have a single, special axis. In cylindrical polar coordinate systems, the coordinates of a point in real space are: r, the distance from that special axis to a particular point; θ, the rotation angle around that axis, relative to an arbitrarily chosen reference direction perpendicular to the axis; and z, the vertical position of the point along the special axis, again relative to an arbitrary origin. If the special axis is chosen as the z-axis for Cartesian coordinate system, and the vector that passes through the origin of the z-axis in the reference direction used for measuring θ is the x-axis, then the relationship between the coordinates of that point in the two coordinate systems will be:

$$r = \left(x^2 + y^2\right)^{1/2}, x = r \cos \theta, y = r \sin \theta, \text{ and } z = z.$$

A similar set of relationships will hold in reciprocal space:

$$R = \left(X^2 + Y^2\right)^{1/2}, X = R \cos \Theta, Y = R \sin \Theta, \text{ and } Z = z.$$

Thus, in cylindrical polar coordinates $\rho(x, y, z)$ becomes $\rho(r, \theta, z)$.

It is clear from these definitions that $\rho(r, \theta, z) = \rho(r, (\theta + 2\pi), z)$ for all values of r and z. Thus:

$$\rho\,(r, \theta, z) = \sum_{-\infty}^{+\infty} g_n\,(r, z) \exp\,(in\theta), \qquad (13.3.1)$$

which is to say that ρ can be expressed as a sum of components, each of which is the product of a function of r and z, multiplied by a sin/cos term that has the appropriate periodicity in θ.

Now the Fourier transform of $\rho(x, y, z)$, is:

$$F\,(X, Y, Z) = \iiint \rho\,(x, y, z) \exp\left[\left(-2\pi i\,(xX + yY + zZ)\right)\right] dx\,dy\,dz.$$

Replacing x, y, z, X, Y, and Z with their polar equivalents, we find:

$$F\,(R, \Theta, Z) = \iiint \rho\,(r, \theta, z) \exp\left[\left(-2\pi i\,(rR\,(\cos\theta \cos\Theta + \sin\theta \sin\Theta) + zZ)\right]\right) r\,dr\,d\theta\,dz.$$

This equation is equivalent to:

$$F(R, \Theta, Z) = \iiint \rho(r, \theta, z) \exp(-2\pi i [rR \cos(\theta - \Theta) + zZ]) \, r \, dr \, d\theta \, dz.$$

Substituting equation (13.3.1) into this expression, and rearranging to separate its angular term from the rest, we obtain:

$$F(R, \Theta, Z) = \sum \iint g_n(r, z) \exp(-2\pi i z Z) \, r \, dr \, dz$$
$$\int \exp[-2\pi i r R \cos(\theta - \Theta) + in\theta] \, d\theta.$$

Now, if a new variable, θ', is defined as being equal to $\theta - \Theta$, this equation can be written:

$$F(R, \Theta, Z) = \sum \exp(in\Theta) \iint g_n(r, z) \exp(-2\pi i z Z) \, r \, dr \, dz$$
$$\int \exp\left(-2\pi i r R \cos\left(\theta'\right) + in\theta'\right) d\theta'.$$

In appendix 9.3, it was pointed out that the definite integral involving angles in the preceding equation was evaluated ages ago, and in integral tables it is usually written:

$$J_n(x) = \left(i^{-n}/2\pi\right) \int \exp(ix \cos\theta) \exp(in\theta) \, d\theta,$$

where $J_n(x)$ is the Bessel function of order n. Thus:

$$\int \exp\left(i\left[2\pi r R \cos\left(\theta'\right) - n\theta'\right]\right) = 2\pi i^n J_n(-2\pi r R),$$

but because $J_n(-x) = (-1)^n J_n(x)$, we can also write:

$$\int \exp\left(i\left[2\pi r R \cos\left(\theta'\right) - n\theta'\right]\right) = 2\pi \exp(-in\pi/2) J_n(2\pi r R).$$

Hence:

$$F(R, \Theta, Z) = 2\pi \sum \exp[in(\Theta - \pi/2)] \iint g_n(r, z) J_n(2\pi r R)$$
$$\exp(-2\pi i z Z) \, r \, dr \, dz. \qquad \textbf{(13.3.2)}$$

Starting with the expression:

$$f(x, y, z) = \iiint F(X, Y, Z) \exp\left[2\pi i \left(xX + yY + zZ\right)\right] dX\, dY\, dZ,$$

and:

$$F(R, \Theta, Z) = \sum G_n(R, Z) \exp\left[in\left(\Theta - \pi/2\right)\right], \tag{13.3.3}$$

and using similar arguments, we can show:

$$f(r, \theta, z) = 2\pi \sum \exp(in\theta) \iint G_n(R, Z)\, J_n(2\pi rR) \exp(-2\pi izZ)\, R\, dR\, dZ.$$

It follows that:

$$g_n(r, z) = \iint G_n(R, Z)\, J_n(2\pi rR) \exp(-2\pi izZ)\, 2\pi R\, dR\, dZ. \tag{13.3.4}$$

INDEX